Groups and Related Topics

MATHEMATICAL PHYSICS STUDIES

VOLUME 13

Groups and Related Topics

Proceedings of the
First Max Born Symposium

edited by

R. Gielerak, J. Lukierski
and Z. Popowicz
University of Wrocław,
Institute of Theoretical Physics,
Wrocław, Poland

SPRINGER SCIENCE+BUSINESS MEDIA, B.V.

Library of Congress Cataloging-in-Publication Data

Max Born Symposium (1st : 1991 : Wojnowice Castle)
    Groups and related topics : proceedings of the First Max Born
Symposium / edited by R. Gielerak, J. Lukierski, and Z. Popowicz.
        p.   cm. -- (Mathematical physics studies ; v. 13)
    Includes bibliographical references and index.
    ISBN 978-0-7923-1924-5   978-94-011-2801-8 (eBook)

    DOI 10.1007/978-94-011-2801-8
    1. Mathematical physics--Congresses. 2. Quantum groups-
-Congresses. 3. Geometry, Differential--Congresses.   I. Gielerak,
R. (Roman), 1951-   . II. Lukierski, Jerzy. III. Popowicz, Z.
IV. Title. V. Series.
QC19.2.M39 1991
530.1'5--dc20   92-25133

ISBN 978-0-7923-1924-5

Foreword

In the seventies and eighties, scientific collaboration between the Theory Section of the Physics Department of Leipzig University and the Institute of Theoretical Physics of the University of Wrocław was established. This manifested itself, among other things, in the organization of regular, twice-yearly seminars located alternatively in Wrocław and Leipzig. These Seminars in Theoretical Physics took place 27 times, the last during November 1990.

In order to continue the traditions of German-Polish contacts in theoretical physics, we decided to start a new series of Seminars in Theoretical Physics and name them after the outstanding German theoretical physicist Max Born who was born in 1883 in Wrocław. We hope that these seminars will continue to contribute to better scientific contacts and understanding between German and Polish theoretical physicists.

The First Max Born Symposium was held in Wojnowice Castle, 20 km west of Wrocław, 27 - 29 September 1991. Wojnowice Castle was built in the 16th century by the noble Boner family, in the Renaissance style, and has been recently adapted as a small conference center.

The preferred subjects at the Symposium were Quantum Groups and Integrable Models. The Symposium was organized by Doctors R. Gielerak and Z. Popowicz under the scientific supervision of the undersigned.

The organizers would like to thank Wrocław University, and especially the Rector, Prof. Józef Ziólkowski, for financial support and for accepting the idea of a new Seminar series. We would also like to express our thanks to the authors, who mostly submitted their paper in camera-ready form, and to Kluwer Academic Publishers for their cooperation in publishing the Proceedings of the First Max Born Symposium.

Jerzy Lukierski

Table of Contents

SECTION I
QUANTUM GROUPS

SUGAWARA CONSTRUCTION AND THE Q-DEFORMATION OF VIRASORO ALGEBRA

M. Chaichian

Department of High Energy, Physics, University of Helsinki
Siltavuorenpenger 20 C, SF-00170 Helsinki, Finland.

and

P. Prešnajder

Department of Theoretical Physics,
Comenius University
Mlynská dolina F2, CS-84215 Bratislava, Czechoslovakia

December 20, 1991

Abstract

The q-deformed Virasoro algebra is obtained using the annihilation and creation operators of the q-deformed infinite Heisenberg algebra, which has the Hopf structure. The generators of the q-deformed Virasoro algebra are expressed as a Sugawara construction in terms of normal ordered binomials in these annihilation and creation operators, and become double-indexed as the reminder of a degeneracy removal. The obtained q-deformed Virasoro algebra with central extension reduces to the standard one in the non-deformed limit.

1. Introduction

For many years great attention has been paid to the Virasoro algebra, the underlying symmetry of the string theory [1]. The Virasoro algebra is the central extension of the Lie algebra of conformal transformations of a complex plane. The generation of the latter are given in terms of annihilation and creation operators satisfying the communication relation

$$[a, a^+] = 1 \tag{1}$$

in the following way

3

R. Gielerak et al. (eds.), Groups and Related Topics, 3–12.
© *1992 Kluwer Academic Publishers.*

$$L_m = (a^+)^{m+1} a \tag{2}$$

They satisfy the Lie algebra relations

$$[L_n, L_m] = (m - n) L_{m+n} \tag{3}$$

During the last few years a growing interest has appeared in the study of quantum groups and algebras [2,3]. In particular, a great deal of attention has been paid to q-deformations of centreless Virasoro algebra [4-7] and to its central extension [8-10]. Several of these investigations [8,9] are based on the use of q-deformed harmonic oscillators satisfying the relations [11]

$$[a, a^+]_q = q^{-N},$$

$$[N, a] = -a, \quad [N. a^+] = a^+, \tag{4}$$

where $[A, B]_q = AB - qAB$, and q is the deformation parameter.

The q-deformed Virasoro generators are expressed by the following generalization of eq. (3):

$$L_m = q^{-N} (a^+)^{m+1} a, \tag{5}$$

and they satisfy the following, algebraic relations

$$[L_n, L_m]_{(q^{m-n}, q^{n-m})} = [m - n] L_{n+m}, \tag{6}$$

where $[A, B]_{(q_1, q_2)} = q_1 AB - q_2 BA$, and $[x] = (q^x - q^{-x})/(q - q^{-1})$. Furthermore, the problem of their central extensions has been investigated by using the q-deformed generalizations of the Jacobi identity [8-10]. However, until now these constructions have not been shown to have a Hopf algebra structure [12-14].

In this letter we shall follow an approach based on the Sugawara construction in the string theory, leading to a Virasoro algebra with central extension (see for example, [1]). The generators L_m are represented by an infinite set of bosonic operators a_n, with n-integer, satisfying the infinite Heisenberg algebra $H(\infty)$, defined by the relations

$$[a_m, a_n] = m H \delta_{m+n,0},$$

$$[H, a_m] = 0, \tag{7}$$

with H as a central element. The formula for L_m reads as follows

$$L_m = \frac{1}{2} \sum_{k,n=-\infty}^{+\infty} : a_k a_n : \delta_{k+n,m}, \tag{8}$$

where the colons denote the normal ordering. Owing to quantization, the operators L_m satisfy the (centrally-extended) Virasoro relations (see ref. [1]):

$$[L_n, L_m] = (m - n) H L_{n+m} + \frac{1}{12} H^2 (n^3 - n) \delta_{n+m,0}. \tag{9}$$

We wish to stress the explicit appearance of the central element H in eqs. (7) and (9), which will be substantial later (by a suitable choice of units we can put $H = 1$, as is usually assumed).

We introduced in [15] the deformation $H(\infty)_q$ of the Heisenber algebra (7). The deformation we use for a_n was inspired by ref. [16], where for one pair of annihilation and creation operators, the quantum algebra $H(1)_q$ was obtained.

By introducing the algebra $H(\infty)_q$, we perform the q-deformation of the Sugawara construction of the Virasoro generators. It appears that the latter become double-indexed L_m^α, and are formed by normal-ordered binomials in q-deformed annihilation and creation operators, and by factors of $q^{\alpha H/2}$, with α-integer. The introduction of α is inevitable in this approach, but for $q \to 1$ the α-dependence disappears: $L_m^\alpha \to L_m$. The q-deformed Virasoro algebra is then realized by the usual commutator among L_m^α as a Lie algebra with central extension.

We generalize the results to the fermionic case and briefly discuss the case of the q-deformed Virasoro superalgebra with central extensions. The infinite Lie (super) algebras associated with our q-deformed Virasoro (super) algebras have trigonometric functions as their structure constants. Infinite Lie (super)algebras, which have been proposed previously [17], have similar trigonometric structure constants.

2 q-Deformation of the Heisenberg algebra

We shall denote by

$$a_n, \quad a_{-n} = a_n^+, \quad n = 1, 2, ...,$$

the infinite set of annihilation and creation operators that satisfy the com munication relations

$$[a_m, a_n] \equiv a_m a_n - a_n a_m = mH\delta_{m+n,0}, \tag{10}$$

where H is a central element commuting with all the operators a_n,

$$Ha_n = a_n H. \tag{11}$$

Equations (10) and (11) define a Lie algebra. As for the Heisenberg algebra $H(\infty)$, we shall denote the corresponding universal enveloping algebra generated by H and a_n.

We define the deformation of $H(\infty)$ by changing the commutators (10) to

$$[a_m, a_n] = \omega_{mn}, \tag{12}$$

where

$$\omega_{mn} = \frac{1}{\varepsilon} \sinh(m\varepsilon H)\delta_{m+n,0}, \tag{13}$$

and the deformation parameter q is taken to be equal to e^ε. The relations (11) remain unchanged.

The associative algebra $H(\infty)_q$, generated by H and a_n, with the defining relations (11)-(13), is a Hopf algebra with

i) co-product

$$\Delta a_n = q^{-|n|H/2} \otimes a_n + a_n \otimes q^{|n|H/2} ,$$

$$\Delta H = 1 \otimes H + H \otimes 1 ,$$

ii) co-unit

$$\varepsilon(a_n) = 0 , \quad \varepsilon(H) = 0 ,$$

iii) co-inverse (antipode)

$$S(a_n) = -a_n , \quad S(H) = -H .$$

Here 1 denotes the unit element in $H(\infty)_q$ for which $\Delta 1 = 1 \otimes 1$, $\varepsilon(1) = 1$, and, $S(1) = 1$. Such a deformation for one pair of annihilation and creation operators was proposed in ref. [16].

For completness, we present the corresponding R−matrix, which is given by the formula

$$R = q^{-(H \otimes N + N \otimes H)} q^{2K} ,$$

where

$$N = \sum_{n=1}^{\infty} \frac{n\varepsilon}{\sinh(n\varepsilon)} a_{-n} a_n ,$$

$$K = \sum_{n=1}^{\infty} q^{nH/2} a_n \otimes q^{-nH/2} a_{-n} .$$

This R−matrix satisfies the Yang-Baxter equation

$$R_{12} R_{13} R_{23} = R_{23} R_{13} R_{12}$$

and the universality condition [18]

$$R \, \Delta A \, R^{-1} = \sigma \, \Delta A,$$

where σ is defined by $\sigma(A \otimes B) = B \otimes A$.

We now extend our approach to the case of q−deformed fermionic annihilation and creation operators (in the Neveu-Schwarz sector, see ref. [1]),

$$b_r, \ b_{-r} = b_r^+ \ , \quad r = 1/2, \ 3/2, ...,$$

satisfying the communication relations

$$[b_r, b_s]_+ \equiv b_r b_s + b_s b_r = \omega_{rs} \ , \tag{14}$$

where

$$\omega_{rs} = \frac{1}{\varepsilon} \sinh(\varepsilon \tilde{H}) \delta_{r+s,0} \tag{15}$$

and $\tilde{H}$ is the central element

$$\tilde{H} b_r = b_r \tilde{H}. \tag{16}$$

The associative superalgebra $s - H(\infty)_q$, generated by $\tilde{H}$ and b_r, defined by the relations (14)-(16), is a Hopf superalgebra (for a definition see, for example , [13]) with

i) co-product

$$\Delta b_r = q^{-\tilde{H}/2} \otimes b_r + b_r \otimes q^{\tilde{H}/2} \ ,$$

$$\Delta \tilde{H} = 1 \otimes \tilde{H} + \tilde{H} \otimes 1 \ ,$$

ii) co-unit

$$\varepsilon(b_r) = 0 \ , \quad \varepsilon(\tilde{H}) = 0 \ ,$$

iii) co-inverse

$$S(b_r) = -b_r \ , \quad S(\tilde{H}) = -\tilde{H} \ .$$

In ref. [20] the $q-$deformed Heisenberg algebra $H(1)_q$ and the super one, $s - H(1)_q$, have been obtained from the contraction of the quantum superalgebra $osp_q(1|2)$. Our annihilation and creation operators used for $H(\infty)$ and $s - H(\infty)_q$ coincide with those of the special case in ref. [20] with the same Hopf structure.

3 q-deformed Sugawara construction for Virasoro algebra

The generators of Virasoro algebra are the elements of $H(\infty)$,

$$L_m = \frac{1}{2} \sum : a_k a_n : \delta_{k+n,m} \ , \quad m \ - \ integer \ , \tag{17}$$

where the summation runs over $k, n = \pm 1, \pm 2, \dots$ (with respect to the string theory, we exclude the zero mode describing the position of the string, which is unessential for us).

In eq. (14) the normal product is defined as

$$: a_k a_n := a_k a_n - 0(k) H \delta_{k+n,0} \ , \tag{18}$$

where $0(k) = 1$ or 0 for k positive or negative, respectively. The operators L_m satisfy the communication relations [1]

$$[L_m, L_n] = (n - m) H L_{m+n} + \frac{1}{12} H^2 (m^3 - m) \delta_{m+n,0} \ . \tag{19}$$

We define the q–deformed Virasoro generators as

$$L_m^\alpha = \frac{1}{2} \sum \cosh(\frac{k-n}{2} \varepsilon \alpha H) \ : a_k a_n : \ \delta_{k+n,m} \ , \tag{20}$$

where α is an additional index - we later discuss the need for its presence. The normal product is now given as

$$: a_k a_n := a_k a_n - 0(k) \omega_{kn} \ , \tag{21}$$

with ω_{kn} defined in eq. (13). We note that $L_m^{-\alpha} = L_m^\alpha$ and that, in the non-deformed limit, from the Sugawara construction (20) we obtain

$$L_m^\alpha \to L_m \ , \quad for \ \varepsilon \to 0 \ (i.e. \ q \to 1) \ , \tag{22}$$

irrespective of the values of the additional index α .

Using eqs. (20) and (21), we obtain the q-deformed Virasoro algebra relations,

$$\begin{aligned}
[L_m^\alpha, L_{m'}^{\alpha'}] &= \frac{1}{2\varepsilon} \sinh(\frac{m - m' - m'\alpha + m\alpha'}{2} \varepsilon H) \ L_{m+m'}^{\alpha+\alpha'+1} \\
&+ \frac{1}{2\varepsilon} \sinh(\frac{m - m' + m'\alpha - m\alpha'}{2} \varepsilon H) \ L_{m+m'}^{-\alpha-\alpha'+1} \\
&+ \frac{1}{2\varepsilon} \sinh(\frac{m - m' + m'\alpha + m\alpha'}{2} \varepsilon H) \ L_{m+m'}^{\alpha-\alpha'-1} \\
&+ \frac{1}{2\varepsilon} \sinh(\frac{m - m' - m'\alpha - m\alpha'}{2} \varepsilon H) \ L_{m+m'}^{-\alpha+\alpha'-1} \\
&+ \frac{1}{16\varepsilon^2} (C_{m-1}^{\alpha,\alpha'} + C_{m-1}^{\alpha,-\alpha'}) \ \delta_{m+m',0} \ .
\end{aligned} \tag{23}$$

where the quantities in the central term are given by

$$\begin{aligned}
C_m^{\alpha,\alpha'} &= \frac{\sinh((\alpha + \alpha' + 1)/2 \ m\varepsilon H)}{\sinh((\alpha + \alpha' + 1)/2\varepsilon H)} \\
&+ \frac{\sinh((\alpha + \alpha' - 1)/2m\varepsilon H)}{\sinh((\alpha + \alpha' - 1)/2\varepsilon H)} \\
&- 2\cosh(m\varepsilon H + \varepsilon H)\frac{\sinh((\alpha + \alpha')/2m\varepsilon H)}{\sinh((\alpha + \alpha')/2\varepsilon H)}
\end{aligned} \tag{24}$$

We see that the odd integer values of the parameter α respect a minimal set for which the relations (23) are closed. Of course, the set of α integers is also admissible.

Using eq. (22), we can show directly that in the limit $q \to 1$ the rela- tions (23) reduce to the non-deformed Virasoro algebra relations (19). We would like to emphasize that owing to the form of the q-deformed Sugawara construction (20), we were unambiguously forced to introduce an additional index α, since the q-deformed Virasoro is realized as a commutator Lie algebra in $H(\infty)_q$ (satisfying the usual Jacobi identity) and spanned by quadratic expressions in a_n, with integer power of the group-like element $Q = q^{H/2}$. In the limit $q \to 1$ the central subgroup Q^α, with α-integer, degenerates to a unit element. We can say that the q-deformation eliminates this degeneracy.

It is remarkable that the structure constants and the central term on the right-hand side of eq. (23) only depend on the central Heisenberg generator H. However in a unitary irreducible representation, H is constant, and we can absorb it into ε, or simply put $H = 1$. In this way we can associate with eq. (23) a usual Lie algebra.

In the fermionic realization we define the q-deformed Virasoro generators by

$$\tilde{L}_m^\beta = \frac{1}{2} \sum \frac{\sinh((s - r)/2\varepsilon\beta\tilde{H})}{2\sinh(\varepsilon\tilde{H}/2)} : b_r b_s : \delta_{r+s,m} , \tag{25}$$

where the summation runs over $r, s = \pm 1/2, \pm 3/2, \dots$. The normal ordered product is defined in the usual way,

$$: b_r b_s := b_r b_s - \theta(r)\omega_{rs}. \tag{26}$$

with ω_{rs} given in eq. (15).

Evidently, $\tilde{L}_m^{-\beta} = -\tilde{L}_m^\beta$, and in the non-deformed limit

$$\tilde{L}_m^\beta \to \beta\tilde{L}_m , \tag{27}$$

where $\tilde{L}_m$ are the standard Virasoro generators in the fermionic realization,

$$\tilde{L}_m = \frac{1}{2} \sum \frac{s - r}{2} : b_r b_s : \delta_{r+s,m} , \tag{28}$$

with b_r and b_s , the usual fermionic operators, satisfying

$$[b_r, b_s]_+ = \tilde{H} \, \delta_{r+s,0} .$$

The generators $\tilde{L}_m$ satisfy the usual, non-deformed, Virasoro algebra relations,

$$[\tilde{L}_m, \tilde{L}_n] = (n - m)\tilde{H}\tilde{L}_{m+n} + \frac{1}{24}\tilde{H}^2(m^3 - m)\delta_{m+n,0} . \tag{29}$$

The q-deformed Virasoro algebra in the fermionic realization is obtained by a direct calculation

$$\begin{aligned}
[\tilde{L}_m^\beta, \tilde{L}_{m'}^{\beta'}] &= \frac{\cosh(\varepsilon\tilde{H}/2)}{\varepsilon} \sinh(\frac{\beta m' - \beta'm}{2}) \, \tilde{L}_{m+m'}^{\beta+\beta'} \\
&- \frac{\cosh(\varepsilon\tilde{H}/2)}{\varepsilon} \sinh(\frac{\beta m' + \beta'm}{2}) \, \tilde{L}_{m+m'}^{\beta-\beta'} \\
&+ \frac{\cosh^2(\varepsilon\tilde{H}/2)}{\varepsilon} (S_m^{\beta+\beta'} - S_m^{\beta-\beta'}) \, \delta_{m+m',0} ,
\end{aligned} \tag{30}$$

where the quantities in the central term are given by

$$S_m^\theta = \frac{\sinh(\beta/2\ m\epsilon\tilde{H})}{\sinh(\beta/2\ \epsilon\tilde{H})} .$$ (31)

By definition we put $\tilde{L}^0 = 0$. The algebra (29) is closed if the parameter β is a multiple of some given natural number. For example, the set of β—even is admissible and is complementary to the minimal set of α—odd appearing in the bosonic realization [cf. the note after eq. (21)]. Another example of an admissible set is given by the β-integer.

Putting $\tilde{H} = 1$ in eqs (30) and (31) we obtain a usual Lie algebra with central extension. The centreless double-indexed Virasoro algebra found in ref. [19] is, after rescaling, given by eq. (30) without the central term. The Lie algebra associated to (30) is closely related to that proposed in ref. [17],

$$
\begin{aligned}
[\hat{K}_m^\beta, \hat{K}'^{\beta'}_{m'}] &= \frac{1}{2\epsilon}\ \sinh(\frac{\beta m' - \beta' m}{2}\epsilon)\ \hat{K}^{\beta+\beta'}_{m+m'} \\
&+ (cm + d\beta)\delta_{m+m',0}\delta_{\beta+\beta',0}\ ,
\end{aligned}
$$ (32)

where c and d are constants specifying the general central term. By putting

$$\tilde{L}_m^\beta = \cosh(\frac{\epsilon}{2})(\hat{K}_m^\beta - \hat{K}_m^{-\beta})\ ,$$ (33)

we obtain the Lie algebra (30), but with a central term linear in m. This indicates that although the algebras (30) and (32) seem to be closely related, they differ considerably [also the relations (33) are not invertible].

To construct the Virasoro superalgebra we have to take bosonic and fermionic annihilation and creation operators simultaneously. We shall assume that they commute,

$$b_r a_n = a_n b_r\ ,\quad i.e.\ n = \pm 1, \pm 2, ...\ ,\quad r = \pm 1/2, \pm 3/2, ...\ ,$$ (34)

and that they satisfy the relations (12) and (14) with the common central element, i.e. $H = \tilde{H}$. Then the even Virasoro generators in the bosonic realization commute with their fermionic partners:

$$[L_m^\alpha, \tilde{L}_n^\beta] = 0\ .$$ (35)

We shall add to them the odd Virasoro generators, which we define as

$$G_r^\gamma = \sum q^{(n-s)\gamma H/2} a_n b_s \delta_{n+s,r}\ ,$$ (36)

where the summation runs over $n = \pm 1, \pm 2, ...$ and $s = \pm 1/2, \pm 3/2, ...$ The commutation relations containing the odd Virasoro generators are given in ref. [15]. They close, together with the relations (23), (30) and (35), the q—deformed Virasoro superalgebra, which in the limit $\epsilon \to 1$ reduces to the usual Virasoro superalgebra [1] (here the case where $N = 1$)

4 Conclusions

In this article we present the q-deformed analogue of infinite Heisenberg algebra $H(\infty)_q$, formed by the central element H and the infinite set of bosonic and fermionic annihilation and creation operators, which has the structure   of the Hopf algebra.

Using the q-analogue of the Sugawara construction, we construct the double -indexed q-deformed Virasoro algebra with central extension, realized by the commutators of the generators L_m^α . These generators are normal-ordered binomials in the introduced q-deformed annihilation and creation operators of $H(\infty)_q$, and contain the factors $q^{\alpha H/2}$ with α-integer, as powers of the group-like element $q^{H/2}$. Thus, the additional index α is inevitably connected with the q-deformation, and for $q \to 1$ the α-dependence disappears.   At present it seems to us that the appearance of an extra index α is a necessity, at least within the framework of a Sugawara construction.

We generalize these results to the fermionic realization - the infinite Heisenberg superalgebra $s - H(\infty)_q$ is formed by a central element $\tilde H$, and fermionic annihilation and creation operators. Assuming simultaneously bosonic and fermionic operators, we perform the Sugawara construction of the q-deformed Virasoro superalgebra, the generators of which are again double-indexed and reduce to the usual Virasoro superalgebra in the non-deformed limits. The situation that emerges in connection with the double-indexing of generators remind us of a kind of degeneracy removal as a result of q-deformation. It is also of interest to study the problem of the corresponding additional symmetry involved and its origin.

References

[1] M.B. Green, J.H. Schwarz and E. Witten, Superstring theory (Cambridge Univ. Press, 1987).

[2] L.D. Fadeev in Les Houches Lectures 1982 (North Holland, Amsterdam, 1984). P.P. Kulish and E.K. Sklyanin, Lecture Notes in Physics (Springer, Berlin, 1982), Vol. 151. P.Kulish and N. Reshetikin, J.Sov.Math 23 (1983) 2435.

[3] W.G.Drinfeld, Sov. Mat. Doklady 32 (1985) 254. Proc. ICM, Berkeley (Academic Press, New York, 1986), vol. 1, p. 793. M.Jimbo, Lett. Math. Phys. 11 (1986) 247 and Commun. Math. Phys. 102 (1986) 537.

[4] M. Chaichian, P.Kulish and J.Lukierski, Phys.Lett B237 (1990) 401.

[5] T.Curtright and C.Zachos, Phys. Lett. B243 (1990) 237.

[6] F.J. Narganes-Quijano, J.Phys. A24 (1991) 593.

[7] A.P. Polychronakos, Phys.Lett. B256 (1991),35.

[8] M. Chaichian, D. Ellinas and Z. Popowicz, Phys.Lett. B248 (1990) 95.

[9] M. Chaichian, A.P. Isaev, J. Lukierski, Z. Popowicz, Phys.Lett. B248(1990) 95.

[10] N. Aizawa and H. Sato, Phys.Lett. B256(1991) 185. H. Sato, Kyoto Univ. preprint
     YITP/K-930, HUPD-9108, May 1991.

[11] A.J. Mcfarlane, J. Phys. A22 (1989) 4581. L.C.Biederharn, J.Phys. A22 (1989),
     L 873. T. Hayashi, Commun. Math. Phys. 127 (1990 129. M. Chaichian and P.
     Kulish, Phys.lett B234 (1990) 72.

[12] E. Abe, Hopf algebras, Cambridge Tracts in Mathematics (Cambridge Univ., Press,
     1980).

[13] Yu.I. Manin, Commun. Math. Phys. 123 (1989) 163.

[14] L.D.Fadeev, N.Y. Reshetikin and L.A.Takhtajan, Advanced Series in Mathema- tical
     Physics (World Scientific, Singapore, 1989), eds. C.N. Yang and M.L.Ge, Vol. 9.

[15] M. Chaichian and P. Presnajder, preprint CERN-TH.6225/91, august 1991.

[16] E. Celeghini, R. Giachetti, E. Sorace and M. Tarlini, J. Math. Phys. 31 (1991) 1155.

[17] D.B. Fairlie, P. Fletcher and C.K. Zachos, Phys. Lett,. B218 (1989) 203

[18] A.N. Kirillov and N.Yu. Reshetikin, LOMI preprint E-9-88, 1988.

[19] H.Hiro-Oka, O. Matsui, T. Naito and S. Saito, Tokyo Metropolitan Univ , preprint
     TMUP-HEL.9004, March 1990. S. Saito in Karpacz 91, XXVII Winter School of
     Theoretical Physics (World Scientific, 1991).

[20] E. Celeghini, R. Giacchetti, P.P. Kulish, E. Sorace and M. Tarlini, preprint Univ.
     Florence, DFF 139/6/91, July 1991.

Complex Quantum Groups and Their Dual Hopf Algebras

Bernhard Drabant[1], *Michael Schlieker* [2], *Wolfgang Weich*[2], *Bruno Zumino*[3]

presented by

Michael Schlieker

Abstract. We construct complexified versions of the quantum groups associated with the Lie algebras of type A_{n-1}, B_n, C_n and D_n. Following the ideas of Faddeev, Reshetikhin and Takhtajan we obtain the Hopf algebras of regular functionals $U_{\mathcal{R}}$ on these complexified quantum groups. In the special example A_1 we derive the q−deformed enveloping algebra $U_q(sl(2, \mathbb{C}))$. In the limit $q \rightarrow 1$ the undeformed $U(sl(2, \mathbb{C}))$ is recovered.

1. Introduction

For quantum groups associated with the Lie algebras g of type A_{n-1}, B_n, C_n and D_n there exist well defined correlations between the quantum group itself and the corresponding q−deformed universal enveloping algebra $U_q(g)$ [Dri, FRT]. Coming from the quantum group, one can construct the algebra of regular functionals which is shown to be the algebra $U_q(g)$ for a certain completion. Though the q-deformed Lorentz group already exists in at least two versions [CSSW, PW], there is not yet such a straightforward procedure like in the case of compact Lie groups to derive the corresponding quantized universal enveloping algebra. However this q−deformed algebra is the very object of interest since it should be fundamental for the construction of a q−deformed relativistic field theory.

In this paper we present the quantized universal enveloping algebra $U_q(sl(2, \mathbb{C}))$ of the q−deformed Lorentz group $Sl_q(2, \mathbb{C})$. In section (2) we construct complex quantum groups for the Lie algebras A_{n-1}, B_n, C_n and D_n. These are complexifications of the original quantum groups. The algebraic relations can be written in a generalized RTT−formulation and the usual determinant or metric relations. Following the ideas of [FRT] this fact is used in section (3) to build up the algebra of regular functionals on the complex quantum groups[†]. In section (4) we derive as a special example the algebra $U_q(sl(2, \mathbb{C}))$[‡]. We investigate the limit $q \rightarrow 1$ in section (5) and recover $U(sl(2, \mathbb{C}))$.

---

[1] *Max-Planck-Institut für Physik, Werner-Heisenberg-Institut, München*
[2] *Sektion Physik der Universität München, Lehrstuhl Professor Wess*
[3] *Department of Physics, University of California, Berkeley*

[†] *The same universal enveloping algebra corresponding to the complex quantum group is constructed by analyzing the algebra of the fundamental bicovariant bicomodule [CW].*
[‡] *This algebra also has been investigated in [SWZ, OSWZ] by an alternative approach.*

R. Gielerak et al. (eds.), Groups and Related Topics, 13–22.
© *1992 Kluwer Academic Publishers.*

2. Complexified Quantum Groups

In the approach of [FRT] the quantum group is a Hopf algebra with comultiplication Φ, counit e and antipode κ [Abe], generated by the matrix elements $t^i{}_j$ ($i, j = 1, \ldots, N$; $N = n$ for A_{n-1} and $N = 2n + 1$ for B_n, $N = 2n$ for C_n, D_n) with the relations

$$I_{t,t}{}^{ij}_{st} := \hat{R}^{ij}_q{}_{kl}\, t^k{}_s\, t^l{}_t - t^i{}_v\, t^j{}_w\, \hat{R}^{vw}_q{}_{st} = 0 \tag{2.1}$$

and

$$\begin{cases} \det(t^i{}_j) = \frac{(-1)^{n-1}}{[n]_q!} q^{-\binom{n}{2}} \varepsilon^{k_1 \ldots k_n}\, t^{l_1}{}_{k_1} \cdot \ldots \cdot t^{l_n}{}_{k_n}\, \varepsilon_{l_1 \ldots l_n} = 1 & \text{for } A_{n-1} \\ t^i{}_s\, (C^{-1})^{sk}\, t^l{}_k\, C_{lj} = (C^{-1})^{ik}\, t^l{}_k\, C_{ls}\, t^s{}_j = \delta^i_j\, 1 & \text{for } B_n, C_n, D_n \end{cases} \tag{2.2}$$

where $\varepsilon_{i_1 \ldots i_n} = (-1)^{n-1} \cdot \varepsilon^{i_1 \ldots i_n} = (-q)^{l(\sigma)}$, $[n]_q!$ is the usual q–factorial [CSWW] and C_{ij} is the usual metric [FRT]. The $\hat{R}$–matrices for the respective quantum groups are taken from [FRT] with $q > 0$ real.

To find the complexified versions of these quantum groups one has to introduce the complex conjugates $t^{*i}{}_j$ of the generators $t^i{}_j$ as additional generators with the complex conjugate versions of the relations (2.1) and (2.2) above [CSWW]. With the definition:

$$\hat{t}^i{}_j := \left(\kappa(t^j{}_i) \right)^* \tag{2.3}$$

we get

$$I_{\hat{t},\hat{t}}{}^{ij}_{st} := (\hat{R}^{-1}_q)^{ij}{}_{kl}\, \hat{t}^k{}_s\, \hat{t}^l{}_t - \hat{t}^i{}_v\, \hat{t}^j{}_w\, (\hat{R}^{-1}_q)^{vw}{}_{st} = 0, \tag{2.4}$$

$$\begin{cases} \det(\hat{t}^i{}_j) = \frac{(-1)^{n-1}}{[n]_q!} q^{-\binom{n}{2}} \varepsilon^{k_1 \ldots k_n}\, \hat{t}^{l_1}{}_{k_1} \cdot \ldots \cdot \hat{t}^{l_n}{}_{k_n}\, \varepsilon_{l_1 \ldots l_n} = 1 & \text{for } A_{n-1} \\ \hat{t}^i{}_s\, (C^{-1})^{sk}\, \hat{t}^l{}_k\, C_{lj} = (C^{-1})^{ik}\, \hat{t}^l{}_k\, C_{ls}\, \hat{t}^s{}_j = \delta^i_j\, 1 & \text{for } B_n, C_n, D_n. \end{cases} \tag{2.5}$$

One still has to define commutation relations between the generators $t^i{}_j$ and their complex conjugates:

$$I_{\hat{t},t}{}^{ij}_{st} := \hat{R}^{ij}_q{}_{kl}\, \hat{t}^k{}_s\, t^l{}_t - t^i{}_v\, \hat{t}^j{}_w\, \hat{R}^{vw}_q{}_{st} = 0. \tag{2.6}$$

With this choice of commutation relations one can identify the function algebra over the unitary group as the quotient $\hat{t}^i{}_j = t^i{}_j$. There is a second possibility interchanging the role of $t^i{}_j$ and $\hat{t}^i{}_j$ in (2.6) which is equivalent to the first.

Summarizing we are considering the following quantum group

$$\mathcal{A} := \mathbb{C} < t^i{}_j, \hat{t}^i{}_j > \Big/ \left(I_{t,t}{}^{ij}_{st}, I_{\hat{t},\hat{t}}{}^{ij}_{st}, I_{\hat{t},t}{}^{ij}_{st}, (2.2), (2.5) \right). \tag{2.7}$$

Proposition 1.

The algebra $\mathcal{A}$ becomes a $*-$Hopf algebra with comultiplication Φ, counit e and antipode κ which are defined on the generators through

$$\Phi(t^i{}_j) = t^i{}_k \otimes t^k{}_j,$$
$$e(t^i{}_j) = \delta^i_j,$$
$$\kappa(t^i{}_j) = \begin{cases} \dfrac{q^{-\binom{n}{2}}}{[n-1]_q!} \varepsilon^{i \ldots k_1 \ldots k_{n-1}} t^{l_1}{}_{k_1} \cdot \ldots \cdot t^{l_{n-1}}{}_{k_{n-1}} \varepsilon_{l_1 \ldots l_{n-1} j} & \text{for } A_{n-1} \\ (C^{-1})^{ik} t^l{}_k C_{lj} & \text{for } B_n, C_n, D_n \end{cases}$$

(2.8)

and

$$\Phi(t^{*i}{}_j) = t^{*i}{}_k \otimes t^{*k}{}_j,$$
$$e(t^{*i}{}_j) = \delta^i_j,$$
$$\kappa(t^{*i}{}_j) = \left(\kappa^{-1}(t^i{}_j)\right)^*.$$

(2.9)

It is convenient to introduce an $RTT-$formulation for this complexified quantum group. Set $(I) := (i, \bar{i})$, $\bar{I} := (\bar{i}, \bar{\bar{i}}) = (\bar{i}, i)$, $(i, \bar{i} = 1, \ldots, N)$ and define the $2N \times 2N-$matrix

$$T^I{}_J := \begin{pmatrix} t & 0 \\ 0 & \hat{t} \end{pmatrix}^I_J.$$

(2.10)

Correspondingly one defines the $\hat{\mathcal{R}}-$matrix

$$\hat{\mathcal{R}}_q^{IJ}{}_{KL} := \begin{pmatrix} \alpha_0 \hat{R}_q & 0 & 0 & 0 \\ 0 & 0 & \alpha_1 \hat{R}_q & 0 \\ 0 & \alpha_2 \hat{R}_q^{-1} & 0 & 0 \\ 0 & 0 & 0 & \alpha_3 \hat{R}_q^{-1} \end{pmatrix}$$

(2.11)

with $\alpha_i \in \mathbb{C}$.

Then the relations (2.1), (2.4) and (2.6) can be written in compact form as

$$\hat{\mathcal{R}}_q^{IJ}{}_{KL} T^K{}_R T^L{}_S = T^I{}_V T^J{}_W \hat{\mathcal{R}}_q^{VW}{}_{RS}$$

(2.12)

and $\hat{\mathcal{R}}_q$ fulfills the Yang-Baxter-Equation:

$$(\mathbf{E} \otimes \hat{\mathcal{R}}_q)(\hat{\mathcal{R}}_q \otimes \mathbf{E})(\mathbf{E} \otimes \hat{\mathcal{R}}_q) = (\hat{\mathcal{R}}_q \otimes \mathbf{E})(\mathbf{E} \otimes \hat{\mathcal{R}}_q)(\hat{\mathcal{R}}_q \otimes \mathbf{E})$$

(2.13)

with $\mathbf{E}^I_J = \delta^I_J$.

There are three further possibilities for the choice of the $\hat{\mathcal{R}}_q-$matrix which we disregard here, since one of them yields equivalent results and the others do not admit a simple involution on the algebra of regular functionals.

3. The Algebra of Regular Functionals

The dual space $\mathcal{A}^*$ of the Hopf algebra $\mathcal{A}$ is an algebra with the convolution product. One can introduce an antimultiplicative involution "†" on $\mathcal{A}^*$: For $f \in \mathcal{A}^*$ one sets

$$\forall a \in \mathcal{A}: \quad f^\dagger(a) := \overline{f(\kappa^{-1}(a^*))}. \tag{3.1}$$

In the following we are working mostly with the multiplicative involution " $-$ ":

$$\bar{f} := f^\dagger \circ \kappa^{-1}. \tag{3.2}$$

It is also possible to consider an involution where κ^{-1} is replaced by κ in (3.1) and (3.2). Since $\kappa((\kappa(a^*))^*) = \kappa^{-1}((\kappa^{-1}(a^*))^*) = a \ \forall a \in \mathcal{A}$ the multiplicative involutions coincide for both cases. This is also true for the antimultiplicative ones for $q \to 1$. We now construct the algebra of regular functionals on $\mathcal{A}$. We define functionals $L^{\pm I}{}_J \in \mathcal{A}^*$ through their action on the generators of $\mathcal{A}$:

$$L^{\pm I}{}_J(1) := \delta^I_J,$$
$$L^{\pm I}{}_J(T^K{}_L) := \hat{\mathcal{R}}_q^{\pm 1 IK}{}_{LJ} \tag{3.3}$$

and their comultiplication

$$\forall a, b \in \mathcal{A}: \quad L^{\pm I}{}_J(ab) = L^{\pm I}{}_K(a) L^{\pm K}{}_J(b). \tag{3.4}$$

This definition is compatible with the algebra relations in $\mathcal{A}$ and it holds

Proposition 2.

$$L^{\pm i}{}_j = L^{\pm j}{}_i = 0 \quad \forall i, j \ ,$$
$$\hat{\mathcal{R}}_q^{JI}{}_{LK} L^{\pm K}{}_V L^{\pm L}{}_W = L^{\pm I}{}_A L^{\pm J}{}_B \hat{\mathcal{R}}_q^{BA}{}_{WV} \ , \tag{3.5}$$
$$\hat{\mathcal{R}}_q^{JI}{}_{LK} L^{+K}{}_V L^{-L}{}_W = L^{-I}{}_A L^{+J}{}_B \hat{\mathcal{R}}_q^{BA}{}_{WV} \ .$$

The equations (2.2) and (2.5) partly determine the coefficients α_i in eq. (2.11):

Proposition 3.

For A_{n-1} one has

$$(\alpha_0)^{-n} = (\alpha_1)^{-n} = (\alpha_2)^n = (\alpha_3)^n = q. \tag{3.6}$$

In the cases of B_n, C_n, D_n one gets

$$(\alpha_0)^2 = (\alpha_1)^2 = (\alpha_2)^2 = (\alpha_3)^2 = 1. \tag{3.7}$$

Definition.

The algebra $U_\mathcal{R}$ of regular functionals on $\mathcal{A}$ is the unital algebra generated by $\{L^{\pm I}{}_J\}$.

Proposition 4.

The algebra $U_\mathcal{R}$ becomes a bialgebra with comultiplication $\Delta : U_\mathcal{R} \to U_\mathcal{R} \otimes U_\mathcal{R}$ and counit $\epsilon : U_\mathcal{R} \to \mathbb{C}$ through the definitions

$$\Delta(L^{\pm I}{}_J) := L^{\pm I}{}_K \otimes L^{\pm K}{}_J,$$
$$\epsilon(L^{\pm I}{}_J) := \delta^I_J \tag{3.8}$$

on the generators of $U_\mathcal{R}$.

Consider now the map $\tilde{S} : \mathcal{A}^* \to \mathcal{A}^*$ defined by

$$\tilde{S} := . \circ \kappa . \tag{3.9}$$

With this definition we get the following

Proposition 5.

$$\tilde{S}(U_\mathcal{R}) = U_\mathcal{R} \tag{3.10}$$

and

$$\tilde{S}(L^{\pm i}{}_j) = \begin{cases} \dfrac{q^{-\binom{n}{2}}}{[n-1]_q!} \varepsilon^{k_{n-1}\ldots k_1\, i}\, L^{\pm l_1}{}_{k_1} \ldots L^{\pm l_{n-1}}{}_{k_{n-1}}\, \varepsilon_{j\, l_{n-1}\ldots i_1} & \text{for } A_{n-1} \\ (C^{-1})^{ki}\, L^{\pm l}{}_k\, C_{jl} & \text{for } B_n,\, C_n,\, D_n \end{cases}$$

$$\tilde{S}(L^{\pm \bar{i}}{}_{\bar{j}}) = \begin{cases} \dfrac{q^{-\binom{n}{2}}}{[n-1]_q!} \varepsilon^{\bar{k}_{n-1}\ldots \bar{k}_1\, \bar{i}}\, L^{\pm \bar{l}_1}{}_{\bar{k}_1} \ldots L^{\pm \bar{l}_{n-1}}{}_{\bar{k}_{n-1}}\, \varepsilon_{\bar{j}\, \bar{l}_{n-1}\ldots \bar{i}_1} & \text{for } A_{n-1} \\ (C^{-1})^{\bar{k}\bar{i}}\, L^{\pm \bar{l}}{}_{\bar{k}}\, C_{\bar{j}\bar{l}} & \text{for } B_n,\, C_n,\, D_n. \end{cases} \tag{3.11}$$

Consequently the algebra $U_\mathcal{R}$ becomes a Hopf algebra with antipode $S := \tilde{S}_{|U_\mathcal{R}}$. And it holds

$$L^{\pm I}{}_J\, S(L^{\pm J}{}_K) = \delta^I_K\, e. \tag{3.12}$$

Proposition 6.

The involution on the generators of $U_\mathcal{R}$ is

$$\overline{L^{\pm J}{}_I} = L^{\pm \bar{I}}{}_{\bar{J}} \tag{3.13}$$

if

$$\overline{\alpha_0} \cdot \alpha_3 = 1 \tag{3.14}$$

and

$$\overline{\alpha_2} \cdot \alpha_1 = 1. \tag{3.15}$$

With the involution "†" $U_\mathcal{R}$ becomes a $*$–Hopf algebra. Nevertheless the coefficients α_i are not yet completely fixed. For further calculations we introduce the so called root–of–unity–homomorphisms $e_{r,s}$ which are elements of $\mathcal{A}^*$ and are defined multiplicatively on the generators of $\mathcal{A}$ as follows:

$$e_{r,s}(\mathbf{1}) := 1,$$
$$e_{r,s}(t^a{}_b) := e^{2\pi i \cdot r/\Theta} \cdot \delta^a_b,$$
$$e_{r,s}(\hat{t}^{\bar{a}}{}_{\bar{b}}) := e^{2\pi i \cdot s/\Theta} \cdot \delta^{\bar{a}}_{\bar{b}}, \tag{3.16}$$

where $r, s \in \mathbb{Z}$, $\Theta := \begin{cases} n & \text{for } A_{n-1} \\ 2 & \text{for } B_n, C_n, D_n \end{cases}$.

One can easily check the following

Proposition 7.

1. $e_{r,s}$ is a well defined algebra homomorphism,

2. $e_{0,0} = (e_{r,s})^{\Theta} = e$,

3. $e_{l,k} \cdot e_{m,n} = e_{l+m,k+n}$,

4. $[e_{r,s}, f] = 0 \quad \forall f \in \mathcal{A}^*$,

5. $\overline{e_{r,s}} = e_{s,r}$.

Using the special form of the $\hat{\mathcal{R}}_q$–matrix and the form of the matrices $\hat{R}_q$ for A_{n-1}, B_n, C_n or D_n, we get

Proposition 8.

1. $L^{+i}{}_j$ is upper–triangular,

   $L^{+\bar{i}}{}_{\bar{j}}$ is lower–triangular.

2. $L^{+i}{}_i \cdot L^{+\bar{i}}{}_{\bar{i}} = L^{+\bar{i}}{}_{\bar{i}} \cdot L^{+i}{}_i = e_{l,l}$

   where $\alpha_0 \cdot \alpha_2 = \alpha_1 \cdot \alpha_3 = e^{2\pi i \cdot l/\Theta}$.

3. $L^{-\bar{i}}{}_{\bar{j}} = L^{-i}{}_j \cdot e_{r,-r}$

   where $\alpha_0 \cdot \alpha_1^{-1} = (\alpha_2 \cdot \alpha_3^{-1})^* = e^{2\pi i \cdot r/\Theta}$.

4. $[L^{+i}{}_i, L^{+j}{}_j] = [L^{+\bar{i}}{}_{\bar{i}}, L^{+\bar{j}}{}_{\bar{j}}] = 0$.

5. $L^{+1}{}_1 \cdot \ldots \cdot L^{+N}{}_N = L^{+\bar{1}}{}_{\bar{1}} \cdot \ldots \cdot L^{+\bar{N}}{}_{\bar{N}} = e$

   for A_{n-1}, C_n, D_n and

   $(L^{+1}{}_1 \cdot \ldots \cdot L^{+N}{}_N)^2 = (L^{+\bar{1}}{}_{\bar{1}} \cdot \ldots \cdot L^{+\bar{N}}{}_{\bar{N}})^2 = e$

   for B_n.

4. The Hopf Algebra $U_q(sl(2, \mathbb{C}))$

To illustrate the above developed formalism we now investigate the easiest example, that is the Hopf algebra $U_q(sl(2, \mathbb{C}))$ with the additional choice $\alpha_0 = \alpha_1$. The other possibility,

$\alpha_0 = -\alpha_1$, would provide the additional algebra homomorphism $e_{1,1}$ in $U_{\mathcal{R}}$. We do not consider this case in this paper. As a consequence of these restrictions we get $\alpha_0 \cdot \alpha_2 = \alpha_0 \cdot (\alpha_1)^{-1} = 1$ and thus the equations in proposition 8 only contain $e_{0,0} = e$. Therefore in the case A_1 we only have to consider the unit e and the generators

$$L^{+1}{}_1, L^{+1}{}_2, L^{+2}{}_{\bar{1}}, L^{-1}{}_1, L^{-1}{}_2, L^{-2}{}_1, L^{-2}{}_2, (L^{+1}{}_1)^{-1}. \tag{4.1}$$

For further considerations we define the element

$$\Delta := L^{-1}{}_2 \cdot L^{-2}{}_1 \in U_{\mathcal{R}}. \tag{4.2}$$

Proposition 9.

1. $\{\Delta^n | n \in \mathbb{N}^0\}$ is a linearly independent set in $\mathcal{A}^*$.

2. $\Delta^n = 0$ for monomials $t^{g_1} \hat{t}^{g_2}$ with $\min(g_1, g_2) < n$.

Property 2 of proposition 9 allows us to handle power series in $\mathcal{A}^*$ of the form

$$\begin{aligned}
\Lambda^1_1 &= L^{-1}{}_1 (e + \sum_{n=1}^{\infty} \alpha_n \Delta^n), \\
\Lambda^2_2 &= L^{-2}{}_2 (e + \sum_{n=1}^{\infty} \beta_n \Delta^n)
\end{aligned} \tag{4.3}$$

where α_n, β_n are arbitrary complex numbers. Because of this fact, property 1 of proposition 9 and (3.12) we obtain

Proposition 10.

$L^{-1}{}_1$ is invertible and $(L^{-1}{}_1)^{-1} = L^{-2}{}_2 \left(e + \sum_{n=1}^{\infty} (-q)^{-n} \Delta^n\right)$ is an element of a certain minimal extension of $U_{\mathcal{R}}$.

Consequently there remain six essential generators since $L^{-2}{}_2$ can now be expressed through Δ and $(L^{-1}{}_1)^{-1}$. Using (3.5) and (3.12) we get the following algebra relations

$$[L^{-1}{}_2, L^{-2}{}_1] = 0,$$

$$[L^{-1}{}_1, L^{+1}{}_1] = 0,$$

$$[L^{+1}{}_2, L^{-1}{}_2] = [L^{-2}{}_1, L^{+\bar{2}}{}_{\bar{1}}] = 0,$$

$$[L^{-2}{}_1, L^{+1}{}_2] = (q - q^{-1})\left\{(L^{+1}{}_1)^{-1} L^{-1}{}_1 - (L^{-1}{}_1)^{-1}(e + q^{-1}\Delta) L^{+1}{}_1\right\},$$

$$[L^{-1}{}_2, L^{+\bar{2}}{}_{\bar{1}}] = (q - q^{-1})\left\{L^{-1}{}_1 L^{+1}{}_1 - (L^{+1}{}_1)^{-1}(L^{-1}{}_1)^{-1}(e + q^{-1}\Delta)\right\},$$

$$[L^{+1}{}_2, L^{+\bar{2}}{}_{\bar{1}}] = (q - q^{-1})\left\{(L^{+1}{}_1)^2 - (L^{+1}{}_1)^{-2}\right\},$$

$$L^{\pm1}{}_1 L^{\pm1}{}_2 = q^{-1} L^{\pm1}{}_2 L^{\pm1}{}_1,$$

$$L^{-1}{}_1 L^{-2}{}_1 = q^{-1} L^{-2}{}_1 L^{-1}{}_1,$$

$$L^{-1}{}_2 L^{+1}{}_1 = q L^{+1}{}_1 L^{-1}{}_2,$$

$$L^{-2}{}_1 L^{+1}{}_1 = q^{-1} L^{+1}{}_1 L^{-2}{}_1,$$

$$L^{-1}{}_1 L^{+1}{}_2 - q L^{+1}{}_2 L^{-1}{}_1 = (q^{-1} - q) L^{-1}{}_2 L^{+1}{}_1,$$

$$L^{-1}{}_1 L^{+\bar{2}}{}_{\bar{1}} - q L^{+\bar{2}}{}_{\bar{1}} L^{-1}{}_1 = (q^{-1} - q) L^{-2}{}_1 (L^{+1}{}_1)^{-1},$$

$$L^{+\bar{2}}{}_{\bar{1}} L^{+1}{}_1 = q^{-1} L^{+1}{}_1 L^{+\bar{2}}{}_{\bar{1}}.$$

(4.4)

In the next step we make an ansatz similar to [FRT] with H_i, $X_i^{\pm}$; $i = 1, 2$. We set

$$L^{+1}{}_1 = \exp(h/2\, H_1),$$

$$L^{-1}{}_1 = \exp(h/2\, H_2),$$

$$L^{+1}{}_2 = -(q - q^{-1}) X_1^-,$$

$$L^{+\bar{2}}{}_{\bar{1}} = (q - q^{-1}) X_1^+,$$

$$L^{-2}{}_1 = (q - q^{-1}) X_2^-,$$

$$L^{-1}{}_2 = -(q - q^{-1}) X_2^+$$

(4.5)

where $q = e^h$.

The equations (4.4) and (4.5) yield the following algebra relations

$$[H_1, H_2] = [X_1^{\pm}, X_2^{\mp}] = [X_2^+, X_2^-] = 0,$$

$$[H_1, X_1^{\pm}] = \pm 2 X_1^{\pm},$$

$$[H_1, X_2^{\pm}] = \mp 2 X_2^{\pm},$$

$$[H_2, X_2^{\pm}] = -2 X_2^{\pm},$$

$$[H_2, X_1^{\pm}] = 2 X_1^{\pm} - 4 X_2^{\mp} \exp(\mp h/2 (H_1 \pm H_2)),$$

$$[X_1^+, X_1^-] = \frac{\exp(h\, H_1) - \exp(-h\, H_1)}{(q - q^{-1})},$$

$$[X_1^{\pm}, X_2^{\pm}] = \frac{\exp(\pm h/2 (H_1 \pm H_2)) - \exp(\mp h/2 (H_1 \pm H_2))}{(q - q^{-1})} +$$

$$+ (1 - q^2) \exp(\mp h/2 (H_1 \pm H_2)) X_2^+ X_2^-.$$

(4.6)

Coming from H_i, $X_i^\pm$ one can argue that this algebra is a certain completion of U_R and a $*-$Hopf algebra with coproduct Δ

$$\Delta(X_1^\pm) = X_1^\pm \otimes \exp(-h/2\ H_1) + \exp(h/2\ H_1) \otimes X_1^\pm\,,$$
$$\Delta(X_2^+) = X_2^+ \otimes \exp(-h/2\ H_2)(e - q^{-2}\mathcal{D}) + \exp(h/2\ H_2) \otimes X_2^+\,,$$
$$\Delta(X_2^-) = X_2^- \otimes \exp(h/2\ H_2) + \exp(-h/2\ H_2)(e - q^{-2}\mathcal{D}) \otimes X_2^-\,,$$
$$\Delta(H_1) = H_1 \otimes e + e \otimes H_1\,,$$

(4.7)

$$\Delta(H_2) = \frac{2}{h}\sum_{k=1}^{\infty}\frac{1}{k}(-1)^{k-1}\Bigg(\exp\left(h/2(H_2 \otimes e + e \otimes H_2)\right) - (q - q^{-1})^2 X_2^+ \otimes X_2^- +$$
$$-\,e \otimes e\Bigg)^k\,,$$

antipode S

$$S(X_1^\pm) = -\exp(-h/2\ H_1)\,X_1^\pm\,\exp(h/2\ H_1)\,,$$
$$S(X_2^\pm) = -\exp(\mp h/2\ H_2)\,X_2^\pm\,\exp(\pm h/2\ H_2)\,,$$
$$S(H_1) = -H_1\,,$$

(4.8)

$$S(H_2) = \frac{2}{h}\sum_{k=1}^{\infty}\frac{1}{k}(-1)^{k-1}\left(\exp(-h/2\ H_2)(e - q^{-2}\mathcal{D}) - e\right)^k$$

and counit ϵ

$$\epsilon(H_1) = \epsilon(H_2) = 0\,,$$
$$\epsilon(X_1^\pm) = \epsilon(X_2^\pm) = 0$$

(4.9)

where $\mathcal{D} := q(q - q^{-1})^2\ X_2^+ X_2^-$.

As a formal power series in h the generators H_1 and H_2 are well defined and unique [Ogi].

After taking the $q \to 1$ limit for the algebra relations (4.6) one has to redefine the algebra in the following way in order to recover the usual $U(sl(2,\mathbb{C}))-$structure

$$\hat{H}_1 := 1/2\,(H_1 - H_2)\,,$$
$$\hat{H}_2 := 1/2\,(H_1 + H_2)\,,$$
$$\hat{X}_1^+ := X_2^-\,,$$
$$\hat{X}_1^- := (X_1^- - X_2^+)\,,$$
$$\hat{X}_2^+ := (X_1^+ - X_2^-)\,,$$
$$\hat{X}_2^- := X_2^+\,.$$

(4.10)

Finally we get the relations

$$[\hat{H}_i, \hat{X}_i^\pm] = \pm 2\,\hat{X}_i^\pm\,,$$
$$[\hat{X}_i^+, \hat{X}_i^-] = \hat{H}_i\,,$$
$$[\hat{H}_1, \hat{H}_2] = 0\,,$$
$$[\hat{H}_1, \hat{X}_2^\pm] = [\hat{H}_2, \hat{X}_1^\pm] = 0\,,$$
$$[\hat{X}_1^\pm, \hat{X}_2^\pm] = [\hat{X}_1^\pm, \hat{X}_2^\mp] = 0$$

(4.11)

and the involution

$$\hat{H}_1^\dagger = \hat{H}_2 \, ,$$
$$(\hat{X}_1^\pm)^\dagger = \hat{X}_2^\mp \, . \tag{4.12}$$

Considering comultiplication and antipode in this limit one recovers the universal enveloping algebra of $sl(2, \mathbb{C})$.

References

[Abe]   E. Abe, Hopf Algebras, Cambridge Tracts in Mathematics, vol. 74, Cambridge Univ. Press (1980).

[CSSW]   U. Carow-Watamura, M. Schlieker, M. Scholl and S. Watamura, *Z. Phys.* **C 48**, 159 (1990).
            U. Carow-Watamura, M.Schlieker, M. Scholl und S. Watamura, *Int. J. Mod. Phys.* **A**, Vol. 6, No. 17, 3081 (1991).

[CSWW] U. Carow-Watamura, M.Schlieker, S.Watamura and W.Weich, preprint KA-THEP-1990-26 (1990), to appear in *Comm. Math. Phys.*.

[CW]   U. Carow-Watamura and S. Watamura, Tohku preprint (1991).

[Dri]   V.G. Drinfel'd, Proceedings of the International Congress of Mathematicians, Berkeley, California, USA, 798 (1986).

[DSWZ]   B. Drabant, M. Schlieker, W. Weich and B. Zumino, in preparation.

[FRT]   L.D. Faddeev, N.Yu. Reshetikhin and L.A. Takhtajan, *Algebra and Analysis* **1**, 178 (1987).

[Jur]   B. Jurco, *Lett. Math. Phys.* **22**, 177 (1991).

[LNRT]   J. Lukierski, A. Nowicki, H. Ruegg, V. N. Tolstoy, Geneva preprints VGVA-DPT 1991/02-710, VGVA-DPT 1991/08-740 to appear in *Phys. Lett. B* (1991).

[Ogi]   O. Ogievetsky, private communication.

[OSWZ]   O. Ogievetsky, W.B. Schmidke, J. Wess and B. Zumino, MPI-Ph/91-51 (1991).

[Pod]   P. Podleś, preprint RIMS 754 (1991).

[PW]   P. Podleś and S.L. Woronowicz, *Comm. Math. Phys.* **130**, 381 (1990).

[SWZ]   W.B. Schmidke, J. Wess and B. Zumino, preprint MPI-Ph/91-15 (1991).

[Wor]   S.L. Woronowicz, *Comm. Math. Phys.* **122**, 125 (1989).

Extremal projector and universal R-matrix for quantized contragredient Lie (super)algebras

S.M. Khoroshkin

Institute of New Technologies, 11, Kirovogradskaya,
113587, Moscow, Russia

V.N. Tolstoy

Institute of Nuclear Physics, Moscow State University,
119899, Moscow, Russia

March 29, 1992

Abstract

Two basic elements of the representation theory of quantized finite-dimensional contragredient Lie (super)algebras g ($U_q(g)$) are presented. These are the universal R-matrix to be an interwining operator, and the extremal projector which gives a powerful method for decomposition of representations. Properties of Cartan-Weyl basis for $U_q(g)$ are discussed. Some Taylor extension of $U_q(g)$ and $U_q(g) \otimes U_q(g)$ are introduced in terms of this basis. The extremal projector p and the universal R-matrix R are described as unique elements of these extensions. Explicit formulae for p and R are given.

1 Introduction

Theory of quantum algebras and quantum groups is the most new and modern field of the mathematics and theoretical and mathematical physics. The quantized simple Lie algebras g (or the quantum universal enveloping algebras $U_q(g)$) [1,2] arose in the study of the algebraical aspects of quantum integrable systems [3-6]. The general definition of the quantized superalgebras was given in [7] (see also [8]). (Here we use the name "(super)algebras" in the following sense. It includes all semisimple finite-dimensional Lie algebras and all finite-dimensional contragredient Lie superalgebras).

As a rule, the definition of quantum (super)algebras is given in terms of Chevalley generators. However, the description of the quantum (super)algebras in terms of

R. Gielerak et al. (eds.), Groups and Related Topics, 23–32.
© 1992 *Kluwer Academic Publishers.*

the Cartan-Weyl generators is also useful. For this we have several reasons: *(i)* the quantum Cartan–Weyl basis generates a complete basis in $U_q(g)$; *(ii)* this basis plays just the same role in the representation theory as usual Cartan–Weyl basis in nonquantized case; *(iii)* the explicit expressions for the extremal projector [7] and the universal R-matrix [8-11] for the quantum (super)algebras have a compact form in this basis.

There are three different procedures for the construction of the quantum Cartan–Weyl basis. The first is founded on a direct quantization of the usual Cartan–Weyl basis [12]. The second is based on the notion of quantum Weyl group [10,11]. Another simple procedure was proposed in [7,8]. It is connected with normal orders in a positive root system for (super)algebra and uses a notion of a q-(super)commutator. It should be noted that the Cartan–Weyl basis introduced for the quantized $sl(n, \mathbb{C})$ in [9,13] is a special case of our basis.

One of basic elements of the theory of representations of quantum (super)-algebras (just the same as non-quantized (super)algebras) is a extremal projector. It gives powerful and universal method for a solution of many problems in the representation theory. For example, the method allows to describe reduced quantum (super)algebras [7], to classify irreducible modules, to decompose modules on submodules, to construct bases of modules and so on. Explicit form of the extremal projector (and some its applications in representation theory) for the quantum (super)algebras was given in [7]. Here we reproduce this result.

Any quantum (super)algebra is a non-cocommutative Hopf (super)algebra which has an intertwining operator called the universal R-matrix . Thus this element is a basic element for the theory representation also and moreover it is trigonometric solution of the Yang-Baxter equation (without a parameter). An implicit form of the universal R-matrix for quantized simple Lie algebras was given by V.G. Drinfeld in terms of a quantum double [1]. However explicit form of the universal R-matrix in terms of $U_q(g)$ is needed. Such a formula for $U_q(sl(2))$ was found in [1]. M. Rosso [9] obtained an analogous expression for $U_q(sl(n))$ by examining the identification of $U_q(sl(n))$ with quantum double of $U_q(b_+)$ where b_+ is a Borel subalgebra of $g = sl(n, \mathbb{C})$. This formula was generalized in [10,11] for all quantized simple Lie algebras g by using of q-Weyl group for $U_q(g)$. In a super case such formula is obtained in our paper [8] (see also [14] for affine case) without using of the quantum double and the quantum Weyl group. Here we reproduce this result shortly.

The paper is organized as follows. In Section 2 we give the definition of any quantized finite-dimensional contragredient Lie (super)algebra g (or a quantum (super)algebra $U_q(g)$) in terms of Chevalley generators and q-(super)commutator and also in terms of the adjoint action.

In Section 3 we present a procedure of the construction of the quantum Cartan–Weyl basis, proposed in [7,8], and give some properties this basis.

In Section 4 we introduce some extensions of the quantum (super)algebras $U_q(g)$ and $U_q(g) \otimes U_q(g)$ in terms of special formal Taylor series of the Cartan-Weyl generators. These extensions are associative algebras with respect to a multiplications of formal series. We construct also special representations for these algebras, which are irreducible and exact.

In Sections 5 and 6 we formulate uniqueness theorems for the extremal projector and the universal R-matrix and give their explicit forms in terms of the Cartan-Weyl generators.

2 Quantized Lie (super)algebras

Let $g(A, \tau)$ be a finite-dimensional contragredient Lie (super)algebra with symmetrizable Cartan matrix A ($A^{sym} = (a_{ij}^{sym})$ is a corresponding symmetrical matrix) and let $\Pi := \{\alpha_1, \ldots, \alpha_r\}$ be a system of simple roots for $g(A, \tau)$. The quantized (super)algebra $g := g(A, \tau)$ is an unital associative (super)algebra $U_q(g)$ with Chevalley generators $e_{\pm \alpha_i}$, $k_{\alpha_i}^{\pm 1} = q^{\pm h_{\alpha_i}}$, ($i \in I := \{1, 2, \ldots, r\}$), and the defining relations

$$[k_{\alpha_i}^{\pm 1}, k_{\alpha_j}^{\pm 1}] = 0, \qquad k_{\alpha_i} e_{\pm \alpha_j} = q^{\pm(\alpha_i, \alpha_j)} e_{\pm \alpha_j} k_{\alpha_i}, \tag{1}$$

$$[e_{\alpha_i}, e_{-\alpha_j}] = \delta_{ij} \frac{k_{\alpha_i} - k_{\alpha_i}^{-1}}{q - q^{-1}}, \tag{2}$$

$$(\overline{ad}_{q'} e_{\pm \alpha_i})^{n_{ij}} e_{\pm \alpha_j} = 0 \qquad \text{for } i \neq j, \ q' = q, q^{-1}, \tag{3}$$

$$\deg(k_{\alpha_i}) = \bar{0}, \qquad \deg(e_{\pm \alpha_i}) = \bar{1} \qquad \text{for } i \notin \tau,$$

$$\deg(e_{\pm \alpha_i}) = \bar{0} \qquad \text{for } i \in \tau \subset I, \tag{4}$$

where

$$n_{ij} = \begin{cases} 1 & \text{if } a_{ii}^{sym} = a_{ij}^{sym} = 0 \\ 2 & \text{if } a_{ii}^{sym} = 0, \ a_{ij}^{sym} \neq 0 \\ 1 - 2(a_{ij}^{sym}/a_{ii}^{sym}) & \text{if } a_{ii}^{sym} \neq 0. \end{cases} \tag{5}$$

Moreover, there are the following additional triple relation

$$[[e_{\pm \alpha_i}, e_{\pm \alpha_j}]_{q'}, [e_{\pm \alpha_i}, e_{\pm \alpha_k}]_{q'}]_{q'} = 0, \qquad q' = q, q^{-1}, \tag{6}$$

if the three simple roots $\alpha_i, \alpha_j, \alpha_k \in \Pi$ satisfy the condition

$$(\alpha_j, \alpha_j) = (\alpha_i, \alpha_k) = (\alpha_j, \alpha_i + \alpha_k) = 0. \tag{7}$$

Here the bracket $[\cdot, \cdot]$ is an usual supercommutator, $\overline{ad}_{q'}$ and $[\cdot, \cdot]_q$ denote a deformed supercommutator (q-supercommutator) in $U_q(g)$:

$$(\overline{ad}_{q'} e_\alpha) e_\beta \equiv [e_\alpha, e_\beta]_{q'} = e_\alpha e_\beta - (-1)^{\theta(e_\alpha)\theta(e_\beta)}(q')^{(\alpha, \beta)} e_\beta e_\alpha, \tag{8}$$

where (α, β) is a scalar product of the roots α and β: $(\alpha_i, \alpha_j) = a_{ij}^{sym}$. In the formula (8) and below we use the short notation

$$\theta(\gamma) := \theta(e_\gamma) \equiv \deg(e_\gamma). \tag{9}$$

Remarks. *(i)* The triple relations (6) may appear only in supercase for the following situation in the Dynkin diagram:

$$\begin{array}{ccc} \alpha_i & \alpha_j & \alpha_k \\[4pt] \bullet \!\!-\!\!-\!\!-\!\! & \otimes & \!\!-\!\!-\!\!-\!\! \bullet \end{array} \tag{10}$$

where α_j is a grey root and the roots α_i and α_k are not connected.

(ii) The outer q-supercommutator in these relations is really a usual one since $(\alpha_i + \alpha_j, \alpha_j + \alpha_k) = 0$.

(iii) The triple relations have evident classical counterpart.

The quantum (super)algebra $U_q(g)$ is a Hopf (super)algebra with respect to a comultiplication $\Delta_{q'}$, an antipode $S_{q'}$ and a counit ϵ defined as

$$\Delta_{q'}(k_{\alpha_i}) = k_{\alpha_i} \otimes k_{\alpha_i}, \qquad \Delta_{q'}(e_{\alpha_i}) = e_{\alpha_i} \otimes 1 + k'_{\alpha_i} \otimes e_{\alpha_i},$$

$$\Delta_{q'}(e_{-\alpha_i}) = e_{-\alpha_i} \otimes (k'_{\alpha_i})^{-1} + 1 \otimes e_{-\alpha_i}, \tag{11}$$

$$S_{q'}(k'_{\alpha_i}) = (k'_{\alpha_i})^{-1}, \quad S_{q'}(e_{\alpha_i}) = -(k'_{\alpha_i})^{-1} e_{\alpha_i}, \quad S_{q'}(e_{-\alpha_i}) = -e_{-\alpha_i} k'_{\alpha_i}, \tag{12}$$

$$\epsilon(k_{\alpha_i}) = \epsilon(e_{\alpha_i}) = \epsilon(e_{-\alpha_i}) = 0, \qquad \epsilon(1) = 1 \tag{13}$$

where $k'_\alpha = (q')^{h_\alpha}$ and q' may be chosen as $q' = q$ or $q' = q^{-1}$.

We may rewrite the defining relation by means of an adjoint action of $U_q(g)$ on itself. For this aim we introduce new Chevalley generators $\tilde{e}_{\pm\alpha_i}$ by the following formulas

$$\tilde{e}_{\alpha_i} = e_{\alpha_i}, \qquad \tilde{e}_{-\alpha_i} = (q')^{-1} e_{-\alpha_i} k'_{\alpha_i}. \tag{14}$$

In this basis the relations (2), (3), (6) take the following form

$$(ad_{q'}\tilde{e}_{\alpha_i})\tilde{e}_{-\alpha_j} = [\tilde{e}_{\alpha_i}, \tilde{e}_{-\alpha_j}]_{q'} = \delta_{ij} \frac{1 - (k'_{\alpha_i})^2}{1 - q'^2}, \tag{15}$$

$$(ad_{q'}\tilde{e}_{\pm\alpha_i})^{n_{ij}} \tilde{e}_{\pm\alpha_j} = 0, \qquad (i \neq j), \tag{16}$$

$$[(ad_{q'}\tilde{e}_{\pm\alpha_i})\tilde{e}_{\pm\alpha_j}, (ad_{q'}\tilde{e}_{\pm\alpha_i})\tilde{e}_{\pm\alpha_k}] = 0. \tag{17}$$

The last relation holds for the condition (10). Here $ad_{q'}$ is an adjoint action (see details in [15]) defined by

$$(ad_{q'} a)x := ((id \otimes S_{q'})\Delta_{q'}(a)) \circ X \tag{18}$$

for all homogeneous elements $a, x \in U_q(g)$, where the operation $\circ$ is defined by the rule

$$(a \otimes b) \circ x = (-1)^{\theta(b)\theta(x)} axb.$$

Below we denote by a symbol $(*)$ an antiinvolution in $U_q(g)$, defined as $(k_{\alpha_i})^* = k_{\alpha_i}^{-1}$, $(e_{\pm\alpha_i})^* = e_{\mp\alpha_i}$, $(q)^* = q^{-1}$. We also use the standard notations $U_q(K)$ and $U_q(b_\pm)$ for the Cartan and Borel subalgebras, generated by $k_{\alpha_i}^{\pm 1}$ and $e_{\pm\alpha_i}$, $k_{\alpha_i}^{\pm 1}$ correspondingly. We put also

$$exp_q(x) := 1 + x + \frac{x^2}{(2)_q!} + \ldots + \frac{x^n}{(n)_q!} + \ldots = \sum_{n \geq 0} \frac{x^n}{(n)_q!}, \tag{19}$$

$$(a)_q := \frac{q^a - 1}{q - 1}, \qquad [a]_q := \frac{q^a - q^{-a}}{q - q^{-1}}, \qquad q_\alpha := (-1)^{\theta(\alpha)} q^{-(\alpha,\alpha)}. \tag{20}$$

Now we proceed to a description of the Cartan-Weyl basis for the quantum (super)algebras $U_q(g)$.

3   Cartan-Weyl basis for $U_q(g)$

Let Δ_+ be the system of all positive roots with respect to Π. We denote by $\underline{\Delta}_+$ the *reduced root system* which is obtained from Δ_+ by removing such roots α for which $\alpha/2$ are roots.

A procedure of a construction of the quantum Cartan-Weyl basis for $U_q(g)$ has to be in agreement with a choice of normal ordering in $\underline{\Delta}_+$. We recall the definition of normal ordering in $\underline{\Delta}_+$ [16,17].

We say that the system $\underline{\Delta}_+$ is in normal ordering if each composite root $\gamma = \alpha + \beta \in \underline{\Delta}_+$ is written between its compon α and β. We say also that $\alpha < \beta$ if α is located on the left side of β in the normal ordering $\underline{\Delta}_+$.

The quantum Cartan-Weyl basis is being constructed by using the following inductive algorithm [7,8].

We fix some normal ordering in $\underline{\Delta}_+$ and put by induction

$$e_\gamma := [e_\alpha, e_\beta]_q, \qquad e_{-\gamma} := [e_{-\beta}, e_{-\alpha}]_{q^{-1}} \tag{21}$$

if $\gamma = \alpha + \beta$, $\alpha < \gamma < \beta$, and $[\alpha; \beta]$ is a minimal segment including γ. The quantum Cartan-Weyl generators are characterized by the following basic properties.

Proposition 3.1 *The root vectors* $e_{\pm\gamma} \in U_q(g)$, $\gamma \in \Delta_+$, *satisfy the following relations:*

$$(e_{\pm\gamma})^* = e_{\mp\gamma}, \qquad k_\alpha^{\pm 1} e_\gamma = q^{\pm(\alpha,\gamma)} e_\gamma k_\alpha^{\pm 1}, \tag{22}$$

$$[e_\gamma, e_{-\gamma}] = a(\gamma) \frac{k_\gamma - k_\gamma^{-1}}{q - q^{-1}}, \tag{23}$$

$$[e_\alpha, e_\beta]_q = \sum_{\alpha < \gamma_1 < \ldots < \gamma_n < \beta} C_{m_i, \gamma_i} e_{\gamma_1}^{m_1} e_{\gamma_2}^{m_2} \cdots e_{\gamma_n}^{m_n}, \tag{24}$$

where $\sum_i k_i \gamma_i = \alpha + \beta$, *and the coefficients* $C \ldots$ *are rational functions of* q *and ones do not depend on the Cartan elements* k_{α_i}, $i = 1, 2, \ldots n$, *and also*

$$[e_\beta, e_{-\alpha}] = \sum C'_{m_i, \gamma_i; m'_j, \gamma'_j} e_{-\gamma_1}^{m_1} e_{-\gamma_2}^{m_2} \cdots e_{-\gamma_p}^{m_p} e_{\gamma'_1}^{m'_1} e_{\gamma'_2}^{m'_2} \cdots e_{\gamma'_s}^{m'_s}. \tag{25}$$

where the sum is taken on $\gamma_1, \ldots, \gamma_p, \gamma_1', \ldots, \gamma_s'$ *and* $m_1, \ldots, m_p, m_1', \ldots, m_s'$ *such that*

$$\gamma_1 < \ldots < \gamma_p < \alpha < \beta < \gamma_1' < \ldots < \gamma_s', \quad \sum_l (m_l' \gamma_l' - m_l \gamma_l) = \beta - \alpha$$

and the coefficients $C' \ldots$ *are rational functions of* q *and* k_α *or* k_β. *The monomials* $e_{\gamma_1}^{n_1} e_{\gamma_2}^{n_2} \cdots e_{\gamma_p}^{n_p}$ *and* $e_{-\gamma_1}^{n_1} e_{-\gamma_2}^{n_2} \cdots e_{-\gamma_N}^{n_p}$, $(\gamma_1 < \gamma_2 < \cdots < \gamma_p)$, *generate (as a linear space over* $U_q(g)$) *subalgebras* $U_q(b_+)$ *and* $U_q(b_-)$ *correspondingly. The monomials*

$$e_{-\gamma_1}^{n_1} e_{-\gamma_2}^{n_2} \cdots e_{-\gamma_p}^{n_p} e_{\gamma_1'}^{n_1'} e_{\gamma_2'}^{n_2'} \cdots e_{\gamma_s'}^{n_s'},$$

where $\gamma_1 < \gamma_2 < \cdots < \gamma_p$ *and* $\gamma_1' < \gamma_2' < \cdots < \gamma_s')$, *generate* $U_q(g)$ *over* $U_q(g)$.

Now we consider some extensions of $U_q(g)$, $U_q(b_+) \otimes U_q(b_-)$ and $U_q(g) \otimes U_q(g)$ since the extremal projector and the universal R-matrix are elements of these extensions.

4   Taylor extensions of $U_q(g)$, $U_q(b_+) \otimes U_q(b_-)$ and $U_q(g) \otimes U_q(g)$

Let $\mathrm{Fract}\,(U_q(K))$ be a field of fractions over $U_q(K)$, i.e. $\mathrm{Fract}\,(U_q(K))$ is an associative algebra of rational functions of the elements $k_{\alpha_i}^{\pm 1}$, $(i = 1, 2, \ldots, r)$. We put

$$\tilde{U}_q(g) = \mathrm{Fract}\,(U_q(K)) \otimes_{U_q(K)} U_q(g). \tag{26}$$

Evidently, the extension $\tilde{U}_q(g)$ is an associative algebra. The algebra $\tilde{U}_q(g)$ is called the Cartan extension of the quantum algebra $U_q(g)$.

Let $\{e_\gamma, e_{-\gamma}\}$, $\gamma \in \underline{\Delta}_+$, be the root vectors of the quantum Cartan-Weyl basis built in accordance with some fixed normal ordering in $\underline{\Delta}_+$. Let us construct a formal Taylor series on the following monomials

$$e_{-\beta}^{n_\beta} \cdots e_{-\gamma}^{n_\gamma} e_{-\alpha}^{n_\alpha} e_\alpha^{m_\alpha} e_\gamma^{m_\gamma} \cdots e_\beta^{m_\beta} \tag{27}$$

with coefficients from $\mathrm{Fract}\,(U_q(K))$, where $\alpha < \gamma < \cdots < \beta$ in a sense of the fixed normal ordering in $\underline{\Delta}_+$ and nonnegative integers $n_\beta, \ldots, n_\alpha, m_\alpha, \ldots, m_\beta$ are subjected to the constraints

$$\left| \sum_{\alpha \in \underline{\Delta}_+} (n_\alpha - m_\alpha) c_i^{(\alpha)} \right| \leq \mathrm{const}, \qquad i = 1, 2, \cdots, r, \tag{28}$$

where $c_i^{(\alpha)}$ are coefficients in a decomposition of the root α with respect to the system Π of simple roots. Let $T_q(g)$ be a linear space of all such formal series. We have the following proposition [7].

Proposition 4.1 *The linear space $T_q(g)$ is an associative algebra with respect to a multiplication of formal series.*

The algebra $T_q(g)$ is called the Taylor extension of $U_q(g)$.

Let $M_q(g) := \tilde{U}_q(g)/\tilde{U}_q(g)n_+$ where n_+ is a linear envelope of the root vectors e_α, $\alpha \in \underline{\Delta}_+$. It is not difficult to see that the vector space $M_q(g)$ is invariant with respect to left-hand multiplication of elements $T_q(g)$ by elements of $M_q(g)$, i.e. we can interpret the elements as a linear operators in $M_q(g)$.

Theorem 4.1 *Left $T_q(g)$-module $M_q(g)$ is irreducible and exact.*

Let Fract $(U_q(K \otimes K))$ be a field of fractions generated by the following elements $1 \otimes k_{\alpha_i}$, $k_{\alpha_i} \otimes 1$ and $q^{h_{\alpha_i} \otimes h_{\alpha_j}}$, $(i,j = 1,2,\ldots,r)$. Let us consider a formal Taylor series of the following monomials

$$e_{-\beta}^{n_\beta} \cdots e_{-\gamma}^{n_\gamma} e_{-\alpha}^{n_\alpha} \otimes e_\alpha^{m_\alpha} e_\gamma^{m_\gamma} \cdots e_\beta^{m_\beta} \tag{29}$$

with coefficients from Fract $(U_q(K \otimes K))$, where $\alpha < \gamma < \cdots < \beta$ in a sense of the fixed normal ordering in $\underline{\Delta}_+$ and nonnegative integers $n_\beta,\ldots,n_\alpha,m_\alpha,\ldots,m_\beta$ are subjected to the constraint (28). Let $T_q(b_+ \otimes b_-)$ be a linear space of all such formal series. Th

Proposition 4.2 *The linear space $T_q(b_+ \otimes b_-)$ is an associative algebra with respect to a multiplication of formal series.*

The algebra $T_q(b_+ \otimes b_-)$ will be called the Taylor extension of $U_q(b_+) \otimes U_q(b_-)$. At least we consider a formal Taylor series of the following monomials

$$e_{-\beta}^{n_\beta} \cdots e_{-\gamma}^{n_\gamma} e_{-\alpha}^{n_\alpha} e_\alpha^{m_\alpha} e_\gamma^{m_\gamma} \cdots e_\beta^{m_\beta} \otimes e_{-\beta}^{n'_\beta} \cdots e_{-\gamma}^{n'_\gamma} e_{-\alpha}^{n'_\alpha} e_\alpha^{m'_\alpha} e_\gamma^{m'_\gamma} \cdots e_\beta^{m'_\beta} \tag{30}$$

with coefficients from Fract $(U_q(K \otimes K))$, where $\alpha < \gamma < \cdots < \beta$ in a sense of the fixed normal ordering in $\underline{\Delta}_+$ and nonnegative integers $n_\beta,\ldots,n_\alpha,m_\alpha,\ldots,m_\beta$ are subjected to the constraints

$$\left| \sum_{\alpha \in \Delta_+} (n_\alpha + n'_\alpha - m_\alpha - m'_\alpha)c_i^{(\alpha)} \right| \leq \text{const}, \qquad i = 1,2,\cdots,r. \tag{31}$$

Let $T_q(g \otimes g)$ be a linear space of all such formal series. The following proposition holds.

Proposition 4.3 *The linear space $T_q(g \otimes g)$ is an associative algebra with respect to a multiplication of formal series.*

The algebra $T_q(g \otimes g)$ will be called the Taylor extension of $U_q(g) \otimes U_q(g)$. Evidently the following embeddings hold

$$T_q(g \otimes g) \supset T_q(b_+ \otimes b_-),$$

$$T_q(g \otimes g) \supset T_q(g) \otimes T_q(g) \supset \Delta_{q'}(T_q(g)). \tag{32}$$

Let $\tilde{U}_q(g \otimes g)$ be a subalgebra of $T_q(g \otimes g)$ consisting from finite series and let $M_q(g \otimes g) := \tilde{U}_q(g \otimes g)/\tilde{U}_q(g \otimes g)(n_+ \otimes n_+)$. We can interpret elements of $T_q(g \otimes g)$ as linear left-hand operators in $M_q(g \otimes g)$.

Theorem 4.2 *Left $T_q(g \otimes g)$-module $M_q(g \otimes g)$ is irreducible and exact.*

5   Extremal projector for $U_q(g)$

By definition, the extremal projector for $U_q(g)$ is a nonzero element $p := p(g)$ satisfying the equations

$$e_{\alpha_i} p = p e_{-\alpha_i} = 0, \qquad \forall\, \alpha_i \in \Pi; \qquad p^2 = p. \tag{33}$$

Acting by the extremal projector p on any highest weight $U_q(g)$-module M we obtain a space $M^0 = pM$ of highest weight vectors for M (if pM has no singularities). An explicit multiplicative formula for p was given in [7].

Fix some normal ordering in Δ_+ and let e_α be the corresponding Cartan-Weyl generators. The following statement holds for any quantized finite-dimensional contragredient (super)algebra g [7].

Theorem 5.1 *The equations (33) have a unique nonzero solution in the space $T_q(q)$ and this solution has the form*

$$p = \overrightarrow{\prod_{\alpha \in \Delta_+}} p_\alpha, \tag{34}$$

where the order in the product coincides with the chosen normal ordering of Δ_+ and the elements p_α are defined by the formulae

$$p_\alpha = \sum_{m \geq 0} \varphi_{\alpha,m} e_{-\alpha}^m e_\alpha^m, \tag{35}$$

$$\varphi_{\alpha,m} = \frac{(-1)^m (q - q^{-1})^m \left(q^{\frac{1}{2}(\alpha,\alpha)} - (-1)^{\theta(\alpha)} q^{-\frac{1}{2}(\alpha,\alpha)} \right)^m}{(a(\alpha))^m q^{m(\rho - \frac{1}{2}\alpha,\alpha)} \prod_{l=1}^m (q^{\frac{l}{2}(\alpha,\alpha)} - (-1)^{m\theta(\alpha)} q^{-\frac{l}{2}(\alpha,\alpha)})} .$$

$$\left\{ \prod_{l=1}^m \left(k_\alpha q^{(\rho,\alpha) + \frac{l}{2}(\alpha,\alpha)} - (-1)^{(m-1)\theta(\alpha)} k_\alpha^{-1} q^{-(\rho,\alpha) - \frac{l}{2}(\alpha,\alpha)} \right) \right\}^{-1}. \tag{36}$$

Here ρ is a linear function such that $(\rho, \alpha_i) = \frac{1}{2}(\alpha_i, \alpha_i)$, $a(\alpha)$ is a factor from the relation (23).

6   Universal R-matrix for $U_q(g)$

By definition, the universal R-matrix for the Hopf (super)algebra $U_q(g)$ is an invertible element of some extension of $U_q(q) \otimes U_q(g)$, satisfying the equations

$$\Delta'(x) = R\Delta(x)R^{-1}, \qquad \forall\, x \in U_q(g), \tag{37}$$

$$(\Delta \otimes id)R = R^{13}R^{23}, \qquad (id \otimes \Delta)R = R^{13}R^{12}, \tag{38}$$

where $\Delta \equiv \Delta_{q^{-1}}$ and Δ' is an opposite comultiplication: $\Delta' = \sigma\Delta$, $\sigma(x \otimes y) = (-1)^{\deg x \deg y} y \otimes x$ for all homogeneous elements $x, y \in U_q(g)$. In (38) we use standard notation $A^{12} = \sum a_i \otimes b_i \otimes 1$, $A^{13} = \sum a_i \otimes 1 \otimes b_i$, $A^{23} = \sum 1 \otimes a_i \otimes b_i$ if A has a form $A = \sum a_i \otimes b_i$.

Fix some normal ordering in $\underline{\Delta}_+$ and let e_α be the corresponding Cartan-Weyl generators. The following statement holds for any quantized finite-dimensional contragredient (super)algebra g [8,19].

Theorem 6.1 *The equation (37) has a unique (up to a multiplicative constant) invertible solution in the space $T_q(b_+ \otimes b_-)$ and this solution (for a certain value of the constant) has the form*

$$R = (\prod_{\alpha \in \underline{\Delta}_+}^{\rightarrow} R_\alpha) \cdot K, \tag{39}$$

where the order in the product coincides with the chosen normal ordering of $\underline{\Delta}_+$ and the elements R_α and K are defined by the formulae

$$R_\alpha = \exp_{q_\alpha}\left((-1)^{\theta(\alpha)}(q - q^{-1})(a(\alpha))^{-1}(e_\alpha \otimes e_{-\alpha})\right) \tag{40}$$

$$K = q^{\sum_{i,j} d_{ij}(h_{\alpha_i} \otimes h_{\alpha_j})} \tag{41}$$

where $a(\alpha)$ is a factor from the relation (23) and d_{ij} is an inverse matrix for a symmetrical Cartan matrix (a_{ij}^{sym}) if (a_{ij}^{sym}) is not degenerated. (In a case of a degenerated (a_{ij}^{sym}) we extend it up to a non-degenerated matrix $(\tilde{a}_{ij}^{sym})$ and take an inverse to this extended matrix (see [8,14])). Moreover the solution (39) is the universal R-matrix, i.e. it satisfies the equations (38) also.

The uniqueness property of the universal R-matrix is proved in [19]. A derivation of the explicit formula (38) (with a detailed description of all rank two (super)algebras and the corresponding R-matrices) is given in [8,14].

References

[1] Drinfeld, V.G. Quantum groups. *Proc. ICM-86 (Berkeley USA) vol.1*, 798-820. Amer. Math. Soc. (1987).

[2] Jimbo, M. A q-difference analogue of $U(g)$ and the Yang-Baxter equation. *Lett. Math. Phys. 10* (1985), 63-69.

[3] Faddeev, L.D. Integrable models in $1 + 1$ dimensional quantum field theory (*Lectures in Les Houches 1982*). Elservier Science Publishers B.V. (1984).

[4] Kulish, P.P., and Sklyanin, E. *Lectures Notes in Physics 151* (1982), 61.

[5] Kulish, P.P., and Reshetikhin, N.Yu. The quantum linear problem for the sine-Gordon equation and higher representations. *Zap. Nauch. Sem. LOMI 101* (1981), 101-110.

[6] Sklyanin, E. Some algebraic structure connected with the Yang-Baxter equation. *Funk. Anal. Priloz. 16*, No.4 (1982), 27-34.

[7] Tolstoy V.N. Extremal projectors for quantized Kac-Moody superalgebras and some of their applications. The Proc. of the Quantum Groups Workshop. Clausthal, Germany (July 1989). *Lectures Notes in Physics 370* (1990), 118-125.

[8] Khoroshkin, S.M., and Tolstoy, V.N. Universal R-matrix for quantized (super)algebras. *Commun. Math. Phys. 141* (1991), 599-617.

[9] Rosso, M. An analogue of PBW theorem and the universal R-matrix for $U_h(sl(n+1))$ *Commun. Math. Phys. 124* (1989), 307-318.

[10] Kirillov, A.N., and Reshetikhin, N. Yu. Q-Weyl group and a multiplicative formula for universal R-matrices. *Comm. Math. Phys. 134* (1990), 421-431.

[11] Levendorskii, S.Z., and Soibelman, Ya.S. Some application of quantum Weyl groups. The multiplicative formula for universal R-matrix for simple Lie algebras. *Geom. and Phys 7:4* (1990), 1-14.

[12] Reshetikhin, N.Yu., Takhtajan, L.A., and Faddeev, L.D. Quantization of Lie groups and Lie algebras. *Algebra and Analisis 1* (1989), 178-206.

[13] Jimbo, M. *Lett. Math. Phys. 11* (1986), 247-252.

[14] Tolstoy, V.N., and Khoroshkin, S.M. The universal R-matrix for quantum nontwisted affine Lie algebras. *Funkz. Analiz i ego pril. 26;1* (1992), 85-88.

[15] Khoroshkin, S.M., and Tolstoy, V.N. The Cartan-Weyl basis and the universal R-matrix for quantum Kac-Moody algebras and superalgebras. Proc. of The Second Wigner Symposium. Goslar, Germany (July 1991) (to appear).

[16] Asherova, R.M., Smirnov, Yu.F., and Tolstoy, V.N. A description of some class of projection operators for semisimple complex Lie algebras. *Matem. Zametki 26* (1979), 15-25.

[17] Tolstoy, V.N. Extremal projectors for reductive classical Lie superalgebras with nondegenerated general Killing form. *Uspechi Math. Nauk 40* (1985), 225-226.

[18] Tolstoy, V.N. Extremal projectors for contragredient Lie algebras and superalgebras of finite growth. *Uspechi Math.Nauk 44* (1989), 211-212.

[19] Khoroshkin, S.M., and Tolstoy, V.N. Uniqueness theorem for universal R-matrix. *Lett. Math. Phys.* (1992) (to appear).

QUANTUM DEFORMATIONS OF $D = 4$ POINCARÉ ALGEBRA

Jerzy LUKIERSKI

Institute for Theoretical Physics, University of Wrocław,
ul. Cybulskiego 36, 50205 Wrocław, Poland

Anatol NOWICKI *[,†]

Laboratoire de Physique Théorique,[‡]
Université de Bordeaux I, rue du Solarium, F-33175 Gradignan, France

1. INTRODUCTION

Recently the formalism of quantum groups and quantum algebras has been applied to the description of quantum deformations of $D = 4$ Lorentz group and $D = 4$ Lorentz algebra [1-5]. In order to obtain the quantum deformation of semisimple Lie algebras describing Minkowski or Euclidean group of motions mostly the contraction techniques have been used. In particular there were obtained:

a) quantum deformation of $D = 2$ and $D = 3$ Euclidean and Minkowski geometries, described as quantum Lie algebra or quantum Lie group [6,7].

b) quantum deformation of $D = 4$ Poincaré algebra [8,9].

c) quantum deformations of $D = 4$ conformal algebra [10,11].

The main aim of this lecture is to present the results of our papers [8,11], where the quantum deformation of $D = 4$ Poincaré algebra is obtained by contracting the Cartan-Weyl basis of a real form $U_q(O(3,2))$ of the quantum complex $D = 4$ de-Sitter algebra $U_q(Sp(4;C))$, [8] as well as by contracting the Cartan-Weyl basis of a real form $U_q(O(4,2))$ of the quantum complex $D = 4$ conformal algebra $U_q(Sl(4;C))$ [11]. The contraction limit (denoted by $\left\{ {R \to \infty \atop q \to 1} \right\}$) which provides finite quantum inhomogeneous algebras is defined as follows:

$$\left\{ {R \to \infty \atop q \to 1} \right\}_l : \qquad R \to \infty; \qquad R^l \, log \, q \to i^\epsilon \kappa^{-l}$$

$$(0 < \kappa < \infty) \tag{1.1}$$

where $l = 1$ for the contraction of $U_q(O(3,2))$ [8] and $l = 2$ for the contraction of

---

*Presented by A. N. at the First Max Born Symposium

†On leave of absence from Institute of Physics, Pedagogical University, Plac Słowiański 6, 65029 Zielona Góra, Poland

‡Unité associée au CNRS URA 764

R. Gielerak et al. (eds.), Groups and Related Topics, 33–44.
© *1992 Kluwer Academic Publishers.*

$U_q(Sl(4;C))$; $\epsilon = 0$ for q real, and $\epsilon = 1$ for $|q| = 1$[1]. We see that the prescription (1.1) provides a particular way of approaching the limit $q \to 1$ in such a way that appears a new mass- like parameter κ. In this lecture we shall show that the limit (1.1) provides for finite κ a nonlinear modification of the Poincaré algebra("κ - deformation") accompanied by the consistent modification of coproduct and antipode formulae.

It should be stressed that for the quantum algebras of rank ≥ 2 physically more important seems to be the Cartan-Weyl basis, which has the following two advantages over the Cartan-Chevalley basis

i) The q-deformed Cartan-Weyl basis describes the quantum deformation of Lie algebra generators (i.e. in the limit $q \to 1$ one obtains the ordinary Lie algebra relations)

ii) The nonlinear (three-linear or fourlinear for classical Lie algebras) Serre relations are replaced by bilinear relations.

2. QUANTUM COMPLEX $D = 4$ DE SITTER ALGEBRA $U_q(Sp(4))$ AND ITS REAL FORMS

The Drinfeld-Jimbo procedure, valid for simple Lie algebras, yields the following q-deformation $U_q(Sp(4))$ of $Sp(4)$ in the Cartan-Chevalley basis ($[x] \equiv (q - q^{-1})^{-1} \cdot (q^x - q^{-x})$):

$$[h_i, h_j] = 0$$
$$[h_i, e_{\pm j}] = \pm a_{ij} e_{\pm j} \qquad\qquad a_{ij} = \begin{pmatrix} 1 & -1 \\ -1 & 2 \end{pmatrix} \qquad\qquad (2.1)$$
$$[e_i, e_{-j}] = \delta_{ij}[h_i]_q$$

where h_i and $e_{\pm\alpha_i} \equiv e_{\pm i}$ ($i = 1, 2$) describe the Cartan subalgebra and the generators corresponding to simple roots, respectively. a_{ij} denotes the symmetrized Cartan matrix given by the scalar product of the simpe roots $a_{ij} = < \alpha_i, \alpha_j >$.

The q-Serre relations take the form

$$(ad_{q'} e_{\pm i})^{\eta_{ij}} e_{\pm j} = 0 \qquad\qquad \text{for} \qquad i \neq j; \quad q' = q, q^{-1}; \qquad \eta_{ij} = 1 - a_{ij} \quad (2.2)$$

where $(ad_{q'} e_\alpha) e_\beta \equiv [e_\alpha, e_\beta]_{q'} \equiv e_\alpha e_\beta - q'^{<\alpha,\beta>} e_\beta e_\alpha$.

In order to get a quantum Lie algebra one has to specify the coproducts

$$\Delta(h_i) = h_i \otimes 1 + 1 \otimes h_i \qquad\qquad \Delta(e_{\pm i}) = e_{\pm i} \otimes q^{\frac{h_i}{2}} + q^{-\frac{h_i}{2}} \otimes e_{\pm i} \qquad (2.3)$$

and antipodes

$$S(h_i) = -h_i \qquad\qquad S(e_{\pm i}) = -q^{\pm\frac{1}{2}d_i} e_{\pm i} \qquad\qquad (2.4)$$

where $d_i = < \alpha_i, \alpha_i >= (1, 2)$.

We now shall complete the Cartan-Chevalley basis by introducing the generators corresponding to the nonsimple roots $\alpha_3 = \alpha_1 + \alpha_2$ and $\alpha_4 = 2\alpha_1 + \alpha_2$. For the normal

---

[1]For $\epsilon = 0$ such limit was firstly introduced by Firenze group [6,7] for studying the contractions of rank one quantum algebras ($U_q(Sl(2;C))$) and its real forms $U_q(SU(2))$ and $U_q(SU(1,1))$.

We now shall complete the Cartan-Chevalley basis by introducing the generators corresponding to the nonsimple roots $\alpha_3 = \alpha_1 + \alpha_2$ and $\alpha_4 = 2\alpha_1 + \alpha_2$. For the normal order [12,13], we get (defining now $[A, B]_q \equiv AB - qBA$)

$$e_3 = [e_1, e_2]_q \qquad\qquad e_{-3} = [e_{-2}, e_{-1}]_{q^{-1}}$$
$$e_4 = [e_1, e_3]_q \qquad\qquad e_{-4} = [e_{-3}, e_{-1}]_{q^{-1}} \qquad\qquad (2.5)$$

With these definitions, the q-Serre relations (2.2) take the linear form reminiscent of the usual Lie-algebra relations

$$[e_1, e_4]_{q^{-1}} = 0, \qquad [e_{-3}, e_{-2}]_q = 0, \qquad [e_3, e_2]_{q^{-1}} = 0$$
$$[e_4, e_3]_{q^{-1}} = 0, \qquad [e_{-4}, e_{-1}]_q = 0, \qquad [e_{-3}, e_{-4}]_q = 0 \qquad\qquad (2.6)$$

(2.1), (2.5) and (2.6) define the Cartan-Weyl basis.

The complete set of commutation relations for $U_q(Sp(4))$ [8] is given by

$$[e_2, e_4] = (1 - q^{-1})e_3^2 \qquad\qquad [e_1, e_{-3}] = -q^{-h_1}e_{-2}$$
$$[e_3, e_{-2}] = q^{-h_2}e_1 \qquad\qquad [e_4, e_{-3}] = q^{-h_3}e_1$$
$$[e_1, e_{-4}] = -e_{-3}q^{-h_1} \qquad\qquad [e_4, e_{-2}] = -(1 - q^{-1})q^{-h_2}e_1^2 \qquad\qquad (2.7)$$
$$[e_3, e_{-3}] = [h_3]_q \qquad\qquad h_3 = h_1 + h_2$$
$$[e_4, e_{-4}] = [h_4]_q \qquad\qquad h_4 = h_1 + h_3$$

The complete Cartan-Weyl basis of $U_q(Sp(4; C))$ is obtained if we add to the relations (2.7) the conjugated ones $(h_i \to h_i, e_{\pm i} \to e_{\mp i}, q \to q^{-1})$.

The coproduct and antipode are calculated using the fact that Δ is an automorphism of the algebra and S an antiautomorphism:

$$\Delta(e_3) = e_3 \otimes q^{\frac{1}{2}h_3} + q^{-\frac{1}{2}h_3} \otimes e_3 + (q^{-1} - q)q^{-\frac{1}{2}h_2}e_1 \otimes e_2 q^{\frac{1}{2}h_1}$$

$$\Delta(e_{-3}) = e_{-3} \otimes q^{\frac{1}{2}h_3} + q^{-\frac{1}{2}h_3} \otimes e_{-3} + (q - q^{-1})q^{-\frac{1}{2}h_1}e_{-2} \otimes e_{-1} q^{\frac{1}{2}h_2}$$

$$\Delta(e_4) = e_4 \otimes q^{\frac{1}{2}h_4} + q^{-\frac{1}{2}h_4} \otimes e_4$$
$$\qquad +(q - q^{-1})\left\{(1 - q^{-1})q^{-\frac{1}{2}h_2}e_1^2 \otimes e_2 q^{h_1} - q^{-\frac{1}{2}h_3}e_1 \otimes e_3 q^{\frac{1}{2}h_1}\right\}$$

$$\Delta(e_{-4}) = e_{-4} \otimes q^{\frac{1}{2}h_4} + q^{-\frac{1}{2}h_4} \otimes e_{-4}$$
$$\qquad +(q - q^{-1})\left\{(q - 1)q^{-h_1}e_{-2} \otimes e_{-1}^2 q^{\frac{1}{2}h_2} + q^{-\frac{1}{2}h_1}e_{-3} \otimes e_{-1} q^{\frac{1}{2}h_3}\right\} \qquad (2.8a)$$

$$S(e_3) = -q^{\frac{1}{2}}e_3 + q^{\frac{1}{2}}(1 - q^2)e_1 e_2$$
$$S(e_{-3}) = -q^{-\frac{1}{2}}e_{-3} + q^{-\frac{1}{2}}(1 - q^{-2})e_{-2}e_{-1}$$
$$S(e_4) = -qe_4 + (1 - q^2)\left\{(q - 1)e_1^2 e_2 + e_1 e_3\right\}$$
$$S(e_{-4}) = -q^{-1}e_{-4} + (1 - q^{-2})\left\{(q^{-1} - 1)e_{-2}e_{-1}^2 + e_{-3}e_{-1}\right\} \qquad (2.8b)$$

Our aim is to define the "physical" basis for $U_q(Sp(4))$ consisting of the 10 generators

will depend on the choice of real forms, which we will now discuss.

For a Hopf algebra $\mathring{A}$ over C with comultiplication Δ and antipode S one can distinguish four types of involutive homomorphisms in $\mathring{A}$:

i) The $+$ involution, which is an antiautomorphism in the algebra sector, and an automorphism in the coalgebra sector, i.e. $(\alpha_i \in \mathring{A})$:

$$(\alpha_1 \cdot \alpha_2)^+ = \alpha_2^+ \alpha_1^+ \qquad (\Delta(\alpha))^+ = \Delta(\alpha^+) \tag{2.9}$$

ii) the $*$ involution, which is an automorphism in the algebra sector, and an antiautomorphism in the coalgebra sector, i.e.

$$(\alpha_1 \cdot \alpha_2)^* = \alpha_1^* \alpha_2^* \qquad (\Delta(\alpha))^* = \Delta'(\alpha^*) \tag{2.10}$$

where $\Delta' = P\Delta = R\Delta R^{-1}$ (P = permutation operator in $(\mathring{A} \otimes \mathring{A})$).
These two involutions are required to satisfy ($\times \equiv +$ or $*$)

$$S((S(a^\times))^\times) = a \Leftrightarrow S \circ \times = \times \circ S^{-1} \tag{2.11}$$

iii) The $\oplus$ involution, which is an antiautomorphism in both algebra and coalgebra sectors, i.e.

$$(\alpha_1 \cdot \alpha_2)^\oplus = (\alpha_2)^\oplus \cdot (\alpha_1)^\oplus \qquad (\Delta(\alpha))^\oplus = \Delta'(a^\oplus) \tag{2.12}$$

iv) The $\circledast$ involution, which is an automorphism in both algebra and coalgebra sectors, i.e.

$$(\alpha_1 \cdot \alpha_2)^\circledast = (\alpha_1)^\circledast \cdot (\alpha_2)^\circledast \qquad (\Delta(\alpha))^\circledast = \Delta(\alpha^\circledast) \tag{2.13}$$

These two involutions are assumed to satisfy ($\otimes \equiv \oplus$ or $\circledast$)

$$S^{-1}((S(\alpha^\otimes))^\otimes = \alpha \Leftrightarrow S \circ \otimes = \otimes \circ S \tag{2.14}$$

These four types of involutions are implemented by their action on the Cartan-Weyl basis of $U_q(Sp(4))$ in the following way (for details see [9])

$$h_i' = \rho h_i \qquad\qquad e_{\pm 1}' = \lambda e_{m1}$$

$$e_{\pm 2}' = \epsilon e_{m2} \qquad\qquad e_{\pm 3}' = \rho \lambda \epsilon q^{\mp n} e_{m3} \qquad\qquad e_{\pm 4}' = \epsilon q^{\mp n} e_{m4} \tag{2.15}$$

The various parameters in (2.15) take the following values:

$+$	$i)$	$\|q\| = 1$	$\rho = -1$	$m = \pm 1$	$n = 1$	$(\Delta_\pm \to \Delta_\pm)$
$*$	$ii)$	$q \in \mathbb{R}$	$\rho = -1$	$m = \mp 1$	$n = -1$	$(\Delta_\pm \to \Delta_\mp)$
$\oplus$	$iii)$	$\|q\| = 1$	$\rho = 1$	$m = \mp 1$	$n = 0$	$(\Delta_\pm \to \Delta_\mp)$
$\circledast$	$iv)$	$q \in \mathbb{R}$	$\rho = 1$	$m = \pm 1$	$n = 0$	$(\Delta_\pm \to \Delta_\pm)$

$$\tag{2.16}$$

Further we shall choose $\oplus$-involution, corresponding to $\rho = \lambda = 1$ and $\epsilon = -1$. From (2.15-16) follows that

$$h_1^\oplus = h_1 \qquad\qquad e_1^\oplus = e_{-1} \qquad\qquad e_2^\oplus = -e_{-2}$$

$$h_2^\oplus = h_2 \qquad\qquad e_3^\oplus = -e_{-3} \qquad\qquad e_4^\oplus = -e_{-4} \tag{2.17}$$

and the $O(3,2)$ generators $(A, B = 0, 1, 2, 3, 4)$

$$M_{12} = h_1 \qquad\qquad M_{23} = \tfrac{1}{\sqrt{2}}(e_1 + e_{-1})$$

$$M_{31} = \tfrac{1}{i\sqrt{2}}(e_1 - e_{-1}) \qquad\qquad M_{04} = h_3 = h_1 + h_2$$

$$M_{34} = \tfrac{1}{\sqrt{2}}(e_3 - e_{-3}) \qquad\qquad M_{03} = \tfrac{1}{i\sqrt{2}}(e_3 + e_{-3}) \qquad (2.18)$$

$$M_{02} = \tfrac{1}{2}(e_4 - e_{-4} + e_2 - e_{-2}) \qquad\qquad M_{01} = -\tfrac{1}{2i}(e_4 + e_{-4} - e_2 - e_{-2})$$

$$M_{24} = \tfrac{1}{2i}(e_4 + e_{-4} + e_2 + e_{-2}) \qquad\qquad M_{14} = \tfrac{1}{2}(e_4 - e_{-4} - e_2 + e_{-2})$$

satisfy the relation $M_{AB}^{\oplus} = M_{AB}$ and describe the $D = 4$ anti-de-Sitter algebra

$$[M_{AB}, M_{CD}] = i\,(\eta_{BC} M_{AD} - \eta_{AC} M_{BD} + \eta_{AD} M_{BC} - \eta_{BD} M_{AC}) \qquad (2.19)$$

where $\eta_{AB} = diag(+, -, -, -, +)$. The formulae describing Cartan-Weyl basis of $U_q(O(3,2))$ in terms of physical generators (2.18) can be found in [8]. We recall that for the real form (2.16) we should assume that $|q| = 1$.

3. CONTRACTION OF $U_q(O(3,2))$ TO κ-DEFORMED POINCARÉ ALGEBRA

The contraction scheme leading from the real form of $U_q(Sp(4))$ to a quantum Poincaré algebra with corresponding Lorentz metric $g_{\mu\nu}$ ($\mu, \nu = 1, 2, 3, 4$) is obtained by contraction (1.1) with $l = 1$. Let us leave the Lorentz generators $M_{\mu\nu}$ unchanged and define the translation generators P_μ by

$$M_{0\mu} = R P_\mu \qquad (3.1)$$

Using the choice (2.17) of the $O(3,2)$ generators and the Cartan-Weyl basis of $U_q(O(3,2))$ one gets the following κ-deformed Poincaré algebra:

a) Lorentz sector ($M_\pm = M_1 \pm i M_2 \equiv M_{23} \pm i M_{31}$; $M_3 = M_{12}$; $L_\pm = M_{14} \pm i M_{24}$; $L_3 = M_{34}$);

$$[M_+, M_-] = 2M_3 \qquad\qquad [M_3, M_\pm] = \pm M_\pm \qquad (3.2)$$

$$[L_-, L_+] = 2M_3 \cos\tfrac{P_0}{\kappa} - \tfrac{1}{\kappa}\{P_3, L_3\} + \tfrac{1}{2\kappa^2} P_3^2$$

$$[L_3, L_\pm] = \mp e^{\mp i P_0/\kappa} M_\pm \pm \tfrac{1}{2\kappa} L_\pm P_3 + \tfrac{i}{2\kappa} L_3 P_\mp \qquad (3.3)$$

$$[M_3, L_3] = 0 \qquad\qquad [M_+, L_3] = -L_+ - \tfrac{i}{2\kappa} M_3 P_-$$

$$[M_3, L_\pm] = \pm L_\pm \qquad\qquad [M_-, L_3] = L_- - \tfrac{i}{2\kappa} P_+ M_3$$

$$[M_\pm, L_\pm] = \tfrac{1}{2i\kappa} M_\pm P_\mp \qquad (3.4)$$

$$[M_+, L_-] = 2L_3 + \tfrac{i}{2i\kappa} P_+ M_+ + \tfrac{1}{\kappa} P_3 M_3$$

$$[M_-, L_+] = -2L_2 + \tfrac{1}{2i\kappa} M_- P_- - \tfrac{1}{\kappa} P_3 M_3$$

b) Translation sector $(P_\pm = P_2 \pm iP_1)$

$$[P_\mu, P_\nu] = 0 \tag{3.5}$$

$$[M_i, P_0] = 0 \qquad [M_i, P_k] = i\epsilon_{ikl}P_l \tag{3.6}$$

$$[L_3, P_0] = iP_3 \qquad\qquad\qquad [L_3, P_3] = i\kappa \sin \tfrac{P_0}{\kappa}$$

$$[L_3, P_2] = \tfrac{1}{2i\kappa}P_1 P_3 \qquad\qquad [L_3, P_1] = -\tfrac{1}{2i\kappa}P_2 P_3$$

$$[L_\pm, P_0] = iP_1 \mp P_2 \qquad\qquad [L_\pm, P_3] = \tfrac{1}{2i\kappa}P_\mp P_3$$

$$[L_\pm, P_2] = \mp\kappa \sin \tfrac{P_0}{\kappa} - \tfrac{1}{2i\kappa}P_3^2 \qquad [L_\pm, P_1] = i\kappa \sin \tfrac{P_0}{\kappa} \pm \tfrac{1}{2\kappa}P_3^2 \tag{3.7}$$

The coproduct is

$$\Delta M_i = M_i \otimes I + I \otimes M_i \qquad\qquad i = 1, 2, 3$$

$$\Delta P_0 = P_0 \otimes I + I \otimes P_0 \qquad\quad \Delta P_i = P_i \otimes e^{\frac{iP_0}{2\kappa}} + e^{\frac{-iP_0}{2\kappa}} \otimes P_i$$

$$\Delta L_3 = L_3 \otimes e^{\frac{iP_0}{2\kappa}} + e^{\frac{-iP_0}{2\kappa}} \otimes L_3 + \tfrac{i}{2\kappa}e^{\frac{-iP_0}{2\kappa}}M_+ \otimes P_+ - \tfrac{i}{2\kappa}P_- \otimes M_- e^{\frac{iP_0}{2\kappa}}$$

$$\Delta L_+ = L_+ \otimes e^{\frac{iP_0}{2\kappa}} + e^{\frac{-iP_0}{2\kappa}} \otimes L_+ + \tfrac{i}{2\kappa}P_- \otimes M_3 e^{\frac{iP_0}{2\kappa}}$$
$$\qquad - \tfrac{i}{2\kappa}e^{\frac{-iP_0}{2\kappa}}M_3 \otimes P_- + \tfrac{1}{\kappa}e^{\frac{-iP_0}{2\kappa}}M_+ \otimes P_3$$

$$\Delta L_- = L_- \otimes e^{\frac{iP_0}{2\kappa}} + e^{\frac{-iP_0}{2\kappa}} \otimes L_- + \tfrac{i}{2\kappa}P_+ \otimes M_3 e^{\frac{iP_0}{2\kappa}}$$
$$\qquad - \tfrac{i}{2\kappa}e^{\frac{-iP_0}{2\kappa}}M_3 \otimes P_+ + \tfrac{1}{\kappa}P_3 \otimes M_- e^{\frac{iP_0}{2\kappa}} \tag{3.8}$$

The antipode is

$$S(P_\mu) = -P_\mu \qquad\qquad S(L_3) = -L_3 - \tfrac{1}{2\kappa}P_3 + \tfrac{i}{2\kappa}(M_+ P_+ - P_- M_-)$$

$$S(M_i) = -M_i \qquad\qquad S(L_+) = -L_+ - \tfrac{i}{\kappa}P_- + \tfrac{1}{\kappa}M_+ P_3$$

$$S(L_-) = -L_- + \tfrac{i}{\kappa}P_+ + \tfrac{1}{\kappa}P_3 M_- \tag{3.9}$$

4. QUANTUM COMPLEX $D = 4$ CONFORMAL ALGEBRA $U_q(SL(4; C))$

In order to describe the q-deformation of real $D = 4$ conformal algebra $SU(2,2) \simeq O(4,2)$ we introduce firstly the Cartan-Chevaley basis for $U_q(SL(4; C))$ describing quantum complexified $D = 4$ conformal algebra $(i, j = 1, 2, 3)$:

$$[h_i, h_j] = 0$$

$$[h_i, e_{\pm j}] = \pm a_{ij} e_{\pm j} \qquad a_{ij} = \begin{pmatrix} 2 & -1 & 0 \\ -1 & 2 & -1 \\ 0 & -1 & 2 \end{pmatrix} \qquad (4.1)$$

$$[e_i, e_{-j}] = \delta_{ij} [h_i]_q$$

The generators corresponding to nonsimple roots are defined as follows (for the general scheme see [12-13]):

$$\begin{aligned}
e_4 &= [e_1, e_2]_q & e_{-4} &= [e_{-2}, e_{-1}]_{q^{-1}} \\
e_5 &= [e_2, e_3]_q & e_{-5} &= [e_{-3}, e_{-2}]_{q^{-1}} \\
e_6 &= [e_1, e_5]_q & e_{-6} &= [e_{-5}, e_{-1}]_{q^{-1}}
\end{aligned} \qquad (4.2)$$

The relations (4.1) are extended to the generators (4.2) in the following way:

$$\begin{aligned}
[e_4, e_{-4}] &= [h_1 + h_2]_q \equiv [h_4]_q \\
[e_5, e_{-5}] &= [h_2 + h_3]_q \equiv [h_5]_q \\
[e_6, e_{-6}] &= [h_1 + h_2 + h_3]_q \equiv [h_6]_q
\end{aligned} \qquad (4.3)$$

and $h_4 = h_1 + h_2$, $h_5 = h_2 + h_3$, $h_6 = h_1 + h_2 + h_3$, as well as $(\alpha = 4, 5, 6)$

$$[h_i, e_{\pm \alpha}] = \pm a_{i\alpha} e_{\pm \alpha} \qquad (4.4a)$$

where

$$a_{i\alpha} = \begin{pmatrix} 1 & -1 & 1 \\ 1 & 1 & 0 \\ -1 & 1 & 1 \end{pmatrix} \qquad (4.4b)$$

The q-Serre relations produce the following collection of bilinear formulae

$$\begin{aligned}
[e_1, e_2]_q &= e_4 & [e_1, e_3] &= 0 & [e_2, e_4]_q &= 0 \\
[e_1, e_5]_q &= e_6 & [e_2, e_6] &= 0 & [e_1, e_4]_{q^{-1}} &= 0 \\
[e_2, e_3]_q &= e_5 & [e_3, e_6] &= 0 & [e_1, e_6]_{q^{-1}} &= 0 \\
[e_4, e_3]_q &= e_6 & [e_3, e_5]_q &= 0 & [e_2, e_5]_{q^{-1}} &= 0
\end{aligned} \qquad (4.5)$$

which can be supplemented by

$$[e_5, e_4] = (q - q^{-1}) e_6 e_2 \qquad [e_4, e_6]_{q^{-1}} = 0 \qquad [e_5, e_6]_q = 0 \qquad (4.6)$$

Further we obtain

$$\begin{aligned}
[e_1, e_{-5}] &= 0 & [e_2, e_{-6}] &= 0 & [e_4, e_{-3}] &= 0 \\
[e_2, e_{-4}] &= e_{-1} q^{h_2} & [e_5, e_{-2}] &= -e_3 q^{h_2} & [e_4, e_{-5}] &= (q - q^{-1}) q^{-h_2} e_{-3} e_1 \quad (4.7) \\
[e_3, e_{-5}] &= e_{-2} q^{h_3} & [e_6, e_{-1}] &= -e_5 q^{h_1} & [e_5, e_{-6}] &= e_{-1} q^{h_2 + h_3} \\
[e_4, e_{-1}] &= -e_2 q^{h_1} & [e_3, e_{-6}] &= e_{-4} q^{h_3} & [e_6, e_{-4}] &= -e_3 q^{h_1 + h_2}
\end{aligned}$$

If we add to the relations (4.5-4.7) the conjugated ones we obtain the q-deformation of the complete Cartan-Weyl basis of $U_q(SL(4; C))$.

In order to describe $U_q(SL(4;C))$ as quantum bialgebra we introduce the formulae for coproduct $(i = 1, 2, 3)$

$$\Delta(e_{\pm i}) = e_{\pm i} \otimes k_i + k_i^{-1} \otimes e_{\pm i} \qquad \Delta(k_i^{\pm 1}) = k_i^{\pm 1} \otimes k_i^{\pm 1} \qquad (4.8a)$$

and further one gets

$$\Delta(e_4) = e_4 \otimes k_4 + k_4^{-1} \otimes e_4 + (q^{-1} - q)k_2^{-1}e_1 \otimes e_2 k_1$$
$$\Delta(e_{-4}) = e_{-4} \otimes k_4 + k_4^{-1} \otimes e_{-4} + (q - q^{-1})k_1^{-1}e_{-2} \otimes e_{-1}k_2$$
$$\Delta(e_5) = e_5 \otimes k_5 + k_5^{-1} \otimes e_5 + (q^{-1} - q)k_3^{-1}e_2 \otimes e_3 k_2$$
$$\Delta(e_{-5}) = e_{-5} \otimes k_5 + k_5^{-1} \otimes e_{-5} + (q - q^{-1})k_2^{-1}e_{-3} \otimes e_{-2}k_3 \qquad (4.8b)$$
$$\Delta(e_6) = e_6 \otimes k_6 + k_6^{-1} \otimes e_6 + (q^{-1} - q) \cdot \left\{ k_5^{-1}e_1 \otimes e_5 k_1 + k_3^{-1}e_4 \otimes e_3 k_4 \right\}$$
$$\Delta(e_{-6}) = e_{-6} \otimes k_6 + k_6^{-1} \otimes e_{-6} + (q - q^{-1}) \cdot \left\{ k_1^{-1}e_{-5} \otimes e_{-1}k_5 + k_4^{-1}e_{-3} \otimes e_{-4}k_3 \right\}$$

where $k_A = q^{\frac{1}{2}h_A}$ $(A = 1 \dots 6)$. The formulae for antipodes of the Cartan-Chevaley basis

$$S(e_{\pm i}) = -q^{\pm 1}e_{\pm i} \qquad S(k_i^{\pm 1}) = k_i^{\mp 1} \qquad (4.9a)$$

are extended to the generators (4.2) as follows:

$$S(e_{\pm 4}) = q^{\pm 2}\tilde{e}_{\pm 4} \qquad S(e_{\pm 5}) = q^{\pm 2}\tilde{e}_{\pm 5} \qquad S(e_{\pm 6}) = -q^{\pm 3}\tilde{e}_{\pm 6} \qquad (4.9b)$$

where

$$\tilde{e}_4 = [e_2, e_1]_q \qquad\qquad \tilde{e}_{-4} = [e_{-1}, e_{-2}]_{q^{-1}}$$
$$\tilde{e}_5 = [e_3, e_2]_q \qquad\qquad \tilde{e}_{-5} = [e_{-2}, e_{-3}]_{q^{-1}} \qquad (4.10)$$
$$\tilde{e}_6 = [e_3, \tilde{e}_4]_q \qquad\qquad \tilde{e}_{-6} = [\tilde{e}_{-4}, e_{-3}]_{q^{-1}}$$

We see therefore that antipodes describe outer automorphism of the Cartan-Weyl basis.

5. CONTRACTION OF $U_q(O(4,2))$ TO κ-DEFORMED POINCARÉ ALGEBRA

In order to describe real quantum conformal algebra $U_q(O(4,2))$ we should restrict the Cartan-Weyl basis of $U_q(SL(4;C))$ by the reality condition. We shall consider here the following $\oplus$-involution [9], describing an antiautomorphism in both algebra and coalgebra sectors:

$\oplus$-involution: $|q| = 1$; $(i = 1, 2, 3)$

$$h_i^{\oplus} = h_i \qquad e_1^{\oplus} = e_{-1} \qquad e_2^{\oplus} = -e_{-2} \qquad e_3^{\oplus} = e_{-3}$$
$$e_4^{\oplus} = -e_{-4} \qquad e_5^{\oplus} = -e_{-5} \qquad e_6^{\oplus} = -e_{-6} \qquad (5.1)$$

Following the techniques presented in [8], we shall consider the real form (5.1) of $U_q(Sl(4;C))$ as an intermediate step in the derivation of κ-deformation of $D = 4$ Poincaré algebra where κ is a mass-like parameter. We assume the generators of the $D = 4$ conformal algebra in the following way:

$$M_{12} = \tfrac{i}{2}(h_1 + h_3)$$

$$M_{23} = \tfrac{1}{2}(e_1 + e_{-1} + e_3 + e_{-3})$$

$$M_{31} = \tfrac{1}{2}(e_1 - e_{-1} + e_3 - e_{-3})$$

$$M_{03} = \tfrac{i}{2}(e_4 - e_{-4} - e_5 + e_{-5})$$

$$M_{02} = \tfrac{1}{2}(e_6 + e_{-6} - e_2 - e_{-2})$$

$$M_{01} = \tfrac{i}{2}(e_6 - e_{-6} + e_2 - e_{-2})$$

$$(5.2)$$

$$M_{43} = \tfrac{i}{2}(h_1 - h_3)$$

$$M_{42} = \tfrac{1}{2}(e_1 - e_{-1} - e_3 + e_{-3})$$

$$M_{41} = \tfrac{i}{2}(e_1 + e_{-1} - e_3 - e_{-3})$$

$$M_{40} = \tfrac{1}{2}(e_4 + e_{-4} + e_5 + e_{-5})$$

$$M_{54} = -\tfrac{i}{2}(e_4 - e_{-4} + e_5 - e_{-5})$$

$$M_{53} = \tfrac{1}{2}(e_4 + e_{-4} - e_5 - e_{-5})$$

$$M_{52} = \tfrac{i}{2}(e_2 - e_{-2} - e_6 + e_{-6})$$

$$M_{51} = \tfrac{1}{2}(e_2 + e_{-2} + e_6 + e_{-6})$$

$$(5.3)$$

$$M_{50} = \frac{i}{2}(h_1 + h_3 + 2h_2)$$

which due to relations (5.1) satisfy the reality condition $M_{AB}^{\oplus} = -M_{AB}$. The $O(4,2)$ q-deformed commutation relations correspond to the following assignment of the signature $g_{AB} = diag(- + + + + -)$ $(A, B = 0, 1, 2, 3, 4, 5)$ and the physical basis is given by the Lorentz generators $M_{\mu\nu}$ $(\mu\nu = 0, 1, 2, 3)$ and $P_{\mu} = M_{4\mu} + M_{5\mu}$, $K_{\mu} = M_{5\mu} - M_{4\mu}$, $D = M_{45}$. The Cartan subalgebra is described by the following three commuting generators: $(M_3, P_0 + K_0, P_3 - K_3)$.

In order to obtain the κ-deformation of the $D = 4$ Poincaré algebra we introduce the limit (1.1) with $l = 2$ and the following rescaling of the generators

$$\tilde{M}_{\mu\nu} = M_{\mu\nu} \qquad \tilde{K}_{\mu} = K_{\mu} \qquad \tilde{P}_{\mu} = \frac{1}{R}P_{\mu} \qquad \tilde{D} = \frac{1}{R}D \qquad (5.4)$$

The rescaling (5.4) corresponds to the contraction of the nonsymmetric and nonreductive coset $K = \frac{G}{H}$, where G is a conformal group $(O(4,2))$ and H is the Poincaré group $(0(3,1) \ni T_4)$. formed by Lorentz group extended by conformal accelerations. Substituting in the formulae of Sect. 4 the relations (5.2-3), or more explicitly

$$e_{\pm 1} = \tfrac{1}{2i}\left[M_{\pm} + \tfrac{1}{2}\left(R\tilde{P}_1 - K_1\right) \pm \tfrac{i}{2}\left(R\tilde{P}_2 - K_2\right)\right]$$

$$e_{\pm 2} = \tfrac{1}{2i}\left[\pm L_{\mp} \pm \tfrac{1}{2}\left(R\tilde{P}_2 + K_2\right) + \tfrac{i}{2}\left(R\tilde{P}_1 + K_1\right)\right]$$

$$e_{\pm 3} = \tfrac{1}{2i}\left[M_{\pm} + \tfrac{1}{2}\left(K_1 - R\tilde{P}_1\right) \mp \tfrac{i}{2}\left(R\tilde{P}_2 - K_2\right)\right]$$

$$e_{\pm 4} = \tfrac{1}{2i}\left[\pm\left(L_3 + R\tilde{D}\right) + \tfrac{1}{2}\left(R\tilde{P}_0 - K_0\right) + \tfrac{i}{2}\left(R\tilde{P}_3 + K_3\right)\right] \qquad (5.5)$$

$$e_{\pm 5} = \tfrac{1}{2i}\left[\pm\left(R\tilde{D} - L_3\right) + \tfrac{1}{2}\left(R\tilde{P}_0 - K_0\right) - \tfrac{i}{2}\left(R\tilde{P}_3 + K_3\right)\right]$$

$$e_{\pm 6} = \tfrac{1}{2i}\left[\pm L_{\pm} \mp \tfrac{1}{2}\left(R\tilde{P}_2 + K_2\right) + \tfrac{1}{2}\left(R\tilde{P}_1 + K_1\right)\right]$$

$$h_1 = -iM_3 + \tfrac{i}{2}\left(R\tilde{P}_3 - K_3\right)$$

$$h_2 = iM_3 - \tfrac{i}{2}\left(R\tilde{P}_0 + K_0\right) \qquad (5.6)$$

$$h_3 = -iM_3 - \tfrac{i}{2}\left(R\tilde{P}_3 - K_3\right)$$

we obtain in the contraction limit the following relations

a) κ-deformation of Lorentz sector (M_i, L_i)

$$[M_+, M_-] = 2iM_3 \qquad [L_+, L_-] = -2iM_3$$

$$[M_3, M_+] = iM_+ \qquad [L_3, L_+] = -iM_+ - \tfrac{3}{8\kappa^2}\tilde{P}_0\left(\tilde{P}_2 - i\tilde{P}_0\right)$$

$$[M_3, M_-] = -iM_- \qquad [L_3, L_-] = iM_- + \tfrac{3}{8\kappa^2}\tilde{P}_0\left(\tilde{P}_0 + i\tilde{P}_1\right)$$

$$[M_3, L_3] = 0 \qquad [M_+, L_+] = \tfrac{1}{8\kappa^2}\left(\tilde{P}_1 + i\tilde{P}_2\right)^2$$

$$[M_3, L_+] = iL_+ \qquad [M_-, L_-] = -\tfrac{1}{8\kappa^2}\left(\tilde{P}_1 - i\tilde{P}_2\right)^2 \qquad (5.7)$$

$$[M_3, L_-] = -iL_- \qquad [M_\mp, L_\pm] = \mp 2iL_3 \mp \tfrac{1}{8\kappa^2}\left(2\tilde{P}_3^2 - \tilde{P}_1^2 - \tilde{P}_2^2\right)$$

$$[M_+, L_3] = -iL_+ - \tfrac{1}{8\kappa^2}\tilde{P}_3\left(\tilde{P}_1 + i\tilde{P}_2\right)$$

$$[M_-, L_3] = iL_- + \tfrac{1}{8\kappa^2}\tilde{P}_3\left(\tilde{P}_1 - i\tilde{P}_2\right)$$

b) κ-deformation of centrally extended Poincaré algebra (Poincaré $\oplus\tilde{D}$).
The limit (1.1) with $l = 2$ implies supplementing of the algebra (5.7) by the relations

$$[\tilde{P}_\mu, \tilde{P}_\nu] = 0 \qquad [M_{\mu\nu}, \tilde{P}_\lambda] = g_{\nu\lambda}\tilde{P}_\mu - g_{\mu\lambda}\tilde{P}_\nu \qquad [M_i, \tilde{D}] = [L_i, \tilde{D}] = 0 \qquad (5.8)$$

It should be added that the contraction limit (1.1) with $l = 2$ we apply only to the generators $(M_{\mu\nu}, P_\mu, D)$, what leads to closed quantum algebra. It appears that in the conformal momenta sector this contraction limit leads to divergences.

The κ-deformations of Poincaré algebra discussed in Sect. 3. and Sect. 5. have the following common features:
i) three dimensional group of motions E_3^κ (space rotations + space translations) remains unchanged,
ii) the energy operator P_0 commutes with all the generators of E_3^κ,
iii) the four momenta operators $P_\nu = (P_i, P_0)$ form abelian subalgebra for κ-deformed $D = 4$ Poincaré algebra,
iv) the κ-deformed Lorentz algebra generators do not form quantum subalgebra.

6. $U_q'(O(4,2))$ AS A q-DEFORMATION OF $D = 4$ CONFORMAL ALGEBRA

It is interesting to consider the following involution $\oplus$ acting on $U_q(SL(4;C))$:
$\oplus$-involution: q real

$$h_1^\oplus = -h_3 \qquad e_{\pm 1}^\oplus = e_{\pm 3} \qquad e_{\pm 2}^\oplus = e_{\pm 2} \qquad (6.1)$$

$$h_2^\oplus = -h_2 \qquad e_{\pm 4}^\oplus = e_{\pm 5} \qquad e_{\pm 6}^\oplus = e_{\pm 6}$$

The q-deformation $U_q'(O(4,2))$ corresponding to the choice (6.1) of reality conditions is given below.

Lorentz quantum subalgebra:
let us introduce the generators of the Lorentz group as follows:

$$M_+ = M_{23} + iM_{31} = e_1 + e_{-3} \qquad\qquad L_+ = M_{20} + iM_{01} = e_1 - e_{-3}$$
$$M_- = M_{23} - iM_{31} = -(e_3 + e_{-1}) \qquad\quad L_- = M_{20} - iM_{01} = e_{-1} - e_3 \qquad (6.2)$$
$$M_3 = \tfrac{i}{2}(h_1 - h_3) \qquad\qquad\qquad\qquad L_3 = M_{03} = \tfrac{1}{2}(h_1 + h_3)$$

We obtain the following commutation relations

$$[M_+, M_-] = [L_3 + iM_3]_q - [L_3 - iM_3]_q$$
$$[M_3, M_\pm] = \pm iM_\pm \qquad\qquad\qquad\qquad\qquad\qquad (6.3a)$$

$$[L_+, L_-] = [L_3 - iM_3]_q - [L_3 + iM_3]_q$$
$$[L_3, L_\pm] = M_\pm \qquad\qquad\qquad\qquad\qquad\qquad\qquad (6.3b)$$

$$[M_\pm, L_\mp] = [L_3 - iM_3]_q + [L_3 + iM_3]_q$$
$$[M_\pm, L_3] = -L_\pm \qquad\qquad [M_3, L_\pm] = \pm iL_\pm \qquad\qquad (6.3c)$$
$$[M_\pm, L_\pm] = 0 \qquad\qquad\qquad [M_3, L_3] = 0$$

Using the reality conditions (6.1), we obtain that $M_{\mu\nu}^\oplus = -M_{\mu\nu}$ and the relations (6.3a-c) describe the q-deformation of the Lorentz algebra.
We see, therefore, that the q-deformation of the Lorentz subalgebra describes Hopf bialgebra[2], i.e. it is a genuine quantum algebra [14].
Let us introduce the four momenta as follows:

$$P_0 = -i(e_2 + e_6) \qquad\qquad P_1 = e_5 - e_4$$
$$P_3 = i(e_2 - e_6) \qquad\qquad\quad P_2 = i(e_5 + e_4) \qquad\qquad (6.4)$$

where $P_\mu^\oplus = -P_\mu$ (see (6.1)). The q-deformed algebra in the four-momentum sector look as follows:

$$[P_0, P_3] = 0 \qquad\qquad\qquad\qquad [P_3, P_2] = i\tfrac{1-q}{1+q}\{P_3, P_1\}$$

$$[P_0, P_2] = i\tfrac{1-q}{1+q}\{P_0, P_1\} \qquad [P_3, P_1] = i\tfrac{q-1}{q+1}\{P_3, P_2\} \qquad (6.5)$$

$$[P_0, P_1] = i\tfrac{q-1}{q+1}\{P_0, P_2\} \qquad [P_1, P_2] = \tfrac{1}{2i}(q^{-1} - q)(P_3^2 - P_0^2)$$

In this way, one can obtain
   i) Quantum deformation of Lorentz algebra (as a quantum subalgebra of $U_q(SL(2;C))$), which is a Hopf algebra.
   ii) The four momenta are nonabelian and form the quadratic relations (6.5), describing closed subalgebra.

---

[2] the coproduct and antipode formulae are given in [11]

Acknowledgements: The author (A.N.) would like to thank Pierre Minnaert for his hospitality at Bordeaux University I, and CNRS for financial support.

References

[1] P. Podleś and S.L. Woronowicz, Comm. Math. Phys. **130**,(1990)381

[2] V. Carow-Watamura, M. Schlieker, M. Scholl and S. Watamura, Z. Phys.**C48**, (1990)159

[3] V. Carow-Watamura, M. Schlieker, M. Scholl and S. Watamura, Int. Journ. Mod. Phys. **6A**, (1991)3081

[4] S.L. Woronowicz, "New Quantum Deformation of $SL(2; C)$ - Hopf Algebra Level", Warsaw University preprint, 1990

[5] W.P. Schmidke, J. Wess and B. Zumino, Max Planck Institute preprint MPI-Ph/91-15, March 1991

[6] E. Celeghini, R. Giacchetti, E. Sorace and M. Tarlini, J. Math. Phys. **32**, (1991)1155; ibid. 1159

[7] E. Celeghini, R. Giacchetti, E. Sorace and M. Tarlini, "Contractions of quantum groups", Proceedings of First EIMI Workshop on Quantum Groups, Leningrad, October - December 1990, ed. P. Kulish, Springer Verlag, 1991

[8] J. Lukierski, A. Nowicki, H. Ruegg and V.N. Tolstoy. Phys. Lett.**264B**, 331(1991)

[9] J. Lukierski, A. Nowicki and H. Ruegg, Phys. Lett.**271B**, 321(1991)

[10] V. Dobrev, "Canonical q-Deformations of Noncompact Lie (Super-)Algebras", Göttingen Univ. preprint, July 1991

[11] J. Lukierski, A. Nowicki,"Quantum Deformations of D=4 Poincaré and Weyl Algebra from q-Deformed D=4 Conformal Algebra", Wrocław University preprint ITP UWr 787/91, October 1991

[12] M. Rosso, Comm. Math. Phys. **124**, (1989)307

[13] S.M. Khoroshkin and V.N. Tolstoy, Comm. Math. Phys. **141**, 599 (1991)

"Quantum group" structure and "covariant" differential calculus on symmetric algebras corresponding to commutation factors on $\mathbf{Z}^n$ *

Rainer Matthes[†]

Institut für Mechanik

Chemnitz

January 10, 1992

Abstract

For any given commutation factor ϵ on $\mathbf{Z}^n$ a first order differential calculus on a certain symmetric algebra C_ϵ^n corresponding to ϵ is constructed. It is shown that there exists a kind of a quantum group structure (ϵ-Hopf algebra) on each C_ϵ^n and that the differential calculus is the unique one being covariant (in an adapted sense) with respect to this "quantum group" structure.

1 Introduction

The most fundamental notion of classical differential geometry is that of a tangent vector. All other notions are defined making reference to this one. However, in a noncommutative situation, it seems to be more appropriate to start with the definition of differential forms (see [1], [4] and [17]). The reason for this is that there may not be "enough" derivations, and if one defines differential forms in the classical spirit as objects "dual" to derivations, one has difficulties to define local objects, because the derivations do in general not form a left module over the algebra in a natural way (see [9] for another approach to locality using ideals). These difficulties disappear for a special class of noncommutative algebras, namely for ϵ-commutative algebras, where ϵ is a commutation factor on an abelian group which defines a grading of the algebra. One has to use ϵ-derivations instead of derivations. They form a left module over the algebra and a so-called ϵ-Lie algebra, which makes it possible to define all notions of differential geometry in a "classical"

---

*Extended version of a talk at the Max Born Symposium, Wojnowice, september 1991

†Current postal address: Fachbereich Physik/Theoretische Physik, Universität Leipzig, Augustusplatz 10/11, O-7010 Leipzig

R. Gielerak et al. (eds.), Groups and Related Topics, 45–54.

way with the only difference that one "takes always into account the commutation rules". These things can be found in [3] (for the $\mathbf{Z}_2$-graded case), [6], [14] and [15]. Here, we consider the special example of the symmetric algebra $\mathbf{C}_\epsilon^n$ corresponding to a general commutation factor ϵ on $\mathbf{Z}^n$ and a $\mathbf{Z}^n$-graded vector space which has a homogeneous basis whose elements have as degrees just the generators of $\mathbf{Z}^n$. First we construct by the "classical" procedure a first order differential calculus on $\mathbf{C}_\epsilon^n$ in the sense of Woronowicz ([17]). Then we show that $\mathbf{C}_\epsilon^n$ has an ϵ-Hopf algebra structure which is a kind of a quantum group structure deforming the additive group structure of $\mathbf{C}^n$ and that the differential calculus is in a certain sense (modifying the definition of [17]) covariant with respect to this "quantum group" structure. The quantum hyperplane $\mathbf{C}_q^n$ of [12], [13] and [16] appears as a special case. The differential calculus given here is different from the calculi of [12] and [16] and therefore not covariant with respect to coactions of standard matrix quantum groups on $\mathbf{C}_q^n$. The results presented here are essentially contained already in [11].

Our conventions are as follows: All algebras are associative with unity I over the field of complex numbers $\mathbf{C}$. Furthermore, we always make use of Einstein's sum convention.

2   The algebra $\mathbf{C}_\epsilon^n$

Definition 1 *([14]). Let Γ be an abelian group. A vector space A is said to be Γ-graded if there exist subspaces $A_\alpha \subset A$, $\alpha \in \Gamma$, such that $A = \bigoplus_{\alpha \in \Gamma} A_\alpha$.*
An algebra A is said to be Γ-graded if in addition $A_\alpha A_\beta \subset A_{\alpha+\beta}$ $\forall \alpha, \beta \in \Gamma$. We put $A^h = \bigcup_{\alpha \in \Gamma} A_\alpha$ (set of homogeneous elements). $g(x)$ denotes the degree of $x \in A^h$. A commutation factor on Γ is a map $\epsilon : \Gamma \times \Gamma \longrightarrow \mathbf{C}$ with
1. $\epsilon(\alpha, \beta)\epsilon(\beta, \alpha) = 1$
2. $\epsilon(\alpha + \beta, \gamma) = \epsilon(\alpha, \gamma)\epsilon(\beta, \gamma)$ $\forall \alpha, \beta, \gamma \in \Gamma$.
A Γ-graded algebra A is said to be ϵ-commutative if $xy = \epsilon(\alpha, \beta)yx$ for $x \in A_\alpha$, $y \in A_\beta$.
Let A, B be Γ-graded vector spaces. A linear map $X : A \longrightarrow B$ is homogeneous of degree $g(X)$ if $X(A_\alpha) \subset B_{\alpha+g(X)}$ $\forall \alpha \in \Gamma$.

We have also $\epsilon(\alpha, \beta + \gamma) = \epsilon(\alpha, \beta)\epsilon(\alpha, \gamma)$ $\forall \alpha, \beta, \gamma \in \Gamma$ as a direct consequence of the definition.

Here, we will consider the case of the abelian group $\Gamma = \mathbf{Z}^n$, a general commutation factor ϵ on Γ, and the $\mathbf{Z}^n$-graded algebra $\mathbf{C}_\epsilon^n$ defined as the quotient

$$\mathbf{C} < x^1, \ldots, x^n > /I_R$$

where $\mathbf{C} < x^1, \ldots, x^n >$ is the free associative unital algebra generated by $x^1, \ldots, x^n$ and I_R is the twosided ideal generated by the elements

$$x^i x^j - \epsilon(\delta_i, \delta_j) x^j x^i$$

where $\delta_k \overset{\text{def}}{=} (0,\ldots,1,\ldots,0)(1$ at the k-th position).
A general commutation factor on $\mathbf{Z}^n$ can be characterized as follows: From the
fact that the δ_i are generators of $\mathbf{Z}^n$ and from the properties of a commutation
factor ϵ one easily concludes

$$\epsilon(\alpha_1,\ldots,\alpha_n,\beta_1,\ldots,\beta_n) = \prod_{i,j} \epsilon(\delta_i,\delta_j)^{\alpha_i\beta_j}$$

Thus, ϵ is determined by

$$\epsilon(\delta_k,\delta_k) = q_{kk} = \pm 1$$

(because in general $\epsilon(\alpha,\alpha) = \pm 1$) and

$$\epsilon(\delta_i,\delta_j) = q_{ij} \in \mathbf{C} \setminus \{0\}$$

with $q_{ij} = q_{ji}^{-1}$ for $i \neq j$. Now, choose any $q \in \mathbf{C} \setminus \{0\}$. Because the complex
exponential function takes any nonzero value we can find $m_{ij} \in \mathbf{C}$ such that
$q_{ij} = q^{m_{ij}}$. We have $m_{ij} = -m_{ji}$ for $i \neq j$ and $m_{ii} = 0$ if $q_{ii} = 1$, $m_{ii} = \frac{\pi\sqrt{-1}}{\ln q}$ if
$q_{ii} = -1$. Thus, we have

$$\epsilon(\alpha,\beta) = q^{\sum_{i,j} m_{ij}\alpha_i\beta_j}$$

and in the case $\epsilon(\alpha,\alpha) = 1$ ϵ can be given by a nonzero complex number q and an
antisymmetric bilinear map $m : \mathbf{Z}^n \times \mathbf{Z}^n \longrightarrow \mathbf{C}$.
The elements $x^{1^{i_1}} \cdots x^{n^{i_n}}, (i_1,\ldots,i_n) \in \mathbf{N}^n, \mathbf{N} = \{0,1,\ldots\}$, form a basis of the
vector space $\mathbf{C}_\epsilon^n$. $\mathbf{C}_\epsilon^n$ is a $\mathbf{Z}^n$-graded algebra with one-dimensional homogeneous
components

$$\mathbf{C}_\epsilon^{n(i_1,\ldots,i_n)} = \begin{cases} \mathbf{C}x^{1^{i_1}} \cdots x^{n^{i_n}} & \text{if } i_1,\ldots,i_n \in \mathbf{N} \\ 0 & \text{otherwise} \end{cases}$$

Notice that we have $x^{k^2} = 0$ for $\epsilon(\delta_k,\delta_k) = -1$ (odd generator), and the basis
elements containing higher than linear terms in x^k are equal to zero in this case.
$\mathbf{C}_\epsilon^n$ is by definition an ϵ-commutative algebra. It could be defined equivalently as
follows([14]): Let V be a $\mathbf{Z}^n$-graded n-dimensional vector space. Denote by $T(V)$
the tensor algebra over V and by $I_\epsilon(V)$ the two-sided ideal of $T(V)$ generated by
the elements

$$x \otimes y - \epsilon(g(x),g(y))y \otimes x$$

for $x,y \in V^h$. Scheunert ([14]) calls this "the ϵ-symmetric algebra of the $\mathbf{Z}^n$-graded
vector space V". If we assume that V has a basis $(x^i)_{i=1,\ldots,n}$ with $g(x^i) = \delta_i$ then
$\mathbf{C}_\epsilon^n$ is isomorphic to $T(V)/I_\epsilon(V)$.

3 ϵ-derivations

Definition 2 *Let A be a Γ-graded algebra, let ϵ be a commutation factor on Γ. $X \in End_C(A)$ is said to be a homogeneous ϵ-derivation of degree $g(X) \in \Gamma$ if X is homogeneous and*

$$X(xy) = X(x)y + \epsilon(g(X), g(x))xX(y)$$

for $x \in A^h$. $Der_\epsilon^\alpha(A)$ denotes the vector space of homogeneous ϵ-derivations of degree α. $Der_\epsilon(A) = \bigoplus_{\alpha \in \Gamma} Der_\epsilon^\alpha(A)$ is the vector space of ϵ-derivations of A. $Der_\epsilon^h(A) = \bigcup_{\alpha \in \Gamma} Der_\epsilon^\alpha(A)$ is the set of homogeneous ϵ-derivations of A.

The following two propositions are well known (see [6], [14]).

Proposition 1 *Let $X_1, X_2 \in Der_\epsilon^h(A)$. Then we have*

$$[X_1, X_2]_\epsilon \overset{\text{def}}{=} X_1 X_2 - \epsilon(g(X_1), g(X_2))X_2 X_1 \in Der_\epsilon^h(A)$$

with $g([X_1, X_2]_\epsilon) = g(X_1) + g(X_2)$.

$[.,.]_\epsilon$ is called ϵ-commutator.

Proposition 2 *For $X, X_1, X_2, X \in Der_\epsilon^h(A), a \in A^h$, we have*
1. $[X, X]_\epsilon = 0$
2. $[X_2, X_1]_\epsilon = -\epsilon(g(X_2), g(X_1))[X_1, X_2]_\epsilon$
3. $\epsilon(g(X_3), g(X_1))[[X_1, X_2]_\epsilon, X_3]_\epsilon + \epsilon(g(X_1), g(X_2))[[X_2, X_3]_\epsilon, X_1]_\epsilon$
$\quad + \epsilon(g(X_2), g(X_3))[[X_3, X_1]_\epsilon, X_2]_\epsilon = 0$
If A is ϵ-commutative then we have
4. $aX \in Der_\epsilon^h(A)$ with $g(aX) = g(a) + g(X)$
5. $[aX_1, X_2]_\epsilon = a[X_1, X_2]_\epsilon - \epsilon(g(X_1) + g(a), g(X_2))X_2(a)X_1$

1.-3. say that $Der_\epsilon(A)$ is an ϵ-Lie algebra.
Let us consider now the ϵ-commutative algebra C_ϵ^n. By 4., $Der_\epsilon(C_\epsilon^n)$ is a left C_ϵ^n- module. There exists a basis of this module consisting of analogues ∂_i of partial derivatives which are defined as follows: Define first operators ∂_i on the free algebra $C < x^1, \ldots, x^n >$ by the recursive prescription

$$\partial_i(I) = 0, \quad \partial_i(x^j) = \delta_i^j I$$

$$\partial_i(x^j x) = \delta_i^j x + \epsilon(g(\partial_i), g(x^j))x^j \partial_i(x)$$

with $g(\partial_i) = -\delta_i$. The operators ∂_i are well defined by this prescription because any homogeneous element of $C < x^1, \ldots, x^n >$ is obtained by successive multiplication from the left of generating elements to one fixed (for the chosen homogeneous element) generating element. Notice that $C < x^1, \ldots, x^n >$ can be considered as a $\mathbf{Z}^n$- graded algebra in a natural way.

Proposition 3 *1.* $\partial_i(xy) = \partial_i(x)y + \epsilon(g(\partial_i), g(x))x\partial_i(y)$ *for homogeneous* $x \in \mathbf{C} < x^1, \ldots, x^n >$.
2. $\partial_i(I_R) \subset I_R$, *where* I_R *is the ideal defining* $\mathbf{C}_\epsilon^n$.

Proof: 1. is proved by induction: The equation is true for $x = x^j$. Let us assume that the equation is true for a homogeneous x. It is sufficient to show that it is then also true for any $x^j x$ in place of x. We have

$$\partial_i(x^j xy) = \delta_i^j xy + \epsilon(g(\partial_i), g(x^j))x^j \partial_i(xy)$$

By the assumption, this is equal to

$$\delta_i^j xy + \epsilon(g(\partial_i), g(x^j))x^j(\partial_i(x)y + \epsilon(g(\partial_i), g(x))x\partial_i(y))$$

By the definition of ∂_i, we have

$$\partial_i(x^j x)y = \delta_i^j xy + \epsilon(g(\partial_i), g(x^j))x^j \partial_i(x)y$$

Using property 2 of ϵ, a comparison of these three formulas gives the desired result.
2. One shows immediately that $\partial_i(x^j x^k - \epsilon(g(x^j), g(x^k))x^k x^j) = 0$. A general homogeneous element of I_R contains one term $x^i x^j - \epsilon(\delta_i, \delta_j)x^j x^i$ as a factor. "Distributing" ∂_i over the factors according to 1. one obtains as summands either zero or an element of I_R.$\square$

With othter words, the ∂_i project to well-defined ϵ-derivations of $\mathbf{C}_\epsilon^n$ which we denote by the same symbol.

Proposition 4 *For any* $X \in Der_\epsilon(\mathbf{C}_\epsilon^n)$ *we have*

$$X = X(x^i)\partial_i$$

Proof: It is sufficient to proof the assertion for homogeneous X. One proceeds inductively: It is obvious that $X(x^i) = X(x^j)\partial_j(x^i)$. Both X and $X(x^i)\partial_i$ are homogeneous ϵ- derivations of the same degree $g(X)$, and it is easy to show inductively that ϵ-derivations that coincide on generating elements must coincide on all the algebra $\mathbf{C}_\epsilon^n$.$\square$

It follows immediately from $\partial_i(x^j) = \delta_i^j I$ that the ∂_i are linearly independent as elements of the left $\mathbf{C}_\epsilon^n$-module $Der_\epsilon(\mathbf{C}_\epsilon^n)$, i.e., $Der_\epsilon(\mathbf{C}_\epsilon^n)$ is a free left $\mathbf{C}_\epsilon^n$-module with basis $(\partial_i)_{i=1,\ldots,n}$.

4 ϵ-differential forms

The following definition is due to Woronowicz([17]):

Definition 3 *Let A be an algebra, and let Γ be an A-bimodule.*
A pair (Γ, d) is called first order differential calculus on A, if $d : A \longrightarrow \Gamma$ is a linear map and the following holds:
1. $d(ab) = d(a)b + ad(b)$ $\forall a, b \in A$ (Leibniz rule).
2. For any $\omega \in \Gamma$ one finds a finite number of $a_k, b_k \in A$ such that $\omega = \sum_k a_k b_k$.

Definition 4 *We denote by $\bigwedge_{0\epsilon}^{1h}(\mathbf{C}_\epsilon^n)$ the set of homogeneous $\mathbf{C}$-linear maps α : $Der_\epsilon(\mathbf{C}_\epsilon^n) \longrightarrow \mathbf{C}_\epsilon^n$ with*

$$\alpha(aX) = \epsilon(g(\alpha), g(a))a\alpha(X)$$

for $a \in \mathbf{C}_\epsilon^{nh}$. The vector space of finite sums of elements of $\bigwedge_{0\epsilon}^{1h}(\mathbf{C}_\epsilon^n)$ is denoted by $\bigwedge_{0\epsilon}^{1}(\mathbf{C}_\epsilon^n)$. Elements of this space are called ϵ -differential forms. We define

$$d : \mathbf{C}_\epsilon^n \longrightarrow \bigwedge_{0\epsilon}^{1}(\mathbf{C}_\epsilon^n)$$

by

$$da(X) = \epsilon(g(a), g(X))X(a)$$

for $a \in \mathbf{C}_\epsilon^{nh}, X \in Der_\epsilon^h(\mathbf{C}_\epsilon^n)$.

Proposition 5 *$(\bigwedge_{0\epsilon}^{1}(\mathbf{C}_\epsilon^n), d)$ is a first order differential calculus on $\mathbf{C}_\epsilon^n$.*

Proof: $\bigwedge_{0\epsilon}^{1}(\mathbf{C}_\epsilon^n)$ has a bimodule structure defined by
$(a\alpha)(X) = a(\alpha(X)) = \epsilon(g(a), g(\alpha))(\alpha a)(X) =$
$\epsilon(g(a), g(\alpha))\alpha(aX) = \epsilon(g(a), g(\alpha) + g(X))\alpha(X)a.$
Using the remark that, obviously, $g(da) = g(a)$, one proves the Leibniz rule :
$d(ab)(X) = \epsilon(g(a) + g(b), g(X))X(ab)$
$= \epsilon(g(a) + g(b), g(X))(X(a)b + \epsilon(g(X), g(a))aX(b))$
$= \epsilon(g(b), g(X))da(X)b + \epsilon(g(b), g(X))aX(b) = (dab)(X) + adb(X).$
Condition 2. is proved by showing the formula $\alpha(X) = \alpha(\partial_i)dx^i(X)$: We have
$dx^i(\partial_j) = \epsilon(g(x^i), g(\partial_j))\delta_j^i = \delta_j^i$. Therefore, $\alpha(\partial_i)dx^i(\partial_j) = \alpha(\partial_j)$. α and $\alpha(\partial_i)dx^i$
are homogeneous ϵ-differential forms with the same degree $g(\alpha)$. Now it is sufficient
to show inductively that homogeneous ϵ-differential forms with the same degree
that coincide on the ∂_i also coincide on any ϵ-derivation : Let α, β be such forms
and assume $\alpha(X) = \beta(X)$ for some ϵ-derivation X. Then we have $\alpha(x^j X) =$
$\epsilon(g(\alpha), g(x^j))x^j\alpha(X) = \epsilon(g(\beta), g(x^j))x^j\beta(X) = \beta(x^j X)$. Due to proposition 4, the
desired formula is proved.□

It follows immediately from $dx^i(\partial_j) = \delta_j^i I$ that the $(dx^i)_{i=1,...,n}$ are a basis of the
$\mathbf{C}_\epsilon^n$-bimodule $\bigwedge_{0\epsilon}^{1}(\mathbf{C}_\epsilon^n)$.

5   Covariance

For $\epsilon = 1$, there is an obvious Hopf algebra structure on C_1^n which is just isomorphic to the Hopf algebra of polynomial functions on the additive group C^n. For $\epsilon \neq 1$, there seems to be no simple way to define a Hopf algebra structure on C_ϵ^n which may be considered as a deformation of this additive group structure. However, C_ϵ^n can be given the structure of a "ϵ-Hopf" algebra: Let us define the algebra structure of $C_\epsilon^n \otimes C_\epsilon^n$ not as usual for tensor products of algebras but by $(x_1 \otimes y_1)(x_2 \otimes y_2) = \epsilon(g(y_1), g(x_2))x_1 x_2 \otimes y_1 y_2$.. Further, define
1. an algebra homomorphism $\Delta : C_\epsilon^n \longrightarrow C_\epsilon^n \otimes C_\epsilon^n$ by $\Delta(x^i) = x^i \otimes I + I \otimes x^i$ and algebra homomorphy (with respect to the algebra structure just defined).
2. an algebra homomorphism $\varepsilon : C_\epsilon^n \longrightarrow C$ by $\varepsilon(I) = 1$, $\varepsilon(x^i) = 0$ and algebra homomorphy.
3. a homomorphism $\kappa : C_\epsilon^n \longrightarrow C_\epsilon^n$ by $\kappa(x^i) = -x^i$ and algebra homomorphy.
It is easy to show that these maps are indeed correctly defined. From $g(\kappa(x)) = g(x)$ follows that $\kappa(xy) = \epsilon(g(x), g(y))\kappa(y)\kappa(x)$. Thus, we could call ε an "ϵ-antihomomorphism". We have

Proposition 6 *1. $id \otimes \Delta \circ \Delta = \Delta \otimes id \circ \Delta$*
2. $id \otimes \varepsilon \circ \Delta = \varepsilon \otimes id \circ \Delta = id$
3. $\mu \circ id \otimes \kappa \circ \Delta = \mu \circ \kappa \otimes id \circ \Delta = \varepsilon(.)I$
*   (μ - the multiplication $C_\epsilon^n \otimes C_\epsilon^n \longrightarrow C_\epsilon^n$)*

Proof: Because $id \otimes \Delta \circ \Delta$ and $\Delta \otimes id \circ \Delta$ are algebra homomorphisms $C_\epsilon^n \longrightarrow C_\epsilon^n \otimes C_\epsilon^n \otimes C_\epsilon^n$ (where the algebra structure of the threefold tensor product is again defined "taking into account the commutation rules") it is sufficient to prove the first equation for the generating elements x^i, which is a simple direct computation. 2. and 3. are proved by the same argument.□

We call $(C_\epsilon^n, \Delta, \varepsilon, \kappa)$ a ϵ -Hopf algebra.
The above differential calculus is in an adapted sense covariant with respect to this "quantum group" structure:
Let us define a $C_\epsilon^n \otimes C_\epsilon^n$- bimodule structure of $C_\epsilon^n \otimes \wedge_{0\epsilon}^1(C_\epsilon^n)$ by
$x \otimes y \cdot z \otimes \alpha = \epsilon(g(y), g(z))xz \otimes y\alpha$ for $y, z \in C_\epsilon^{nh}$
$z \otimes \alpha \cdot x \otimes y = \epsilon(g(\alpha), g(x))zx \otimes \alpha y$ for $x \in C_\epsilon^{nh}, \alpha \in \wedge_{0\epsilon}^{1h}(C_\epsilon^n)$.
Further, define $\Delta_l^1 : \wedge_{0\epsilon}^1(C_\epsilon^n) \longrightarrow C_\epsilon^n \otimes \wedge_{0\epsilon}^1(C_\epsilon^n)$ by $\Delta_l^1(dx^i) = I \otimes dx^i$ and $\Delta_l^1(x\alpha) = \Delta(x)\Delta_l^1(\alpha)$ (homomorphy with respect to the left $C_\epsilon^n \otimes C_\epsilon^n$-module structure of $C_\epsilon^n \otimes \wedge_{0\epsilon}^1(C_\epsilon^n)$ just defined). Analogously, define a $C_\epsilon^n \otimes C_\epsilon^n$ -bimodule structure of $\wedge_{0\epsilon}^1(C_\epsilon^n) \otimes C_\epsilon^n$ and a map $\Delta_r^1 : \wedge_{0\epsilon}^1(C_\epsilon^n) \longrightarrow \wedge_{0\epsilon}^1(C_\epsilon^n) \otimes C_\epsilon^n$. It is easy to show that these definitions are correct.

Proposition 7 *We have*

1. $\Delta_l^1(\alpha x) = \Delta_l^1(\alpha)\Delta(x)$ $\qquad$ 1.' $\Delta_r^1(x\alpha) = \Delta(x)\Delta_r^1(\alpha)$

2. $\Delta \otimes id \circ \Delta_l^1 = id \otimes \Delta_l^1 \circ \Delta_l^1$ $\qquad$ 2.' $\Delta_r^1 \otimes id \circ \Delta_r^1 = id \otimes \Delta \circ \Delta_r^1$

3. $\varepsilon \otimes id \circ \Delta_l^1 = id$ $\qquad\qquad\qquad$ 3.' $id \otimes \varepsilon \circ \Delta_r^1 = id$

4. $\Delta_l^1 \circ d = id \otimes d \circ \Delta$ $\qquad\qquad$ 4.' $\Delta_r^1 \circ d = d \otimes id \circ \Delta$

$\qquad\qquad$ 5. $\Delta_l^1 \otimes id \circ \Delta_r^1 = id \otimes \Delta_r^1 \circ \Delta_l^1$

The proof is reduced to simple calculations for the x^i and the dx^i using the homomorphy properties of the maps.

A comparison with [17] shows that these are just the conditions for bicovariance of a first order differential calculus. In spite of the fact that we have not used the usual structures of Hopf algebras and bimodules we say that $\bigwedge_{0\varepsilon}^1(C_\varepsilon^n)$ is a bicovariant bimodule and $(\bigwedge_{0\varepsilon}^1(C_\varepsilon^n), d)$ is a bicovariant first order differential calculus. Notice that the $(dx^i)_{i=1,\dots,n}$ are a basis of (left and right) invariant differential forms. The calculus is unique in the following sense: Let (Λ, d') be another left covariant calculus on C_ε^n such that $(d'x^i)_{i=1,\dots,n}$ is a basis of Λ. "Left covariant" means that a map $\Delta_l^{\prime 1} : \Lambda \longrightarrow C_\varepsilon^n \otimes \Lambda$ is defined which fulfils 1.-4. of proposition 7 and $\Delta_l^{\prime 1}(x\alpha) = \Delta(x)\Delta_l^{\prime 1}(\alpha)$. Here, the $C_\varepsilon^n \otimes C_\varepsilon^n$- bimodule structure of $C_\varepsilon^n \otimes \Lambda$ is defined taking into account both the commutation rules of C_ε^n and the C_ε^n-bimodule structure of Λ. It follows immediately from 4. that $\Delta_l^{\prime 1}(d'x^i) = I \otimes d'x^i$, and one shows easily that Λ is isomorphic to $\bigwedge_{0\varepsilon}^1(C_\varepsilon^n)$ as a C_ε^n-bimodule writing out 4. on quadratic terms and using homomorphy properties and the fact that $(d'x^i)_{i=1,\dots,n}$ is a basis. Because of this it is also clear that (Λ, d') is isomorphic to $(\bigwedge_{0\varepsilon}^1(C_\varepsilon^n), d)$ as a first order differential calculus. The same reasoning can be applied for the case of right covariance.

Remarks

1. The quantum hyperplane C_q^n of [12], [13] and [16] appears here as the special case

$$\epsilon(\alpha, \beta) = q^{m(\alpha,\beta)}$$

with

$$m(g(x), g(y)) = (i_1 + \dots + i_{n-1})k_n + (i_1 + \dots + i_{n-2} - i_n)k_{n-1} + \dots + (-i_2 - \dots - i_n)k_1$$

for $g(x) = (i_1, \dots, i_n)$, $g(y) = (k_1, \dots, k_n)$. The differential calculus constructed in the present note is different from those of [12] and [16] and therefore not covariant with respect to left or right coactions of matrix quantum groups on the comodule C_q^n.

2. Having defined algebraic analogues of vector fields and differential 1-forms on C_ε^n there is no essential problem to introduce other notions of differential geometry (tensor calculus, calculus of differential forms, connection, curvature, compare

[6]). The bases $(\partial_i)_{i=1,...,n}$ and $(dx^i)_{i=1,...,n}$ play a role fully analogous to their role in classical differential geometry. This has been made explicite for the case of the algebra C_q^n in [10] (without knowing the papers of Scheunert and Marcinek).

Following the ideas of [17] one can define coactions of C_ε^n on the differential forms of higher degree that are compatible with exterior product and differentiation ("covariance" of the higher order calculus).

3. It seems to be possible to obtain the above results also in the more general situation of symmetric algebras defined by Yang-Baxter operators with square 1 (cf. [2] and [5]). In particular, this is possible for symmetric algebras related to commutation factors on general abelian grading groups, because one only needs the general properties of a commutation factor and the fact that the symmetric algebra is always ε-commutative. (Besides [14] and [15] see also [7] and [8] for information about commutation factors and various related algebraic constructions.)

Acknowledgements

I am indebted to J. Lukierski and M. Scheunert who drew my attention to the articles [6], [14] and [15] and to W. Marcinek and Z. Oziewicz for valuable remarks and for sending me their manuscripts. Moreover, I would like to thank K. Schmüdgen, G. Rudolph and H. D. Doebner for their interest and the latter for generous support.

References

[1] Connes, A.: Noncommutative differential geometry, Publ. Math. IHES **62** (1986), 41-144

[2] Gurevich, D. J.: Quantum Yang-Baxter equation and a generalization of the formal Lie theory, Seminar on supermanifolds No. 4 (D. Leites ed.), Reports Department of Mathematics University of Stockholm No. 24, 1986, pp. 34-123

[3] Jadczyk, A. and D. Kastler: Graded Lie-Cartan pairs, University of Wrocław, preprint No 677, 1987; Graded Lie-Cartan pairs II, Preprint IHES, 1987

[4] Kastler, D.: Introduction to Alain Connes' non- commutative differential geometry, Preprint CPT-86/PE.1929 Marseille 1986

[5] Lyubashenko, V. V.: Vectorsymmetries, Seminar on supermanifolds No. 15 (D. Leites ed.), Reports Department of Mathematics, University of Stockholm No. 19, 1987

[6] Marcinek, W.: Generalized Lie-Cartan pairs, Rep. Math. Phys. **27** (1989), 385-400

[7] Marcinek, W.: Algebras based on Yang-Baxter operators, Preprint ITP UWr 731/89, Wroclaw November 1989

[8] Marcinek, W.: Graded algebras and geometry based on Yang-Baxter operators, Preprint ITP UWr 745/90, Wroclaw June 1990

[9] Matthes, R.: A general approach to connections: algebra and geometry, preprint KMU-NTZ 89-03, to appear in Rep. Math. Phys.

[10] Matthes, R.: Vector fields and differential forms on C_q^n, unpublished manuscript 1990

[11] Matthes, R.: A covariant differential calculus on the "quantum group" C_q^n, to appear in the Proceedings of the Wigner Symposium Goslar 1991

[12] Pusz, W. and S. L. Woronowicz: Twisted second quantization, Rep. Math. Phys. **27** (1989), 231-257

[13] Reshetikhin, N. Yu., Takhtajan, L. A. and L. D. Faddeev: Kvantovanie grupp Li i algebr Li, Algebra i analiz **1** (1989), 178-206

[14] Scheunert, M.: Generalized Lie algebras, J. Math. Phys. **20** (1979) 712-720

[15] Scheunert, M.: Graded tensor calculus, J. Math. Phys. **24** (1983), 2658-2670

[16] Wess, J. and B. Zumino: Covariant differential calculus on the quantum hyperplane, preprint 1990

[17] Woronowicz, S. L.: Differential calculus on compact matrix pseudogroups (quantum groups), Comm. Math. Phys. **122** (1989), 125-170

Remarks on the Use of R-matrices

Arne Schirrmacher

Max–Planck–Institut für Physik
München*

Abstract

Examples for the application of the R-matrix formalism on problems related to $GL(n)$ quantum groups are given. (1) The Yang-Baxter property of the R-matrix provides a simple means to replace the explicit use of the diamond lemma and thus eliminates lengthy calculations. (2) The R-matrix description induces a pair of bicovariant differential calculi on the group. Reasons are given that these may be the only suitable ones.

---

*postal address: Föhringer Ring 6, W-8000 München 40, Germany; e-mail: ars at dm0mpi11.

R. Gielerak et al. (eds.), Groups and Related Topics, 55–65.
© 1992 Kluwer Academic Publishers.

1   Introduction

These remarks are initiated by some recent work by Yu. I. Manin introducing the quantum de Rahm complex of $GL_{p,q}(2)$. It is completely determined by its automorphism structure and the so-called Wess-Zumino condition, i.e. that the tensor product of the algebra of 'basic variables', the generators of the functions on the group, with the algebra of their differentials (with constant coefficients) gives the full de Rahm complex. In order to archieve this for $GL_{p,q}(2)$ the commutation relations as required by the transformation of the quantum spaces had to be supplemented and its coherence was assured by explicit use of the diamond lemma. The way in which the appropriate 'missing relations' are recognized is not very straight-forward and the proof of the consistency of the relations with help of the diamond lemma rather lengthy.

In order to make a contribution to this discussion we should like to dwell on two observations. (a) The relations for the two quantum de Rahm complexes related to $GL_{p,q}(2)$ can easily be written down as $\hat{R}$-matrix relations that generalize immediately. (b) This allows us to demonstrate the coherence of the relations since they can be solved such that unordered monomials are related to ordered ones.

Working out these points permits to exhibit many links among the different approaches to quantum groups. To shed some more light on the topic, we first give a brief review of standard material which the remarks will refer to. [1]

1.1. (Manin) A *q-plane* is the linear space of all monomials in two variables x and y subject to the conditions $xy = q\,yx$, $x^2 \neq 0$, $y^2 \neq 0$, where $q \in \mathbb{C}\backslash\{0\}$. (This generalizes immediately to *quantum spaces* of n indeterminates with $x^i x^j = q_{ij}\,x^j x^i$, $1 \leq i < j \leq n$.) A *quantum group* G given by a quantum matrix T transforms a q-plane into a q-plane, i.e. for $x' = Tx$ we also find $x'y' = q\,y'x'$. This does not completely determine the commutation structure of the quantum matrix. Requiring that T in addition transforms a *second* $\tilde{q}$-plane $\xi\eta = \tilde{q}\,\eta\xi$ fixes all relations if one allows $\xi^2 = \eta^2 = 0$, $\tilde{q} \neq q$. Such a plane is called *exterior (quantum) plane.* —Cf.[1].

1.2. (Wess, Zumino) The second plane can be identified with the differentials of the first: $\xi \equiv dx$, $\eta \equiv dy$. For convenience write $\tilde{q} \equiv -q/r^2 \equiv -p$ which is a parameter independent from q. The transformation is recognized as that from the two-parameter quantum group $GL_{p,q}(2)$. —Cf.[2,3].

1.3. (Faddeev *et.al.*) The quantum group relations can be written with help of a matrix $\hat{R}$ that acts on the square tensor space of the group/space variables and is a solution of the Yang-Baxter equation (YBE):

$$\boxed{\hat{R}\,(T \otimes T) = (T \otimes T)\,\hat{R}} \tag{1}$$

This relation is invariant under the replacement $\hat{R} \leftrightarrow \hat{R}^{-1} = \hat{R} - (r - r^{-1})$ where we omit the unit matrix. $\hat{R}$ has a decomposition into orthogonal projectors. $\hat{R} = r^{-1}P_A + rP_S$ where $P_A \propto \hat{R} - r = \hat{R}^{-1} - r^{-1}$ and $P_S \propto \hat{R} + r^{-1} = \hat{R}^{-1} + r$ for GL. The plane relations are thus (for x and ξ now vectors with components x^i or ξ^i):

---

[1]Sections with an asterix (*) have not been lectured in detail or contain new material.

$$P_A(x \otimes x) = 0, \quad P_S(\xi \otimes \xi) = 0 . \tag{2}$$

The relation (1) ensures the transformation properties for planes. For $z' = Tz$, $z = x$ or ξ, $P = P_A$ or P_S one finds:

$$P(z' \otimes z') = P(T \otimes T)(z \otimes z) = (T \otimes T) P(z \otimes z) = 0 . \tag{3}$$

—Cf.[4].

1.4. (Wess, Zumino) The two planes can be combined to a *differential graded quantum algebra* (DGQA) of the q-plane with a differential operator d that is nilpotent and obeys the graded Leibniz rule ($\hat{f} = deg\ f$):

$$d(fg) = df.g + (-1)^{\hat{f}} f.dg . \tag{4}$$

Commutation relations among plane variables and differentials are to be added. The only possibilities are:

$$\xi \otimes x = r\hat{R} x \otimes \xi \quad \text{and} \quad \xi \otimes x = r^{-1}\hat{R}^{-1} x \otimes \xi . \tag{5}$$

These relations respect transformations as in (3). Here the unique structure splits into a pair of "differential calculi"; in the following we will stick to the first choice. E.g. for $GL_{p,q}(2)$ this reads $\xi x = pq\ x\xi$, $\xi y = q\ y\xi + (pq - 1)x\eta$, $\eta x = p\ x\eta$, $\eta y = pq\ y\eta$. It is easy to check that the relations of (2) and (5) are consistently linked by the application of d. —Cf.[2].

1.5. (Bergman) The *diamond lemma* is the general answer in ring theory to the question under which conditions one can show that an ordering prescription brings every expression in a unique irreducible form. While mathematicians seem to think that one cannot help using this lemma, physicists do usually not go beyond asking whether the $\hat{R}$-matrix satisfies the YBE.—Cf.[5].
— For quadratic algebras this general lemma may be replaced by some simpler methods.

1.6. (Maltsiniotis/Manin) The DGQA of a q-plane is understood as a *quantum de Rahm complex* of a q-plane. The quantum group induces a algebra (auto-)morphism of DGQAs, i.e. the tensor product of group and plane algebras becomes a *differential graded tensor product* of DGQAs:

$$T : \Omega(A_q) \longrightarrow \Omega(GL_{p,q}(2)) \overset{.}{\otimes} \Omega(A_q) \tag{6}$$

This requirement, however, does not give all commutation relations for $\Omega(G)$ (i.e. the Poincare series does not match that of the classical case, the deformation is not *flat*). The missing relations are determined by consideration of special cubic monomials (Manin) and the desired properties of the coproduct (Maltsiniotis), respectively. The diamond lemma is used to demonstrate the coherence of the relations. This makes a lengthy calculation. Manin formulated the need of a "more intelligent approach" and so does Maltsiniotis[2]. —Cf.[6,7,8,9].

---

[2] "Il est en tout cas souhaitable de trouver une preuve plus conceptionelle."[9]

— We will demonstrate that the problem of "missing relations" can easily be evaded and the coherence of the relations can be proven within a couple of lines.

1.7. (Corrigan *et.al.*) A quantum matrix τ of nilpotent entries $(\tau_k^i)^2 = 0$ can be described by a matrix $\hat{R}_\Lambda$:

$$\hat{R}_\Lambda (\tau \otimes \tau) = -(\tau \otimes \tau) \hat{R}_\Lambda . \tag{7}$$

It was a puzzle that the relations are coherent while $\hat{R}_\Lambda$ is *not* a solution of the YBE. —Cf.[10].
— As in the considerations for the planes in **1.2.** we can identify here $\tau \equiv dT$. We will see that it is still the usual YB $\hat{R}$-matrix that gives relation (7).

1.8.* (Alekseev, Faddeev) The 'quantized' phase space T^*G_r for the quantum group $G_r = SL_r(2)$ is defined with help of physical considerations. Classically, an element of the phase space T^*G is a pair (g, ω), $g \in G$ and $\omega \in \mathcal{G}^*$ with a basis $\{t_a\}$ dual to that of the Lie algebra $\mathcal{G}$ $\{t^a\}$. The phase space has a symplectic structure relating one-forms and vector fields. The relations of the phase space can be written down in Poisson brackets:

$$\{\omega^a, g\} = t^a g , \qquad \{\omega^a, \omega^b\} = -f^{ab}{}_c \omega^c \tag{8}$$

where $\omega = \omega^a t_a$ and $f^{ab}{}_c$ the structure constants with respect to t^a. These relations are rewritten using the two-dimensional representation of $SL(2)$ employing σ-matrices and a change of variables (towards canonical ones). With help of matrices u and v we can diagonalize g and ω:

$$\omega = \frac{1}{2}u \begin{pmatrix} -ip & 0 \\ 0 & ip \end{pmatrix} u^{-1}, \qquad g = u \begin{pmatrix} e^{-iq} & 0 \\ 0 & e^{iq} \end{pmatrix} v \equiv uQv \tag{9}$$

$$\begin{aligned} \{u \otimes 1, 1 \otimes u\} &= (u \otimes u) r_0(p) \\ \{v \otimes 1, 1 \otimes v\} &= -r_0(p)(v \otimes v) \end{aligned} \tag{10}$$

where $r_0(p) = -i/p\,(\sigma_+ \otimes \sigma_- - \sigma_- \otimes \sigma_+)$. There are additional relations e.g. $\{p, q\} = 1$ exhibiting that p and q are a canonical pair of variables. This quadratic Poisson algebra (10) is the motivation for the following proposal for a deformed one:

$$\begin{aligned} \hat{R}(u \otimes u) &= (u \otimes u)\hat{R}(p) \\ \check{R}(p)(v \otimes v) &= (v \otimes v)\check{R} \end{aligned} \tag{11}$$

where $\hat{R} = PR$ and $\check{R} = RP$. $\hat{R}(p)$ are the deformed 6j-symbols of $SL_r(2)$(i.e. the solution of a quasi-YBE with associator $Q \otimes 1 \otimes 1$). For

$$\Omega = u \begin{pmatrix} e^{-ip} & 0 \\ 0 & e^{ip} \end{pmatrix} u^{-1} \tag{12}$$

and g as before, relations were derived from (11) where in the end the 6j-symbols disappear. It is supposed that the result can be used in general to define T^*G. — Cf.[11,12].

— We will show that relations of the same type come out automatically from a R-matrix formulation of tensor products of DGQAs without any 'physics' and $6j$-symbols at all.

1.9.[*] (Woronowicz)The *(first order) differential calculus on a matrix bialgebra* $(\mathcal{A}, \Delta, \epsilon)$ with comultiplication Δ and counit ϵ can be defined using

- an $\mathcal{A}$-bimodule $\quad \mathcal{A}^2 := \{ \sum a_k \otimes b_k \in \mathcal{A}^{\otimes 2} : \sum a_k b_k = 0 \}$

- an (algebraic) derivation $\quad \mathrm{D} : \mathcal{A} \longrightarrow \mathcal{A}^2 : a \longmapsto 1 \otimes a - a \otimes 1$

- the canonical epimorphism π with $\mathcal{N}$ a subbimodule of $\mathcal{A}^2$, $\pi : \mathcal{A}^2 \longrightarrow \Gamma \equiv \mathcal{A}^2/\mathcal{N} : \sum a_k \otimes b_k \longmapsto \sum a_k \, db_k$, i.e. $\mathrm{d} = \pi \mathrm{o} \mathrm{D} : \mathcal{A} \longrightarrow \Gamma$ is the differential operator acting on $\mathcal{A}$. Any such differential calculus that obeys the Leibniz rule can be characterized by a bimodule $\mathcal{N} = ker \, \pi$. For any *bicovariant* (first order) differential calculus $\mathcal{N} \subset ker \, \epsilon$ is given by a right ideal $\mathcal{R}$ of $\mathcal{A}$ satisfying

$$r(\mathcal{N}) = \mathcal{A} \otimes \mathcal{R} , \qquad r(a \otimes b) = (a \otimes 1)\Delta(b)$$

$$s(\mathcal{N}) = \mathcal{R} \otimes \mathcal{A} , \qquad s(a \otimes b) = (1 \otimes a)\Delta(b) \qquad (13)$$

The higher order differential calculus is then also fixed — Cf.[13]
— $\mathcal{R}$ will be provided for $GL_{p,q}(2)$ to contrast this with other approaches. The question arises whether or not the construction of Woronowicz is complete to determine the appropriate differential calculi that meet the general understanding of a deformation.

2   Remarks

2.1. A naive consideration of differential grading. — If we want to render the differential grading of plane and group DGQAs unified we expect for the combination $dT \dot\otimes x \equiv \tau x = \tilde\xi$ degree 1, i.e. it should be an exterior plane:

$$\begin{aligned} 0 = P_S \, (\tilde\xi \otimes \tilde\xi) &= (\hat{R} + r^{-1})(\tau \otimes \tau)(x \otimes x) \\ &= \left[\hat{R}(\tau \otimes \tau) + (\tau \otimes \tau)\hat{R}^{-1} \right] (x \otimes x) \end{aligned} \qquad (14)$$

since $(\hat{R}^{-1} - r^{-1})(x \otimes x) = 0$. Thus we find the commutation relations for the group differentials $dT \equiv \tau$:

$$\boxed{\hat{R}(\tau \otimes \tau) = -(\tau \otimes \tau)\hat{R}^{-1}} \qquad (15)$$

This now solves the puzzle of Corrigan *et.al.* in **1.7.** Since $1(\tau \otimes \tau) = (\tau \otimes \tau)1$ is a redundant relation it can freely be added to (15). $\hat{R}$ obeys the relation $\hat{R} - \hat{R}^{-1} = r - r^{-1} \equiv 2\rho$ and thus $\hat{R} - \rho = \hat{R}^{-1} + \rho = \hat{R}_\Lambda$. Like (1) also (15) is invariant under the replacement $\hat{R} \leftrightarrow \hat{R}^{-1}$.

2.2. Tensor product of DGQAs; the $\hat{R}$-matrix approach. — In order to unfold the content of the differential graded tensor product of DGQAs we have to exploit the Leibniz rule that becomes extended over the tensoring, i.e.

$$\mathrm{d} : \quad \Omega(G_{p,q}) \dot\otimes \Omega(A_q) \longrightarrow \Omega(G_{p,q}) \dot\otimes \Omega(A_q)$$

$$\mathrm{d}(a \dot\otimes b) = da \dot\otimes b + (-1)^{\hat{a}} a \dot\otimes db \qquad (16)$$

such that the automorphism property (6) of the group does not get spoiled. (In the following we will omit the sign $\otimes$ between group and space variables; the sign $\otimes$, however, will still be used to indicate the embeddings in the tensor square on which $\hat{R}$ acts.) Thus, we have to require for $\xi' = d(x')$ the basic relations (2) and (5). For the latter we calculate:

$$
\begin{aligned}
0 &= x' \otimes \xi' - r^{-1}\hat{R}^{-1}(\xi' \otimes x') \\
&= Tx \otimes (T\xi + \tau x) - r^{-1}\hat{R}^{-1} (T\xi + \tau x) \otimes Tx \\
&= (T \otimes T)(x \otimes \xi - r^{-1}\hat{R}^{-1}\xi \otimes x) \\
&\quad + (T \otimes \tau)(x \otimes x) - r^{-1}\hat{R}^{-1}(\tau \otimes T)(x \otimes x) \\
&= r^{-1}\left[(T \otimes \tau)\hat{R} - \hat{R}^{-1}(\tau \otimes T)\right](x \otimes x)
\end{aligned}
\tag{17}
$$

since $(r^{-1}\hat{R} - 1)(x \otimes x) = 0$. We conclude

$$
\boxed{\hat{R}^{-1}(\tau \otimes T) = (T \otimes \tau)\hat{R}}
\tag{18}
$$

which gives the commutation relations of the basic variables and its differentials. For the second choice of (5) another differential calculus emerges inducing the replacement $\hat{R} \leftrightarrow \hat{R}^{-1}$ in (18). In addition (2) gives

$$
\begin{aligned}
0 = P_S \,(\xi' \otimes \xi') &= (\hat{R} + r^{-1})\big((\tau \otimes \tau)(x \otimes x) + (T \otimes T)(\xi \otimes \xi) \\
&\qquad\qquad + (\tau \otimes T)(x \otimes \xi) - (T \otimes \tau)(\xi \otimes x)\big) \\
&= (\hat{R} + r^{-1})(\tau \otimes \tau)(x \otimes x)
\end{aligned}
\tag{19}
$$

since the second term in the middle equality clearly vanishes and the third and last cancel due to (5) and (18). Thus we arrive just at the same problem where in **2.1.** we naively started from and which yielded (15). The equations (1), (15), and (18) give the entire commutation structure for the quantum de Rahm complex of a quantum group.

2.3. Example: $\Omega(GL_{p,q}(2))$. — The $\hat{R}$-matrix of $GL_{p,q}(2)$ reads

$$
\hat{R} = \begin{pmatrix} r & 0 & 0 & 0 \\ 0 & r - \frac{1}{r} & \frac{q}{r} & 0 \\ 0 & \frac{r}{q} & 0 & 0 \\ 0 & 0 & 0 & r \end{pmatrix}
\tag{20}
$$

For $T = \left(\begin{smallmatrix} a & b \\ c & d \end{smallmatrix}\right)$ relation (1) gives the well-known quantum group relations ($p \equiv r^2/q$):

$$
ab = p\,ba, \qquad ac = q\,ca, \qquad bc = \frac{q}{p}\,cb,
$$

$$
cd = p\,dc, \qquad bd = q\,db, \qquad ad - da = \left(p - \frac{1}{q}\right)bc\,.
\tag{21}
$$

From (15) the the relations of the differentials follow which coincide with those for the odd quantum variables of Corrigan *et.al.*[10], $\tau \equiv \left(\begin{smallmatrix} \alpha & \beta \\ \gamma & \delta \end{smallmatrix}\right)$:

$$
\begin{aligned}
q\,\alpha\beta &= -\beta\alpha, & p\,\alpha\gamma &= -\gamma\alpha, & \beta\gamma &= -\frac{q}{p}\gamma\beta - (q - 1/p)\alpha\delta, \\
q\,\gamma\delta &= -\delta\gamma, & p\,\beta\delta &= -\delta\beta, & \alpha\delta &= -\delta\alpha, \\
\alpha^2 &= \beta^2 = \gamma^2 = \delta^2 = 0\,.
\end{aligned}
\tag{22}
$$

and (18) provides the mixed relations

$$
\begin{aligned}
\alpha a &= pq\, a\alpha, & \beta b &= pq\, b\beta, & \gamma c &= pq\, c\gamma, & \delta d &= pq\, d\delta, \\
\beta a &= q\ a\beta, & \delta c &= q\ c\delta, & \gamma a &= p\, a\gamma, & \delta b &= p\, b\delta, \\
\alpha b &= (pq-1)a\beta + p\, b\alpha, & & & \gamma d &= (pq-1)c\delta + p\, d\gamma, \\
\alpha c &= (pq-1)a\gamma + q\, c\alpha, & & & \beta d &= (pq-1)b\delta + q\, d\beta, \\
p\,\beta c &= (pq-1)a\delta + q\, c\beta, & & & q\,\gamma b &= (pq-1)a\delta + p\, b\gamma, \\
pq\,\alpha d &= pq\, d\alpha + (pq-1)^2 a\delta + (pq-1)\left[p\, b\gamma + q\, c\beta\right], & & & \delta a &= a\delta .
\end{aligned}
\tag{23}
$$

The other calculus where $\hat{R}$ and $\hat{R}^{-1}$ are interchanged has different mixed relations that simply replace T and τ. (Manin and Maltsiniotis use the latter with reciprocal p and q.) No relations have to be added since consistency is already 'built in' by the $\hat{R}$-matrix.[3]

2.4. Ordering with the $\hat{R}$-matrix. — The standard $\hat{R}$-matrices of the quantum groups of type A_n, B_n, C_n, D_n ([4], for multiparameter solutions [14]) have the following properties for nonvanishing entries:

$$
\hat{R}^{ij}{}_{ab} \neq 0 \quad\Longrightarrow\quad
\begin{aligned}
i &> j \Rightarrow a < b \\
i &< j \Leftarrow a > b
\end{aligned}
$$

and since $\hat{R}^{-1}_{r;q_{ij}} = P\hat{R}_{1/r;1/q_{ij}}P$ for $\hat{R}^{-1}$ > and < are interchanged. For $GL(n)$ we have in addition $\hat{R}^{\pm 1}{}^{ii}{}_{ab} \neq 0 \Leftrightarrow a = b = i$. Exploiting this we can define an ordering operator $\mathcal{O}$ on quadratic monomials of the DGQA of $GL_{r;q_{ij}}(n)$:

$$
\mathcal{O}(T_k^i T_l^j) = \left\{
\begin{array}{ll}
T_k^i T_l^j & : \ (i < j) \text{ or } (i = j,\ k < l) \\
\hat{R}^{ij}{}_{ab}\, T_c^a T_d^b\, (\hat{R}^{-1})^{cd}_{kl} & : \ i > j \\
(\hat{R}^{-1})^{ij}{}_{ab}\, T_c^a T_d^b\, \hat{R}^{cd}_{kl} & : \ i = j,\ k > l
\end{array}
\right.
$$

$$
\mathcal{O}(T_k^i \tau_l^j) = T_k^i \tau_l^j
$$
$$
\mathcal{O}(\tau_k^i T_l^j) = \hat{R}^{ij}{}_{ab}\, T_c^a \tau_d^b\, \hat{R}^{cd}_{kl}
$$
$$
\mathcal{O}(\tau_k^i \tau_l^j) = \left\{
\begin{array}{ll}
\tau_k^i \tau_l^j & : \ (i < j) \text{ or } (i = j,\ k > l) \\
\hat{R}^{ij}{}_{ab}\, \tau_c^a \tau_d^b\, \hat{R}^{cd}_{kl} & : \ i > j \\
(\hat{R}^{-1})^{ij}{}_{ab}\, \tau_c^a \tau_d^b\, (\hat{R}^{-1})^{cd}_{kl} & : \ i = j,\ k < l
\end{array}
\right.
\tag{24}
$$

The idea to demonstrate that this ordering operator $\mathcal{O}$ provides a unique reduction of any expression to an ordered form is to reduce it to the YB property. Ambiguities can arise if we reorder a cubic monomial. Define $A \leq B :\Leftrightarrow (\mathcal{O} - id)\, AB = 0$. Let $t = T$ or τ and assume $t_1 > t_2 > t_3$. Since we can apply either $\mathcal{O}_{12}\mathcal{O}_{23}\mathcal{O}_{12}$ or $\mathcal{O}_{23}\mathcal{O}_{12}\mathcal{O}_{23}$ we find (α etc. $= \pm 1$ according to (24))

$$
\begin{aligned}
t_1 t_2 t_3 &= (\hat{R}^{\alpha}_{12}\hat{R}^{\beta}_{23}\hat{R}^{\gamma}_{12}) t_1 t_2 t_3 (\hat{R}^{\lambda}_{23}\hat{R}^{\mu}_{12}\hat{R}^{\nu}_{23}) \\
&= (\hat{R}^{\gamma}_{23}\hat{R}^{\beta}_{12}\hat{R}^{\alpha}_{23}) t_1 t_2 t_3 (\hat{R}^{\nu}_{12}\hat{R}^{\mu}_{23}\hat{R}^{\lambda}_{12})
\end{aligned}
\tag{25}
$$

---

[3]Interestingly enough, the determinant of the quantum group $\mathcal{D} = ad - p\,bc$ is not central in the full differential algebra even if one takes $p = q$. This observation is due to Bruno Zumino.

Ambiguities could arise for $T_1T_2T_3$, $\tau_1T_2T_3$, $\tau_1\tau_2T_3$, $\tau_1\tau_2\tau_3$ with the different cases $i > j$ and $i = j$, $k \lessgtr l$. We conclude that the ordering is unique due to the YBE

$$\hat{R}_{12}\hat{R}_{23}\hat{R}_{12} = \hat{R}_{23}\hat{R}_{12}\hat{R}_{23} ,\tag{26}$$

and its variations by multiplications with $\hat{R}^{-1}$.

Clearly, one additional condition has to be added: For $A < B, C < D$ and $\mathcal{O}(BC) = B_i'C_i'$ we need $A < B_i' < C_i' < D$ for all i. Otherwise we cannot restrict the argument to cubic monomials since B_i' or C_i' could jump over several factors of a string of variables and the termination of this procedure is not clear. For $GL_{r;q_{ij}}(n)$ this 'strong ordering condition' is satisfied. (For other groups additional quadratic relations from orthogonality or symplecticity conditions have to be taken into consideration. Possibly, a similar construction can be found.)

2.5. Right invariant one-forms.— From (15) and (18) we can derive the commutation structure for right (and left) invariant one-forms. In doing this we omit the $\otimes$ sign and use leg numbering notation to indicate the summations ($T_1 = T \otimes 1, T_2 = 1 \otimes T$ etc.). Let $\omega = \mathrm{d}T \cdot T^{-1}$, i.e. $\tau = \omega T$. Using (18) we get

$$T_1\omega_2 = \hat{R}^{-1}\omega_1 T_1 T_2 \hat{R}^{-1} T_2^{-1} = \hat{R}^{-1}\omega_1 \hat{R}^{-1} T_1\tag{27}$$

and thus (15) can be rewritten as

$$\begin{aligned}
0 &= \hat{R}\tau_1\tau_2 + \tau_1\tau_2\hat{R}^{-1}\\
&= \hat{R}\omega_1 T_1 \omega_2 T_2 + \omega_1 T_1\omega_2 T_2\hat{R}^{-1}\\
&= \hat{R}\omega_1\hat{R}^{-1}\omega_1\hat{R}^{-1}T_1 T_2 + \omega_1\hat{R}^{-1}\omega_1\hat{R}^{-1}T_1 T_2\hat{R}^{-1}\\
&= \left[\hat{R}\omega_1\hat{R}^{-1}\omega_1 + \omega_1\hat{R}^{-1}\omega_1\hat{R}^{-1}\right]\hat{R}^{-1}T_1 T_2
\end{aligned}$$

yielding

$$\hat{R}\omega_1\hat{R}^{-1}\omega_1 = -\omega_1\hat{R}^{-1}\omega_1\hat{R}^{-1} .\tag{28}$$

These relations (27) and (28) for T and ω are very similar to those which Alekseev and Faddeev derived for $SL_q(2)$ using g and Ω [11].

2.6. Derivatives, vector fields etc. — Manin considers *quantum vector fields* as the dual objects to the differentials [7]:

$$\mathrm{d} \equiv \mathrm{d}(T_k^i) \cdot \frac{\partial}{\partial T_k^i} = \tau : \partial\tag{29}$$

where ':' indicates summation over a pair of indices. Assuming for the dual objects commutation relations of similar type $\hat{S}(\partial \otimes \partial) = (\partial \otimes \partial)\hat{S}$ from

$$\begin{aligned}
0 = \mathrm{d}^2 &= \mathrm{d}(\tau : \partial)\\
&= \tau : (\tau : \partial)\,\partial\\
&\equiv (\tau \otimes \tau) :: (\partial \otimes \partial)_{21}\\
&= -(\hat{R}(\tau \otimes \tau)\hat{R}) :: (\hat{S}(\partial \otimes \partial)\hat{S})_{21}\\
&\propto (\tau \otimes \tau)(\hat{R}\hat{S}_{21}) :: (\partial \otimes \partial)_{21}
\end{aligned}\tag{30}$$

we find

$$\hat{S} = \hat{R}_{21}^{-1} = P\hat{R}_{r;q_{ij}}^{-1} P = \hat{R}_{1/r_{i}1/q_{ij}} \tag{31}$$

i.e. the relations for the derivatives are the same as those of the differentials except that the deformation parameters are the reciprocal. To find the algebra of the left invariant vector fields $\nabla \equiv T \cdot \partial$, i.e. the deformed Lie algebra relations one has to analyse $d^2 = 0$ using $d = \omega : \nabla$.

2.7.* The R-matrix calculi in Woronowicz description. — In order to translate our results in the language of **1.9.** consider

$$\rho = r^{-1}(1 \otimes \sum_{\alpha} c_{\alpha} T_{ka}^{ia} T_{la}^{ja}) = \sum_{\alpha} c_{\alpha} \left[(T_2^{-1} T_1^{-1}) \dot\otimes (T_1 T_2) \right]^{iaja}_{kala} \tag{32}$$

Such that $m \rho \equiv \sum_{\alpha} c_{\alpha} \, \delta_{ka}^{ia} \delta_{ka}^{ja} = 0$ i.e. $\rho \in \mathcal{N} \subset \mathcal{A}^2$ (m is the multiplication map). We have to compute $ker \, \pi$ using (18) and $\Theta = T^{-1} \cdot dT$ the left invariant one-forms (suppressing indices)

$$
\begin{aligned}
\pi(\rho) &= \sum_{\alpha} c_{\alpha} \, T_2^{-1} T_1^{-1} \, d(T_1 T_2) \\
&= \sum_{\alpha} c_{\alpha} \, (T_2^{-1} T_1^{-1} \, \hat{R}^{-1} T_1 \tau_2 \, \hat{R}^{-1} + T_2^{-1} \tau_2) \\
&= \sum_{\alpha} c_{\alpha} \, (\hat{R}^{-1} \Theta_2 \hat{R}^{-1} + \Theta_2) \\
&= \left[\sum_{\alpha} c_{\alpha} (\hat{R}^{-1})^{iaja}_{sm} \hat{R}^{sn}_{kala} + \delta_{ka}^{ia} \delta_{m}^{ja} \delta_{la}^{n} \right] \Theta^{m}_{n} \tag{33}
\end{aligned}
$$

The square bracket must vanish for $\rho \in ker \, \pi$ and for all m and n due to the fact that all N^2 one-forms of $GL(N)$ are linear independent. Going through all quadratic monomials we find the following combinations from $\mathcal{A}^2$ in the case of $GL_{p,q}(2)$ that already generate the ideal:

$$\mathcal{R} = \left\{ b^2, c^2, b(a-d), c(a-d), bc - q \, \frac{r^2 - 1}{r^2 + 1}(d^2 - 1), a^2 + d^2 - (r^2 + 1)ad \right\} \tag{34}$$

For the other differential calculus the last two relations read $cb - \frac{1}{q}\frac{r^2-1}{r^2+1}(d^2 - 1)$ and $a^2 + d^2 - (r^2 - 1)da$. The classical limit of $\mathcal{R}$ clearly is $ker \, \epsilon \otimes ker\epsilon$ ($ker \, \epsilon = \{b, c, a-d\}$).

2.8.* Other differential calculi? — The pair of $\hat{R}$-matrix induced bicovariant differential calculi fits in Woronowicz's general scheme. Also for SO and SU groups differential calculi have been worked out using various constructions with $\hat{R}$-matrices [15,16]. The reversed problem, i.e. the question of the "most general bicovariant calculus" allowed by Woronowicz's construction has recently been addressed by Müller-Hoissen [17]. Checking all consistency conditions for the most general ansatz for the commutation relations in the case of $GL_{p,q}(2)$ one degree of freedom evades fixing. It is thus claimed that there exist two one-parameter families of calculi. For a particular value of the additional parameter things become comparatively simple. It turns out that these are the two $\hat{R}$-matrix calculi we discussed above. For all other values of the additional parameter, however, we cannot help pointing out a number of drawbacks:

- For $p = q = 1$ the additional parameter survives. We prefer to reject such "non-standard" calculi of classical groups and consider the differential geometry of undeformed groups uniquely defined. Thus, some condition is missing to guarantee an appropriate limit.

- The relations for the left invariant one-forms θ^i produce ordering circles on some cubic monomials, e.g. if one mechanically tries to order (omitting numerical factors)

$$\theta^2\theta^1\theta^3 \rightarrow \theta^1\theta^2\theta^3 + \theta^2\theta^4\theta^3 \rightarrow \theta^1\theta^2\theta^3 + \theta^2\theta^3\theta^4 + \theta^2\theta^1\theta^3 \rightarrow \dots \tag{35}$$

  Such circles are cut only for the two $\hat{R}$-matrix calculi.

- As a consequence the YB or braiding property does not entail the coherence of the relations, this remains to be demonstrated.

These points may open a discussion whether all properties of a deformed differential calculus are already formulated in Woronowicz's approach or if some "natural" condition should be added. It might turn out that the simple and rigid $\hat{R}$-matrix formulation (1), (15), and (18) can be considered as the 'canonical' one.

Acknowledgements
Fruitful discussions are acknowledged with J. Schwenk, M. Pillin and W. Weich.

References

[1] Yu. I. Manin: *Quantum groups and non-commutative geometry*, CRM University of Montreal (1988).

[2] J. Wess, B. Zumino: *Covariant differential calculus on the quantum hyperplane*, Nucl. Phys. B (Proc. Suppl.)**18B**, 302-312, (1990).

[3] A. Schirrmacher, J. Wess, B. Zumino: *The two-parameter deformation of GL(2) its differential calculus, and Lie algebra*, Z. Phys. C **49**, 317, (1991).

[4] N. Yu. Reshetikhin, L. A. Takhtadzhyan, L. D. Faddeev,: *Quantization of Lie groups and Lie algebras*, Leningrad Math. J. **1**, 193–225, (1990).

[5] G. M. Bergman: *The diamond lemma for ring theory*, Adv. Math., **29**, (1978) 178.

[6] Yu. I. Manin: *Quantum groups and non-commutative de Rahm complexes*, Bonn preprint MPI/91–47 (1991).

[7] Yu. I. Manin: *Notes on quantum groups and quantum de Rahm complexes*, Bonn preprint MPI/91–60 (1991).

[8] G. Maltsiniotis: *Groupes quantiques et structures differentilles*, C.R. Acad. Sci. Paris, **331**, (1990), 831.

[9] G. Maltsiniotis: *Calcul differentiell sur le groupe linéaire quantique*, exposé, ENS (1990).

[10] E. Corrigan, D. B. Fairlie, P. Fletcher,R. Sasaki: *Some aspects of quantum groups and supergroups* J. Math. Phys., **36**, (1990), 776.

[11] A. Alekseev, L. Faddeev: $(T^*G)_t$; *A toy model for conformal field theory*, Commun. Math. Phys. **141**, (1991), 413.

[12] L. Faddeev: Cargése Lectures 1991, to appear at Plenum press.

[13] S. L. Woronowicz: *Differential calculus on compact matrix pseudogroups*, Commun. Math. Phys., **122**, (1989), 125.

[14] A. Schirrmacher: *Multiparameter R-matrices and their quantum groups*, J. Phys. A. **50**, 321, (1991).

[15] U. Carow-Watamura, M. Schlieker, S. Watamura, W. Weich: *Bicovariant differential calculus on Quantum groups $SU_q(N)$ and $SO_q(N)$*, Commun. Math. Phys. **142**, 605 (1991).

[16] D. Bernard: *Quantum Lie algebras and differential calculus on quantum groups*, Saclay preprint SPhT-90-124 (1990).

[17] F. Müller-Hoissen *Differential calculi on the quantum group $GL_{p,q}(2)$*, Göttingen preprint GOET-TP 55/1991.

CONSTRUCTION OF SOME HOPF ALGEBRAS.

E. Sorace [1]

[1]Dipartimento di Fisica, Università di Firenze and INFN–Firenze,

**Talk given at the First Max Born German–Polish
Symposium in Theorethical Physics.**

1.   The contraction method for quantum groups.

It is a lucky opportunity – and I thank the organizers for the invitation to
a citizen of a State not involved in the meeting – to speak in this Symposium,
devoted to such an outstanding physicist, about a set of quantum groups which
may have immediate relevance in physics. I mean those q-deformed Hopf algebras
whose $q = 1$ limits are, besides the algebra of Heisenberg, the usual kinematical
symmetry groups, namely, Euclides, Poincaré, Galilei and Lorentz ones, which are
noncompact or inhomogeneous or both.

Many successful efforts have been devoted to exploit the knowledge of the
complex semisimple q–algebras by searching for their real forms, between the results of this analysis there are the quantized real versions of the Lorentz group [1,2].
Less efforts seem to have been devoted to the quantization of the inhomogeneous
groups, the main difficulty being their non semisemplicity, owing to the lacking
of a mathematical theory in that situation. Our group in Firenze has been able
to find a constructive way to match together q-deformation and non semisimple
$q = 1$ limit. By this method an exhaustive treatment of many important algebras
of this kind has been done from our group [3,4,5,6] and [7] with P.Kulish, and
in this year by Lukiersky et al [8,9,10]. The method we have introduced is an
extension to the q–deformed Hopf structures of the contraction procedure defined
many years ago on the Lie algebras [11], effective on their representations also
[12]. The q–deformed Hopf strucures recovered at the end of this process are associated by the $q = 1$ limit to non semisimple Lie algebras, so that it is reasonable

E-mail: SORACE@FI.INFN.IT

R.-Gielerak et al. (eds.), Groups and Related Topics, 67–81.
© *1992 Kluwer Academic Publishers.*

to call them *nonsemisimple* quantum groups. In some cases it has been possible
to obtain, in the same framework, the universal R–matrix associated to the non
semisimple quantum algebra. But a general theory about the existence and the
properties of these peculiar limits still is lacking, so the fiability of the method is
demonstrated by the internal consistence of the final results. As a matter of fact
there are problems even in the usual contraction of the Lie algebras realized by a
linear transformation of the basis, depending on a parameter ϵ, singular for $\epsilon \rightarrow$
0. Taking the dominant terms a new "more abelian" algebra may produce: the
method is clearly basis dependent and the goal cannot be prefixed. Neverthless all
the nonsemisimple Lie algebras we are interested in can be obtained in this way.
Of course the same contraction can be extended without problems to the Hopf
algebra associated to the given semisimple Lie algebra.

Thus by requiring the commutativity of the following diagram[3]

$$
\begin{array}{ccc}
 & \epsilon \rightarrow 0 & \\
SQG & \longrightarrow & CQG \\
q \rightarrow 1 \;\; \downarrow & & \downarrow \;\; q \rightarrow 1 \\
SG & \longrightarrow & CG \\
 & \epsilon \rightarrow 0 &
\end{array}
$$

(where SQG and CQG, SG and CG respectively mean simple and contracted
quantum group, simple and contracted classical Hopf algebra, and the limits on q
mean the nondeformed limits) we have tried to extend the contraction technique
to q–algebras. The limiting procedure must be done on all the Hopf relations:
if the results converge we have got a constructive definition of SQG. Between
the algebraic and coalgebraic defining relations of SQG there are some involving
the elements $k_j = e^{zH_j}$ and k_j^{-1}, with H_j in Cartan subalgebra, and $q = e^z$.
Therefore, to avoid exponential singularities, the factors zH_j must be of order 0 in
ϵ during the contraction: this implies the new crucial prescription that z too must
be contracted, and in a reciprocal way to H_j. The introduction of the rescaling
of the deformation parameter [3] has been the essential ingredient to make the
contractions an useful tool in quantum groups. Moreover when ϵ has dimension
the new parameter w derived by contraction from z acquires dimension. Thus in
these contracted structures there will be nothing corresponding to q, but only new
elements derived from k_i, k_i^{-1} and homogeneous functions of w. As it stands this
method needs the starting algebra to be described by means of commutator-like

relations,as e.g.in [13] because till now it doesn't exist a formulation containing Serre relations.

Following this method, by putting $SQG = SU(2)_q$, we recovered $H(1)_q$ and $E(2)_q$ [3,4]. From $SU(2)_q \otimes U(2)_q \equiv O(4)_q$, $E(q)_3$, $G(2)_q$ (Galilei bidimensional) [5,6] and from $Osp(1|2)_q$ the new quantum graded algebra $s - -H(1)_q$ [7] have been obtained. Starting from the pseudo-orthogonal groups $O(3,2)$ and $O(4,2)$ J.Lukierski et al. [8,9] succeeded in recovering various versions of 4-dimensional q–Poincaré.

With the exception of the even-dimension Euclides or Poincaré, the method generates the universal R–matrices satisfying the $QYBE$.

Now I will give you some concrete examples and applications.

2. The quantum Heisenberg group.

The 1-dimensional Heisenberg Lie algebra $H(1)$ of $A, A^\dagger, H, N$:

$$[A, A^\dagger] = H, \quad [N, A] = -A, \quad [N, A^\dagger] = A^\dagger, \quad [H,.] = 0, \qquad (2.1)$$

can be obtained from $SU(2) \otimes U(1) = U(2)$,generated from J_+, J_-, J_3, H and

$$[J_3, J_\pm] = \pm J_\pm, \quad [J_+, J_-] = J_3, \quad [H,.] = 0, \qquad (2.2)$$

by equating the dominant terms after the ϵ–change of basis

$$A = \epsilon J_+, \quad A^\dagger = \epsilon J_-, \quad N = \frac{H}{2\epsilon^2} - J_3, \quad H = H, \qquad (2.3)$$

It is clear that to apply the contraction method to find $U(2)_q$ one have to do the same change of basis (2.3) together with the rescaling of $z = \log(q) : z = w\epsilon^2$. Following the described procedure one then gets straightforward a new Hopf structure whose algebraic relations different from (2.1) are only

$$[A, A^\dagger] = \frac{2}{w} \sinh(wH/2) \qquad (2.4)$$

while for the coproducts one has H, N primitives

$$\Delta(A) = e^{-wH/4} \otimes A + A \otimes e^{wH/4}, \Delta(A^\dagger) = e^{-wH/4} \otimes A^\dagger + A^\dagger \otimes e^{wH/4}, \quad (2.5)$$

Antipode and counity result from the same limit and are all trivials.

The central elements are given by H and $C = AA^\dagger - \frac{\sinh(wH/2)}{w/2} N$.

In this simple case it is evident a general feature of these kind of structures: a basis transformation can be found that trivializes the algebraic sector but it is impossible to do this for all the Hopf algebra.

To search for the R-matrix we start from the well known R-matrix of $SU(2)_q$, multiplied for an arbitrary function r of $H \otimes H$, $r(z, \frac{H \otimes H}{\lambda})$, where λ is a free scale, thus realizing an R-matrix for $U(2)_q$.

Thereof from

$$R = r(z, \frac{H \otimes H}{\lambda})e^{zJ_3 \otimes J_3} \sum_{k \geq 0} \frac{(1 - e^{-z})^k}{[k]!} e^{-zk(k-1)/4}$$

$$\left(e^{zkJ_3/2}(J_+)^k \otimes e^{-zkJ_3/2}(J_-)^k \right)$$

we have easily after the substitution (2.3) and $w = \frac{z}{\epsilon^2}$, by searching for the limit $\epsilon \to 0$:

$$R_H = \lim_{\epsilon \to 0} r(w\epsilon^2, \frac{H \otimes H}{\lambda})e^{w \frac{H \otimes H}{\epsilon^2}} e^{-w/2(N \otimes H + H \otimes N)}{}_e B \otimes B^\dagger$$

where:

$$B \otimes B^\dagger = we^{w\Gamma/4} A \otimes A^\dagger, \quad \Gamma = H \otimes 1 - 1 \otimes H . \tag{2.6}$$

By choosing $r = e^{-zH \otimes H/\lambda}$ and $\lambda = \epsilon^4$ the diverging factors drop out and the finite R_H is recovered

$$R_H = e^{-w/2(N \otimes H + H \otimes N)}{}_e B \otimes B^\dagger \tag{2.7}$$

Of course R_H must satisfy in the Hopf algebra of $H(1)_q$ the relations corresponding to those R satisfies in $U(2)_q$. Anyway it is a tedious but not difficult task to verify directly at the general algebraic level that R_H is the universal R-matrix for $H(1)_q$, that it satisfies the Q.Y.B.E. and equivalently the quasitriangularity conditions.

To complete this example let us consider the lowest dimensional faithful realization of $H(1)_q$ - non hermitean and coinciding with that one of $H(1)$ - given by

$$A = \begin{pmatrix} 0 & 1 & 0 \\ 0 & 0 & 0 \\ 0 & 0 & 0 \end{pmatrix} \qquad A^\dagger = \begin{pmatrix} 0 & 0 & 0 \\ 0 & 0 & 1 \\ 0 & 0 & 0 \end{pmatrix}$$

$$H = \begin{pmatrix} 0 & 0 & 1 \\ 0 & 0 & 0 \\ 0 & 0 & 0 \end{pmatrix} \qquad N = \begin{pmatrix} 0 & 0 & 0 \\ 0 & 1 & 0 \\ 0 & 0 & 0 \end{pmatrix}$$

Therefore, R_H is represented by the 9×9 matrix

$$R_H = \begin{pmatrix} I_3 & wA^\dagger & -\frac{w}{2}N) \\ 0 & I_3 - \frac{w}{2}H & 0 \\ 0 & 0 & I_3 \end{pmatrix},$$

I_3 being the 3×3 identity matrix.

In this representation moreover the group elements are the 3-dimensional triangular matrices T depending on four group coordinates:

$$T = \begin{pmatrix} 1 & \alpha & \beta \\ 0 & 1+\gamma & \delta \\ 0 & 0 & 1 \end{pmatrix}$$

If we now apply the defining relations [14]

$$R_H T_1 T_2 = T_2 T_1 R_H$$

i.e. explicitly

$$\begin{pmatrix} T & \alpha T + wA^\dagger(1+\gamma)T & \beta T + wA^\dagger \delta T - \frac{w}{2}NT \\ 0 & (1+\gamma)T - \frac{w}{2}H(1+\gamma)T & \delta T - \frac{w}{2}H\delta T \\ 0 & 0 & T \end{pmatrix} =$$

$$\begin{pmatrix} T & wTA^\dagger + T\alpha - \frac{w}{2}T\alpha H & -\frac{w}{2}TN + T\beta \\ 0 & T(1+\gamma) - \frac{w}{2}T(1+\gamma)H & T\delta \\ 0 & 0 & T \end{pmatrix}.$$

we find immediatiately that $\alpha, \beta, \gamma, \delta$ as functions on the quantum group $H(1)_q$ satisfy the C^*-algebra relations

$$[\alpha, \beta] = \frac{w}{2}\alpha, \qquad [\alpha, \delta] = 0, \qquad [\delta, \beta] = \frac{w}{2}\delta, \qquad [\gamma, .] = 0,$$

which constitute the dual Hopf algebra structure together with

$$\Delta(T) = T \otimes T, \quad \gamma(T) = T^{-1}, \quad \epsilon(T) = I_3$$

3.   A contracted Hopf superalgebra and an R-matrix for fermions.

An important extension of the previous example is obtained if we start from the supergraded version of the Heisenberg group.

Let us consider the superalgebra $Osp(1|2)$, whose q-deformation has been discovered and studied in [15], the real forms and contractions of which have been recently analised in [10]. It is composed of 3 even generators $X_\pm$, H (which close an $SU(2)$) and 2 odd generators $v_\pm$ satisfying the following graded commutation relations:

$$[H, X_\pm] = \pm X_\pm, \qquad [X_+, X_-] = 2H,$$

$$[H, v_\pm] = \pm \frac{1}{2}v_\pm, \qquad [X_\pm, v_\mp] = v_\pm, \qquad [X_\pm, v_\pm] = 0, \qquad (3.1)$$

$$[v_+, v_-] = -\frac{1}{2}H, \qquad [v_\pm, v_\pm] = \pm \frac{1}{2}X_\pm.$$

In analogy with the H(1) case we add a central even generator $Q : [Q, .] = 0$, and transform the basis as follows:

$$a_\pm = \epsilon X_\pm, \quad \mathcal{N} = -H + \epsilon^{-2}Q, \quad h = 2Q, \quad \text{for even elements,}$$

$$b_\pm = v_\pm, \quad \text{for odd elements,} \qquad (3.2)$$

with ϵ even parameter, and contract by $\epsilon \to 0$. Finally one gets the algebra of bosonic and fermionic oscillators together:

$$[\mathcal{N}, a_\pm] = \pm a_\pm \ , \qquad\qquad [a_-, a_+] = h \ ,$$

$$[\mathcal{N}, b_\pm] = \pm \frac{1}{2} b_\pm \ , \qquad\quad [a_\pm, b_\pm] = [a_\pm, b_\mp] = 0 \ , \qquad (3.3)$$

$$[b_+, b_-] = -\frac{1}{4} h \ , \qquad\qquad [b_\pm, b_\pm] = 0 \ ,$$

h remains a central element and the Casimir c_2 results

$$c_2 = -h\mathcal{N} + \frac{1}{2}(a_- a_+ + a_+ a_-) + (b_- b_+ - b_+ b_-) \ . \qquad (3.4)$$

It is noteworthy that $\mathcal{N}$ appears to be the sum of the bosonic number N_b plus one half of the fermionic one N_F.

In [15] it is shown that the q-deformation $Osp(1|2)_q$ of (3.3) is determined from the graded commutators

$$[v_+, v_-] = -\frac{\sinh(\eta H)}{\sinh(2\eta)}, \quad [H, v_\pm] = \pm\frac{1}{2} v_\pm \qquad [3.5]$$

with η even, $X_\pm$ defined as $\pm 2[v_\pm, v_\pm]$ in the graded envelope and from

$$\Delta v_\pm = v_\pm \otimes e^{\eta H/2} + e^{-\eta H/2} \otimes v_\pm \ ,$$

$$\Delta H = H \otimes 1 + 1 \otimes H \ ,$$

$$\qquad\qquad\qquad\qquad\qquad\qquad\qquad\qquad\qquad\qquad (3.6)$$

$$\gamma(H) = -H \ , \qquad \gamma(v_\pm) = -e^{\pm\eta/4} v_\pm \ ,$$

$$\epsilon(1) = 1, \qquad \epsilon(H) = \epsilon(v_\pm) = 0 \ ,$$

where the tensor product is also graded.

Now to search for the quantum algebra associated to (3.30) we contract $Osp(1|2)_q$ by doing the same change of basis given in (3.2). Moreover to have ηH finite we must rescale η and we put $\eta = \frac{\epsilon^2 w}{2}$. In the new algebra one gets by $\epsilon \to 0$ the only changes are given by:

$$[b_-, b_+] = -\frac{1}{w} \text{sh}(wh/4) \ , \qquad [a_-, a_+] = \frac{2}{w} \text{sh}(wh/2) \ .$$

$$\Delta b_\pm = b_\pm \otimes e^{wh/8} + e^{-wh/8} \otimes b_\pm, \Delta a_\pm = a_\pm \otimes e^{wh/8} + e^{-wh/8} \otimes a_\pm \quad (3.7)$$

$\mathcal{N}$ and h are primitives,antipode and counity trivial.

The set $\mathcal{N}, a_+, a_-, h$ generates $H(1)_q$ and $\mathcal{N}, b_+, b_-, h$ the analogous graded structure $F(1)_q$ for fermions.Moreover doing the same algorithm on the Casimir of $Osp(1|2)_q$ given from [15]

$$c_2(\eta) = \text{ch } 2\eta(H + \frac{1}{4}) - e^{\eta/4} \text{ ch } \eta(H + \frac{1}{2}) \, fe - \frac{e^{\eta/2}}{8} \text{ sh}^{-2}(\eta/4) \, f^2 e^2 \,,$$

$$e = x \, e^{-\eta H/2} \, v_+ \,, \qquad f = x \, e^{\eta H/2} \, v_- \,, \qquad x = \Big(4 \, \text{sh}(\eta/2) \, \text{sh}(2\eta) \Big)^{1/2} \qquad (3.8)$$

we get

$$c_2(w) = -\frac{2}{w} \text{ sh}\Big(\frac{wh}{2}\Big) \, \mathcal{N} + \frac{1}{2}\,(a_- a_+ + a_+ a_-) + \text{ch}\Big(\frac{wh}{4}\Big)\,(b_- b_+ - b_+ b_-) \,. \quad (3.9)$$

The R-matrix of this new deformed graded algebra has been recovered again by contraction from that of $Osp(1|2)$,but as the calculation is rather complex and the method completely analogous to that of the previous bosonic case, I only sketch here a simple heuristic argument to guess it.

The algebra (3.7) is composed by an $H(1)_q$ and its fermionic twin $F(1)_q$: they share the central element h and the particle number $\mathcal{N} = N_B + \frac{1}{2}N_F$, with $[N_B, N_F] = 0$. It is therefore possible to factorize the R-matrix in bosonic and fermionic submatrices $R = R_B \, R_F = R_F \, R_B$, trying to recollect $\mathcal{N}$ at the end. From the previous analysis of $H(1)_q$ we know R_B (2.6),(2.7):

$$R_B = e^{-\frac{w\Omega_B}{2}} e^{w\Lambda_B}; \quad \Omega_B = N_B \otimes h + h \otimes N_B; \Lambda_B = e^{\Gamma w/4}; \quad \Gamma = h \otimes 1 - 1 \otimes h.$$

For the fermionic factor, the analogous structure of $F(1)_q$ suggest an R_F structurally similar to R_B and we guess

$$R_F = e^{-\frac{w\Omega_F}{2}} e^{w\Lambda_F}; \quad \Omega_F = N_F/2 \otimes h + h \otimes N_F/2; \quad \Lambda_F \propto e^{\zeta \Gamma w/4} b_+ \otimes b_-.$$

By inserting this R_F in the equations it must satisfy in $F(1)_q$, one finds that this happens if ζ and the proportionality factor are put equal $1/2$ and 2 respectively. So for the algebra (3.7) the R-matrix satisfying the graded Q.Y.B.E. is

$$R_G = e^{-\frac{w\Omega}{2}} e^{w\Lambda_B} e^{w\Lambda_F} \tag{3.10},$$

with $\Omega = \Omega_B + \Omega_F = N \otimes h + h \otimes N$, and $\Lambda_F = 2e^{\Gamma w/8} b_+ \otimes b_-$.

4.   Inhomogeneous quantum groups and a physical application.

The smallest inhomogeneous group of physical relevance is probably the euclidean group of the plane $E(2)$. Its 3 generators Lie algebra

$$[P_x, P_y] = 0, \quad [J, P_x] = iP_y, \quad [J, P_y] = -iP_x, \tag{4.1}$$

is recovered by means of contractions from the $SU(2)$ one: these same contractions on $SU(2)_q$ give rise to two different quantum versions of $E(2)$.

Starting from the symmetrical change of basis :

$$P_x = \varepsilon J_1, \quad P_y = \varepsilon J_2, \quad J = J_3,$$

one must not rescale the deformation parameter and one finds the $E(2)_q$ studied from Vaksman and Korogodskii [16], in which the algebraic sector coincides with (4.1). By means of the non symmetrical contraction

$$P_y = \varepsilon J_1, \quad J = J_2, \quad P_x = \varepsilon J_3, \quad w = \frac{z}{\varepsilon} \tag{4.2}$$

we find [3,6] the new interesting Hopf algebra, given by:

$$[P_x, P_y] = 0, \qquad [J, P_x] = iP_y, \qquad [J, P_y] = -\frac{i}{w} \text{sh}(wP_x). \tag{4.3}$$

$$\Delta J = e^{-wP_x/2} \otimes J + J \otimes e^{wP_x/2}, \qquad \Delta P_x = 1 \otimes P_x + P_x \otimes 1,$$
$$\Delta P_y = e^{-wP_x/2} \otimes P_y + P_y \otimes e^{wP_x/2}, \tag{4.4}$$

$$\gamma(J) = -J + \frac{i}{2} wP_y, \qquad \gamma(P_x) = -P_x, \qquad \gamma(P_y) = -P_y,$$

and trivial counit. The substitution (4.2) on the invariant of $SU(2)_q$ followed from the limit $\varepsilon \to 0$ gives rise to the Casimir operator which reads

$$C = P_y^2 + \frac{4}{w^2} \text{sh}^2(wP_x/2).$$

It is possible therefore to realize (4.3) in the plane of the eigenvalues p_x, p_y of the diagonal momenta. The boost is then associated to the field with respect to which the Lie derivative of the Casimir curve is zero:

$$ J \equiv i p_y \frac{\partial}{\partial p_x} - (i/w)\mathrm{sh}(w p_x)\frac{\partial}{\partial p_y}; \quad P_x \equiv p_x, \quad P_y \equiv p_y; $$

Moreover it is worth noticing that the 1–D Poincaré group $E(1,1)_q$ has two versions equally admissible, depending on the identification of time generator with $i P_x$ or $i P_y$. But their physical implications are very different, as one gives rise to a wave equation in which there is a fixed spacing in the imaginary time, while the other generates an equation in spatial lattice which is just that of the physical phonons in a crystal [17]. In this last case the non trivial coproduct has produced correct physical predictions about composed states. Anyway it results that the form of this Casimir is the same for the euclidean groups of upper dimension: one has only to add to P_y^2 the squares of the further momenta. On the other hand, if we try to get the R–matrix of $E(2)_q$ by means of the contraction (4.2) applied to the R–matrix of $SU(2)_q$, we find divergences we are unable to control. One can conjecture this difficulty to be connected with the impossibility to get non trivial extension of $E(2)$ by contracting $U(2)$. The contraction from $SU(2)_q$ to $E(2)_q$ recently obtained by Woronowicz, directly at the C^*–level of the functions on the group [18] appears to be an important achievement.

When you pass to 3–D it is possible to find all the properties of $E(3)_q$, in perfect similarity to what has been shown for $H(1)_q$. The peculiar points are the following: we construct an $SO(4)_q$ as the direct product $SU(2)_{q_1} \otimes SU(2)_{q_2}$. By summing and subtracting the homologous generators (the simplest choice, but not the only which works) we get Hopf algebras both for $q_1 = q_2$ and $q_1 = \frac{1}{q_2} = e^z$.

It is a crucial fact that only the last choice gives rise to a structure for which the contraction generating $E(3)$ from $SO(4)$, implemented with the rescaling of the Cartan generators, is a well defined operation recovering a new Hopf algebra.

The rescaling we introduced is

$$ P_s = \varepsilon N_s, \quad , J_s = J_s \quad 2w = z\varepsilon^{-1}, \quad N = J^1 - J^2, \quad J = J^1 + J^2, \quad (s = \pm, 3) , $$

so that the final commutation relations read

$$[J_3, J_\pm] = \pm J_\pm \ ,$$
$$[J_3, P_\pm] = [P_3, J_\pm] = \pm P_\pm \ ,$$
$$[J_s, P_s] = [P_l, P_s] = 0 \ , \tag{4.5}$$
$$[J_+, J_-] = 2 \ J_3 \ \text{ch}(w P_3) \ ,$$
$$[J_\pm, P_\mp] = \pm(2/w) \ \text{sh}(w P_3) \ .$$

while the nontrivial coproducts are

$$\Delta P_\pm = e^{-w P_3/2} \otimes P_\pm + P_\pm \otimes e^{w P_3/2} \ .$$

$$\Delta J_\pm = e^{-w P_3/2} \otimes J_\pm + J_\pm \otimes e^{w P_3/2} - \frac{w}{2}\left(e^{-w P_3/2} \ J_3 \otimes P_\pm - P_\pm \otimes e^{w P_3/2} \ J_3\right) \tag{4.6}$$

with the antipode given by

$$\gamma(J_3) = -J_3 \ , \qquad \gamma(J_\pm) = -(J_\pm \pm w P_\pm) \ ; \tag{4.7}$$
$$\gamma(P_s) = -P_s \ , \qquad (s = \pm, 3)$$

In agreement with general results there is no inclusion of $SO(3)_q$.

In this case the Casimir operators are two, we obtain both by analizing the different orders in ε of the invariants of the two $SU(2)_q$:

$$\left[\frac{J_3 \pm \varepsilon^{-1} P_3}{2}\right]_{e^{2\varepsilon w}} \left[\frac{J_3 \pm \varepsilon^{-1} P_3}{2} + 1\right]_{e^{2\varepsilon w}} + \left(\frac{J_- \pm \varepsilon^{-1} P_-}{2}\right)\left(\frac{J_+ \pm \varepsilon^{-1} P_+}{2}\right) \ .$$

The terms of order ε^{-2} produce the first Casimir of $E(3)_q$

$$C_1 = P_1^2 + P_2^2 + \frac{4}{w^2} \ \text{sh}^2\left(\frac{w P_3}{2}\right) \ , \tag{4.8}$$

and the terms of order ε^{-1} give the second one

$$C_2 = J_1 P_1 + J_2 P_2 + \frac{1}{w} \ J_3 \ \text{sh}(w P_3) \ . \tag{4.9}$$

By means of rather complex algorithms it has been possible [5,6] , by following the same procedure , to recover the R–matrix of $E(3)_q$. The result is the following:

$$R = \exp\{w(J_3 \otimes P_3 + P_3 \otimes J_3)\} \ \exp\{2B \ \mathrm{arcsh}(wA)/(wA)\}(1 + w^2 A^2)^{-1/2}$$

where

$$A = \frac{w}{2}\, Q_+ \otimes Q_- \,,$$

$$B = \frac{w}{2}\, (L_+ \otimes Q_- + Q_+ \otimes L_-)$$

$$- \frac{w^2}{2}\left(Q_+ \otimes Q_- - \frac{1}{2}\,(J_3 Q_+ \otimes Q_- - Q_+ \otimes J_3 Q_-)\right),$$

with

$$Q_\pm = e^{\pm w P_3/2}\, P_\pm\,, \qquad\qquad L_\pm = e^{\pm w P_3/2}\, J_\pm\,.$$

To get quantization of the group we introduce the 4-dimensional representation of $E(3)_q$, so the R-matrix becomes a 16-dimensional one and the generic matrix T can be written

$$T = \begin{pmatrix} \exp(i\Sigma) & \mathbf{v} \\ 0 & 1 \end{pmatrix} \tag{4.10}$$

with $\Sigma \in LieSO(3)$ and we can impose the relation $R\, T_1\, T_2 \ = \ T_2\, T_1\, R$, with $\exp(i\Sigma)$ expressed by the Euler angles θ, ϕ, χ and $\mathbf{v} \ = \ (x, y, z)^T$. To get the Hopf algebra of the representative functions its relations are presented in terms of the angles $(\theta,\ \psi = \phi - \chi,\ \omega = \phi + \chi)$ and of the coordinates $(\tilde{x} = x\cos\phi + y\sin\phi,\ \tilde{y} = -x\sin\phi + y\cos\phi,\ z)$. The representative functions are entire functions of the Euler angles, so we can express the relations coming from the defining relations as

$$[\theta, \tilde{x}] = w\ \sin\theta\ \mathrm{tg}(\theta/2)\,,\qquad [\theta, z] = w\ \sin\theta\,,\qquad [\omega, \tilde{y}] = -2w\ \mathrm{tg}(\theta/2)\,,$$
$$[z, \tilde{x}] = w\ z\,,\qquad\qquad [\tilde{y}, \tilde{x}] = w\ \tilde{y}\,,$$

all other commutators being vanishing. To get a complete formulation of the dual Hopf structure the coalgebra of the representative functions must be given. It is determined, as usual, by the multiplication and the inverse of the representative matrix,

$$\Delta M_{ij} = M_{ik} \otimes M_{kj} , \qquad \gamma(M_{ij}) = (M^{-1})_{ij} .$$

I want to conclude this talk by showing you a simple use [19] of an inhomogeneous quantum group to get new results about the everyday physics. Let us recall the two Casimir of $E(3)_q$ given in (4.8),(4.9) and generalize the first one to the 4–dimensional case by adding P_z^2 and changing the signa for Minkowsky space. In this way you guess the first Casimir of $E(3,1)_q$, confirmed for the different regimes of q in ref. [8,9]. Let us call it M^2, you get the deformed expression for the energy:

$$E = \frac{2}{w} \operatorname{arcsh} \left(\left(\frac{w^2 M^2}{4} + \frac{w^2}{4}(\mathbf{p}^2) \right)^{1/2} \right). \qquad (4.11)$$

When $w \to 0$ you get the usual relativistic relation between mass M, energy E and impulse **p**. But when you search for the q-Galilei limit of your deformed Poincaré (by means of a contraction) you find [6] that your energy is expressed as

$$E = 2/u \ \operatorname{arcsh} \frac{uM}{2} + \frac{\mathbf{p}^2}{2M_{ph}} \qquad (4.12)$$

where u is derived by contraction from w and $M_{ph} = M\sqrt{M^2 u^2/4 + 1}$. So we see that the new kinematics determines the ground classical energy, coinciding with the q–relativistic rest mass, different from the physical mass M_{ph}, and that only the rest mass can be tranformed in energy during a reaction. As it is reasonable to think that something observable, if any, could be found only in interacting microscopic systems, we recover, by means of simple analytical mechanics, from the previous energy *vs* momentum relation an energy *vs* angular momentum relation, getting for the quantized hamiltonian of such systems:

$$\mathcal{H} = \frac{2}{w} \operatorname{arcsh} \left(\left(\frac{w^2 M^2}{4} + \frac{w^2}{4a^2}(\mathbf{J}(\mathbf{J}+1)) \right)^{1/2} \right). \qquad (4.13)$$

where a is proportional to the classical giration radius of the system and J is the quantized angular momentum ($\hbar = 1$). We have fitted by means of (4.13) the energy spectra of the ground rotational bands of a large sample of heavy even-even nuclei,rare earths and actinides. The physical masses of the various isotopes and the rotational energy levels have been used as data, the deformation variable w and the size measure a are the fit parameters. The issues are very good as the average value of the mean quadratic error on 29 deformed bands - each composed of states of even J with J from 2 to a minimum of 16 and a maximum of 28 - is 3.8 Kev over an average value of the fitted energies of about 1750 Kev. The same fit, done on the $SU(2)_q$–rotator model proposed in [20] gives 7.2 Kev.

The values of a, of order of $3 \div 4$ fm, are mainly increasing with the the atomic number A and their distribution could be accomodated into the phenomenological trend $a \propto A^{1/3}$. Moreover the formula (4.13) implies a critical value of J and of the corresponding angular velocity ω at which the plot of angular momentum vs angular velocity ω degenerates. This value of ω is in agreement with the experimental points at which the nuclear backbendings and upbendings have rise. It must be stressed that the maximal velocities at hand are few thousanths of the velocity of the light, as required from the nuclear physics knowledge. Nevertheless the deviations from the classical rotator are rather strong. This is a peculiar effect of the deformed relativity: you can have strong corrections at non relativistic speed, purvue the deformation parameter w is sufficiently high. On the other hand the best fitting values of w are in a range of times scales characteristic of strong processes. This example and the case of unidimensional crystal lattice [17] support the hope that the domain of direct physical applications of quantum inhomogeneous algebras may be not so far from the physics of today.

References.

[1] S.L.Woronowicz and and S.Zakrzewski,“*Quantum deformations of Lorentz group. Hopf∗– algebra level* ”, Preprint.

[2] A.Schirrmacher ,“*Aspects of Quantizing Lorentz Symmetry* ” MPI–PTh/91–77 ,Talk given at Cargèse Institut,July 15–27,1991. Celeghini E., Giachetti R., Reyman A., Sorace E. and Tarlini M., Lett.in Math. Phys. **23** (1991) 45.

[3] E.Celeghini, R.Giachetti, E.Sorace and M.Tarlini, J. Math. Phys. **31** (1990) 2548.

[4] E. Celeghini, R.Giachetti, E.Sorace and M.Tarlini, J. Math. Phys. **32** (1991) 1155.

[5] E. Celeghini, R.Giachetti , E.Sorace and M.Tarlini, J. Math. Phys. **32** (1991) 1159.

[6] E. Celeghini, R. Giachetti, E. Sorace and M. Tarlini, “*Contractions of quantum groups*”, Proceedings of the first semester on quantum groups, Eds. L.D. Faddeev and P.P. Kulish, Leningrad October 1990, Springer-Verlag, in press.

[7] E. Celeghini, R. Giachetti, P.Kulish, E. Sorace and M. Tarlini, J.Physics A:Math. Gen. **24** (1991) 5675.

[8] J. Lukierski, H. Ruegg, A. Nowicki and V.N. Tolstoy, Phys. Lett B **264**, 331 (1991).

[9] J. Lukierski, A. Nowicki, H.Ruegg, "*Quantum deformations of $D = 4$ Poincaré and conformal algebras*, Preprint.

[10] J. Lukierski, A. Nowicki, "*Real forms of $U_q(OS_p(1|2))$ and quantum $D = 2$ Supersymmetry algebras* , Preprint, June 1991 ITP UWr 777/91

[11] R. Gilmore, "*Lie groups, Lie algebras and some of their applications*" (New York, N.Y., 1974, Wiley).

[12] E. Celeghini and M.Tarlini , Nuovo Cim. **B61** (1981) 265, **B65** (1981) 172 and **B68** (1982) 133.

[13] S.M.Khoroskin and V.N.Tolstoy, "*Universal R-Matrix for Quantized (Super)Algebras*, Preprint, Moscow 1991.

14] Faddeev L., Reshetikhin N.Yu. and Takhtajan L., Algebra i Analiz, **1**, (1989),178

[15] P.P. Kulish, N.Yu. Reshetikhin, Lett. Math. Phys. **18**.

[16] L.L. Vaksman and L.I.Korogodskii , Soviet Math. Dokl. **39** (1989) 173.

[17] F. Bonechi, E. Celeghini, R. Giachetti, E. Sorace and M. Tarlini, "*Inhomogeneous Quantum Groups as Symmetry of Phonons.*", University of Florence Preprint, DFF 152/12/91

[18] S.L.Woronowicz private communication. S.L.Woronowicz "*Quantum $SU(2)$ and $E(2)$ groups.Contraction procedure* ", Preprint, Lyon I, December 7, 1991.

[19] E. Celeghini, R. Giachetti, E.Sorace and M.Tarlini , *Quantum groups of motion and rotational spectra of heavy nuclei*, Preprint DFF 151/11/91 Firenze, to be published in Phys.Lett. B.

[20] D. Bonatsos, E.N. Argyres, S.B. Drenska, P.P. Raychev, R.P. Roussev and Yu.F. Smirnov, Phys. Lett. B **251**, 477 (1990).

Realifications
of complex quantum groups

S. Zakrzewski[*†]

Inst. f. Theoretische Physik, TU Clausthal
Leibnizstr. 10, W–3392 Clausthal–Zellerfeld, Germany

Abstract

A passage from a complex group to a real group analogical to the realification procedure is investigated in context of Hopf ($*$-)algebras. As an application, a $(2N^2 - 5N + 5)$–parameter quantum deformation of $SL(N, \mathbf{C})$ as a real group is described.

Introduction

In [1] several (presumably all) quantum deformations of $SL(2, \mathbf{C})$ as a real group have been found. The method consisted in finding all quantum groups of 2×2 matrices satisfying characteristic for $SL(2, \mathbf{C})$ properties of tensor products involving the fundamental representation and its complex conjugate. Practically, the procedure turned out to split into two steps. The first step consisted in picking up a quantum deformation of $SL(2, \mathbf{C})$ as a *complex* group. The second step resembled the realification procedure: along with holomorphic polynomials in the matrix elements we had to introduce anti-holomorphic polynomials and provide consistent commutation relations between polynomials of different type.

It turns out that the consistency conditions concerning the commutation relations, needed for such a realification can be easily formulated in much more general case, including complex quantum groups of the series A_n, B_n, C_n, D_n introduced in [2]. To stress this fact and present some examples is just the aim of this paper.

The paper is organized as follows. Sect. 1 contains basic notions and useful constructions. In Sect. 2 we define twisted realifications by characteristic properties, then we show an equivalent constructive definition in terms of a map defining the commutation rules between 'holomorphic' and 'antiholomorphic' elements. The constructive approach is specialized in Sect. 3 to the case of matrix groups given

---

[*]Supported by Alexander von Humboldt Foundation
[†]On leave from: Department of Mathematical Methods in Physics, Warsaw University

R. Gielerak et al. (eds.), Groups and Related Topics, 83–100.
© 1992 *Kluwer Academic Publishers*.

in terms of generators and relations. We assume there that the product of an antiholomorphic generator by a holomorphic one is a linear combination of products of holomorphic generators by antiholomorphic ones. It turns out that then there is a matrix X governing the commutation rules between holomorphic and antiholomorphic generators, playing the role similar as matrix R in the R-matrix approach. The matrix X has to satisfy conditions analogical to (6), (50) and (51) in [1]. Our main theorem says that having a matrix X satisfying those conditions, we can construct a twisted realification. The method is then applied in Sect. 4 to obtain examples of realifications of known complex matrix quantum groups.

Due to different choices of the matrix X we have different twisted realifications of the same complex quantum group. In Sect. 4 we describe a $(N-1)^2$-parameter family (parameters are real) of twisted realifications of the standard $SL_q(N, \mathbf{C})$. The same family of matrices X provides also twisted realifications of each member of the $(N^2 - 3N + 4)$-parameter family of deformations of complex $SL(N, \mathbf{C})$ described in [3]. In total, this gives a $(2N^2 - 5N + 5)$-parameter family of deformations of $SL(N, \mathbf{C})$ as a real group.

A remarkable example of a twisted realification has been recently worked out in [4], generalizing considerably the result of [5].

In this paper, quantum groups are considered on the level of Hopf $(*$-$)$algebras. In other words, we adopt the following terminology:

I. A *real quantum group* G is given by a Hopf $*$-algebra $(A, *, \Delta)$. We write $(A, *, \Delta) = \mathcal{P}ol\,(G)$ to denote this correspondence (we treat A as the algebra of 'polynomials' on G).

II. A *complex quantum group* H is given by a Hopf algebra (A, Δ) with bijective antipode. Again we write $(A, \Delta) = \mathcal{P}ol\,(H)$ to denote this correspondence (A is treated as the algebra of 'holomorphic polynomials' on H).

In both cases, by a *representation* we mean an element u of $\mathrm{End}\,(K) \otimes A$, where K is a finite-dimensional complex vector space, such that

$$(\mathrm{id} \otimes \Delta)u = u \oplus u, \qquad (\mathrm{id} \otimes e)u = \mathrm{id}_K,$$

where e is the counit. We refer to [6] for the definition of symbols ① and ⊤. The set of all linear maps intertwining [6] two representations u and v is denoted by $\mathrm{Mor}\,(u, v)$.

In order to avoid complicated formulas we shall use a universal notation for the flip map acting in tensor products of vector spaces:

$$\sim : Y \otimes Z \to Z \otimes Y, \qquad y \otimes z \mapsto z \otimes y.$$

The notation does not indicate to what spaces the flip is applied. However, we shall use it only when it is clear what are Y and Z in a particular case. A similar remark applies to the identity map 'id '.

In the next section we recall necessary definitions and facts concerning Hopf $(*$-$)$algebras.

1   Basic notions and constructions

We use the following terminology. By an algebra we mean always an algebra over
C with unit. By a morphism of algebras we mean a homomorphism preserving
the unit. An *opposite algebra* A^{op} of a given algebra A is defined to be equal A as
a vector space and equipped with the opposite multiplication: $m^{\mathrm{op}} = m \sim$, where
$m: A \otimes A \to A$ is the multiplication map in A.

1.1   Hopf (∗-)algebras

A pair (A, Δ) is said to be a *Hopf algebra* if A is an algebra, $\Delta: A \to A \otimes A$ is a
morphism satisfying $(\Delta \otimes \mathrm{id})\Delta = (\mathrm{id} \otimes \Delta)\Delta$ and there exists a morphism $e: A \to C$
and a linear map $k: A \to A$ such that $(e \otimes \mathrm{id})\Delta = \mathrm{id} = (\mathrm{id} \otimes e)\Delta$ and

$$m(k \otimes \mathrm{id})\Delta = \imath e = m(\mathrm{id} \otimes k)\Delta,$$

where $m: A \otimes A \to A$ is the multiplication map and $\imath: C \to A$ is given by $\imath(\lambda) = \lambda I$,
I being the unit of A (e is said to be the *counit* and k is said to be the *antipode*;
both are unique). The antipode k is bijective if and only if $(A^{\mathrm{op}}, \Delta)$ is a Hopf
algebra. If this is the case then the antipode of $(A^{\mathrm{op}}, \Delta)$ equals k^{-1}.

A *morphism* from a Hopf algebra (A_1, Δ_1) to a Hopf algebra (A_2, Δ_2) is a
morphism $f: A_1 \to A_2$ of algebras such that $\Delta_2 f = (f \otimes f)\Delta_1$. Then $e_2 f = e_1$ and
$k_2 f = f k_1$, where e_j, k_j are the counit and the antipode in (A_j, Δ_j), $j = 1, 2$ (see
Lemma 6.8 in [7]).

A Hopf algebra (A_1, Δ_1) is a *Hopf subalgebra* of a Hopf algebra (A, Δ) if $A_1 \subset A$
and the inclusion map is a morphism of Hopf algebras.

Consider a Hopf algebra (A, Δ) and a subalgebra $A_1 \subset A$ containing the unit
and such that $\Delta(A_1) \subset A_1 \otimes A_1$. Then $(A_1, \Delta|_{A_1})$ is a Hopf (sub-)algebra if and
only if $k(A_1) \subset A_1$ (k is the antipode in (A, Δ)).

A triple $(A, *, \Delta)$ is said to be a Hopf ∗-algebra if (A, Δ) is a Hopf algebra,
$(A, *)$ is a ∗-algebra and $\Delta* = (* \otimes *)\Delta$. In this case we have $*k * k = \mathrm{id}$ (cf.
Lemma 6.2 in [7]). In particular, k is bijective.

1.2   Double algebras

A triple of algebras $(B; A_1, A_2)$ is said to be a *double algebra* if A_1, A_2 are subalgebras of B (containing the unit I of B) and the linear maps

$$\phi: A_1 \otimes A_2 \to B, \qquad \psi: A_2 \otimes A_1 \to B \tag{1}$$

defined by $\phi(a_1 \otimes a_2) = a_1 a_2$, $\psi(a_2 \otimes a_1) = a_2 a_1$ for $a_1 \in A_1$, $a_2 \in A_2$, are bijections.

In the above situation one can use ϕ to transfer the algebra structure from B
to the vector space $A_{12} = A_1 \otimes A_2$. The multiplication map $m_{12}: A_{12} \otimes A_{12} \to A_{12}$,

defined by $m_{12} = \phi^{-1}m(\phi \otimes \phi)$, where $m: B \otimes B \to B$ is the multiplication in B, can be written as

$$m_{12} = (m_1 \otimes m_2)(\mathrm{id} \otimes s \otimes \mathrm{id}), \tag{2}$$

where $m_k: A_k \otimes A_k \to A_k$, $k = 1, 2$, are the multiplications in A_1, A_2 and $s = \phi^{-1}\psi$. The unit $I_{12} = \phi^{-1}I$ is given by

$$I_{12} = I_1 \otimes I_2, \tag{3}$$

where I_k is the unit of A_k, $k = 1, 2$. Clearly, all information about the algebra structure induced on A_{12} by ϕ is given by the map s (and the algebraic structure of A_1 and A_2). Note the following properties of s:

$$s(I_2 \otimes a_1) = a_1 \otimes I_2 \qquad \text{for } a_1 \in A_1 \tag{4}$$

$$s(a_2 \otimes I_1) = I_1 \otimes a_2 \qquad \text{for } a_2 \in A_2 \tag{5}$$

$$s(\mathrm{id} \otimes m_1) = (m_1 \otimes \mathrm{id})(\mathrm{id} \otimes s)(s \otimes \mathrm{id}) \tag{6}$$

$$s(m_2 \otimes \mathrm{id}) = (\mathrm{id} \otimes m_2)(s \otimes \mathrm{id})(\mathrm{id} \otimes s) \tag{7}$$

(two last equalities follow easily from the associativity of m_{12}).

Above conditions are actually sufficient to construct a double algebra containing given pair of algebras. Given two algebras A_1, A_2 and a linear bijection $s: A_2 \otimes A_1 \to A_1 \otimes A_2$ satisfying (4)–(7), formulae (2), (3) define on the vector space $A_{12} = A_1 \otimes A_2$ a structure of an algebra, which we denote by $A_1 \circledS A_2$, and $(A_1 \circledS A_2; A_1, A_2)$ is a double algebra (the embedding of A_k in $A_1 \circledS A_2$ is obvious).

In case when $s = \sim$, $A_1 \circledS A_2$ coincides with the usual tensor product $A_1 \otimes A_2$ of two algebras. We refer to $A_1 \circledS A_2$ as to a *twisted tensor product of A_1 and A_2*.

1.3   Double Hopf algebras

A triple of Hopf algebras $((B, \Delta); (A_1, \Delta_1), (A_2, \Delta_2))$ is said to be a *double Hopf algebra* if (A_1, Δ_1), $(A_2, \Delta_2))$ are Hopf subalgebras of (B, Δ) and the linear maps

$$\phi: A_1 \otimes A_2 \to B, \qquad \psi: A_2 \otimes A_1 \to B$$

defined as in (1), are bijections.

Remark 1.1 By the definition of ϕ and ψ we have

$$k\phi = \psi \sim (k_1 \otimes k_2), \tag{8}$$

where k, k_1, k_2 are the antipodes in the considered Hopf algebras. If k, k_1, k_2, ϕ are bijective, then ψ is also bijective.

Suppose we have a double Hopf algebra as in above definition. Let us reformulate its structure in terms of the vector space $A_{12} = A_1 \otimes A_2$ using the map ϕ. Clearly, $(B; A_1, A_2)$ is a double algebra (cf. Sect.1.2) and its image by ϕ is given by the twisted tensor product $A_1 \circledS A_2$ (for a suitable s). The coalgebra structure in A_{12} (i.e. the comultiplication $\Delta_{12} = (\phi^{-1} \otimes \phi^{-1})\Delta\phi$ and the couinit $e_{12} = e\phi$) turns out to be that of the (usual) tensor product of coalgebras:

$$\Delta_{12} = (\mathrm{id}\otimes \sim \otimes\mathrm{id})(\Delta_1 \otimes \Delta_2), \tag{9}$$

$$e_{12} = e_1 \otimes e_2, \tag{10}$$

where e, e_1, e_2 are counits in the considered Hopf algebras. The expression for the antipode $k_{12} = \phi^{-1}k\phi$ follows from (8):

$$k_{12} = s \sim (k_1 \otimes k_2). \tag{11}$$

Note that s is a morphism of coalgebras:

$$\Delta_{12}s = (s \otimes s)\Delta_{21}, \tag{12}$$

where $\Delta_{21} = (\mathrm{id}\otimes \sim \otimes\mathrm{id})(\Delta_2 \otimes \Delta_1)$. Indeed, since Δ_{12} is a morphism of algebras, we have

$$\Delta_{12}(s(a_2 \otimes a_1)) = \Delta_{12}((I_1 \otimes a_2)(a_1 \otimes I_2)) = (s \otimes s)(\mathrm{id}\otimes \sim \otimes\mathrm{id})(\Delta_2(a_2) \otimes \Delta_1(a_1))$$

for $a_1 \in A_1$, $a_2 \in A_2$. By property (12), the counit (10) is automatically preserved by s: $(e_1 \otimes e_2)s = e_2 \otimes e_1$. This can be proved by applying $e_2 \otimes e_1 \otimes e_1 \otimes e_2$ to the both sides of the following version of (12):

$$(\mathrm{id} \otimes s)\Delta_{21} = (s^{-1} \otimes \mathrm{id})\Delta_{12}s.$$

The analysis we have done so far is sufficient to proceed now with a synthesis. One can easily check (diagrams!) that, given two Hopf algebras (A_1, Δ_1), (A_2, Δ_2) and a linear bijection $s: A_2 \otimes A_1 \to A_1 \otimes A_2$ satisfying (4)–(7) and (12), formula (9) defines on $A_1 \circledS A_2$ a structure of a Hopf algebra (with counit (10) and antipode (11)) and

$$((A_1 \circledS A_2, \Delta_{12}); (A_1, \Delta_1), (A_2, \Delta_2))$$

is a double Hopf algebra.

1.4   Conjugation

We remind that with each complex vector space V one can associate its *complex conjugate* space $\overline{V}$. It is defined to coincide with V as an abelian group. The multiplication by scalars $\lambda \in \mathbf{C}$ is however modified:

$$\lambda \bar{v} = \overline{\bar{\lambda}v},$$

for $v \in V$, where $\bar{v}$ denotes the element v when treated as an element of the vector space $\overline{V}$. The parametrization $v \mapsto \bar{v}$ of $\overline{V}$ by V is thus an anti-linear bijection. The construction is completed by a definition of the *complex conjugate map* $\bar{f}: \overline{V} \to \overline{W}$ of a linear map $f: V \to W$, given by $\bar{f}(\bar{v}) = \overline{f(v)}$ for $v \in V$ (leading to a covariant functor whose square is naturally isomorphic to identity).

With each algebra A one can associate its *complex conjugate algebra* $\overline{A}$, by applying the complex conjugation functor both to A (as a vector space) and its multiplication map. The opposite algebra $\overline{A}^{\mathrm{op}}$ of $\overline{A}$ is said to be the *complex conjugate opposite algebra* of A. The map $a \mapsto \bar{a}$ is an anti-linear anti-multiplicative bijection from A to $\overline{A}^{\mathrm{op}}$.

If (A, Δ) is a Hopf algebra, then the pair $(\overline{A}^{\mathrm{op}}, \overline{\Delta})$ is a Hopf algebra if and only if the antipode k of (A, Δ) is bijective. In this case $(\overline{A}^{\mathrm{op}}, \overline{\Delta})$ is said to be the *conjugated Hopf algebra* of (A, Δ) (cf. [4,10]). Its antipode equals $\overline{k}^{-1}$.

2 Twisted realifications

Definition. A real quantum group G, $\mathcal{P}ol(G) = (B, *, \Delta_B)$, is said to be a *twisted realification* of a complex quantum group H, $\mathcal{P}ol(H) = (A, \Delta)$, if (A, Δ) is a Hopf subalgebra of (B, Δ_B) and the linear map

$$\phi: A \otimes A^* \to B$$

defined by $\phi(a \otimes b^*) = ab^*$ for $a, b \in A$, is bijective. Here $A^* = \{a^*: a \in A\} \subset B$.

Assume the situation as in above definition. Let k and k_B denote the antipodes in (A, Δ) and (B, Δ_B). Clearly, A^* is a subalgebra (with unit) in B. We have $\Delta(A^*) \subset A^* \otimes A^*$ and $k_B(A^*) \subset A^*$ (because $k_B(a^*) = (k_B^{-1}(a))^* = (k^{-1}(a))^*$ for $a \in A$). It follows that $(A^*, \Delta_B|_{A^*})$ is a Hopf algebra and the bijective map $a \mapsto a^*$ is anti-linear, anti-multiplicative and comultiplication preserving, as far as the Hopf algebra structures of (A, Δ) and $(A^*, \Delta_B|_{A^*})$ are concerned. Composing this bijection with the complex conjugation (cf. Sect.1.4) we get an isomorphism of Hopf algebras $(A^*, \Delta_B|_{A^*})$ and $(\overline{A}^{\mathrm{op}}, \overline{\Delta})$.

On the other hand, the bijectivity of ϕ implies that (A, Δ) and $(A^*, \Delta_B|_{A^*})$ define in (B, Δ_B) the structure of a double Hopf algebra (cf. Remark 1.1). The Hopf $*$-algebra structure on the vector space $A \otimes A^*$ induced from B by ϕ is therefore given by a twisted tensor product of A and A^* such that condition (12) is satisfied and the star operation $*_{12} = \phi^{-1} * \phi$ such that $(a \otimes b^*)^{*_{12}} = b \otimes a^*$ for $a, b \in A$. Using the former identification of $(A^*, \Delta_B|_{A^*})$ and $(\overline{A}^{\mathrm{op}}, \overline{\Delta})$, we obtain a structure of a Hopf $*$-algebra on the vector space $A \otimes \overline{A}^{\mathrm{op}}$. It is given again by a twisted tensor product $A \circledS \overline{A}^{\mathrm{op}}$ with the star operation characterized as follows:

$$(a \otimes \bar{b})^{*_{12}} = b \otimes \bar{a} \tag{13}$$

for $a, b \in A$. Since

$$((I \otimes \bar{b})(a \otimes \bar{I}))^{*_{12}} = (a \otimes \bar{I})^{*_{12}}(I \otimes \bar{b})^{*_{12}},$$

we obtain the following condition for s:

$$\sim \bar{s} \sim = s. \tag{14}$$

It is easy to see that (13) defines a structure of a $*$-algebra in $A \circledS \bar{A}^{\mathrm{op}}$ if and only if (14) is satisfied.

The following proposition shows that conditions on the map s written so far are sufficient to construct a twisted realification.

Proposition 2.1 *Let (A, Δ) be a Hopf algebra with bijective antipode. Let $s \colon \bar{A}^{\mathrm{op}} \otimes A \to A \otimes \bar{A}^{\mathrm{op}}$ be a linear bijection satisfying*

$$s(\bar{a} \otimes I) = I \otimes \bar{a}, \qquad s(\bar{I} \otimes a) = a \otimes \bar{I} \tag{15}$$

for $a \in A$ (I is the unit of A),

$$s(\mathrm{id} \otimes m) = (m \otimes \mathrm{id})(\mathrm{id} \otimes s)(s \otimes \mathrm{id}) \tag{16}$$

$$s(\bar{m} \sim \otimes \mathrm{id}) = (\mathrm{id} \otimes \bar{m} \sim)(s \otimes \mathrm{id})(\mathrm{id} \otimes s), \tag{17}$$

(m is the multiplication in A), condition (14) and condition (12) with $\Delta_{12} = (\mathrm{id} \otimes \sim \otimes \mathrm{id})(\Delta \otimes \bar{\Delta})$, $\Delta_{21} = (\mathrm{id} \otimes \sim \otimes \mathrm{id})(\bar{\Delta} \otimes \Delta)$.
Then $(A \circledS \bar{A}^{\mathrm{op}}, *_{12}, \Delta_{12})$ with $*_{12}$ defined by (13) is a Hopf $*$-algebra and

$$((A \circledS \bar{A}^{\mathrm{op}}, \Delta_{12}); (A, \Delta), (A, \bar{\Delta}))$$

is a double Hopf algebra.

Proof:
The double Hopf algebra structure follows from Sect. 1.3. It is easy to check that $\Delta_{12}*_{12} = (*_{12} \otimes *_{12})\Delta_{12}$.

$$\text{Q.E.D.}$$

Proposition 2.1 describes all (up to isomorphisms) realifications of a complex quantum group H, $\mathcal{P}ol(H) = (A, \Delta)$. They are in one-to-one correspondence with maps s satisfying required conditions. The realification of H corresponding to a particular s will be denoted by $H^{\mathrm{R}(s)}$:

$$\mathcal{P}ol(H^{\mathrm{R}(s)}) = (A \circledS \bar{A}^{\mathrm{op}}, *_{12}, \Delta_{12}). \tag{18}$$

One can easily see that $s = \sim$ satisfies the required conditions. We set $H^{\mathrm{R}(\sim)} = H^{\mathrm{R}}$. This particular case is said to be the *trivial realification* of H. The trivial realification defines a functor from complex to real quantum groups.

A nontrivial realification of complex quantum groups belonging to the series A_n, B_n, C_n, D_n has been constructed recently in [4] by taking into account their real forms (and requiring the latter to be subgroups of these realifications).

Let $G = H^{\mathbf{R}(s)}$ be a twisted realification of a complex quantum group H as in (18). Each representation $u \in \operatorname{End}(K) \otimes A$ of H can be treated as a representation of G (by the embedding of A into $A \circledS \overline{A}^{\mathrm{op}}$). Moreover, the complex conjugation $\overline{u} \in \operatorname{End}(\overline{K}) \otimes \overline{A}^{\mathrm{op}}$ of u is also a representation of G. One can easily obtain analogues of the last statement in point 1) and the first statement in point 2) of Theorem 3.1 of [4] using arguments therein:

1. If $(u^\alpha)_{\alpha \in \mathcal{J}}$ is a set of irreducible pairwise disjoint representations of H, then

$$(u^\alpha \oplus \overline{u^\beta})_{\alpha, \beta \in \mathcal{J}} \tag{19}$$

is a set of irreducible pairwise disjoint representations of G.

2. If each representation of H is a direct sum of irreducible representations, then each representation of G has the same property, and, for $\mathcal{J}$ containing representatives of all irreducible representations of H, (19) contains representatives of all irreducible representations of G.

Let us only mention the line of reasoning showing that $u \oplus \overline{v}$ is irreducible if u, v are irreducible representations of H (we shall use it in the next section). It uses the following chain of equivalent statements (cf. Lemma 4.8 in [6]):

(i) u, v irreducible,

(ii) matrix elements of u linearly independent, matrix elements of v linearly independent,

(iii) matrix elements of $u \oplus \overline{v}$ linearly independent,

(iv) $u \oplus \overline{v}$ irreducible.

In the next section we shall use also the following lemma.

Lemma 2.2 *Let v, w, r be representations of H and let r be irreducible. Then each linear map T intertwining $v \oplus \overline{r}$ with $w \oplus \overline{r}$ is of the form $T = T_1 \otimes \operatorname{id}$ with $T_1 \in \operatorname{Mor}(v, w)$.*

Proof: The relation of intertwining can be written in matrix components as follows:

$$\sum_{bjk} T^{ai}_{bj} v^b_c \overline{r^j_k} \delta^k_l = \sum_{djk} w^a_d \overline{r^j_k} \delta^i_j T^{dk}_{cl} .$$

Since $\overline{r^j_k}$ are linearly independent we obtain

$$\sum_b T^{ai}_{bj} v^b_c \delta^k_l = \sum_d w^a_d \delta^i_j T^{dk}_{cl} . \tag{20}$$

Applying the counit e to both sides we get

$$T^{ai}_{cj} \delta^k_l = \delta^i_j T^{ak}_{cl} .$$

If we set $k = l = 1$, we obtain $T = T_1 \otimes \operatorname{id}$. Then we use (20) again.

                                                                Q.E.D.

With the assumptions of the above lemma we have therefore

$$\mathrm{Mor}\,(v \oplus \bar{r}, w \oplus \bar{r}) = \mathrm{Mor}\,(v, w) \otimes \mathrm{id}, \tag{21}$$

$$\mathrm{Mor}\,(\bar{r} \oplus v, \bar{r} \oplus w) = \mathrm{id} \otimes \mathrm{Mor}\,(v, w) \tag{22}$$

(the second equality follows in an analogous way as the first one).

3 Realification of matrix groups

Consider a pair (H, u), where H is a complex quantum group, $\mathcal{P}ol\,(H) = (A, \Delta)$, and $u \in \mathrm{End}\,(K) \otimes A$ is a representation of H in K. We assume that the subspace $A^1 = \{(\rho \otimes \mathrm{id})u : \rho \in \mathrm{End}\,(K)'\}$ spanned by the matrix elements of u, generates A algebraically (i.e. (H, u) is a complex *matrix* quantum group). We assume also that u is irreducible.

We shall study realifications of $G = H^{\mathrm{R}(s)}$ of H satisfying the following assumption:

$$s(\overline{A^1} \otimes A^1) = A^1 \otimes \overline{A^1}. \tag{23}$$

This assumption means that the subspaces spanned in $A \circledS \overline{A}^{\mathrm{op}}$ by matrix elements of representations $u \oplus \bar{u}$ and $\bar{u} \oplus u$ are the same. Since these representations are irreducible (see Sect.2), it follows that they are equivalent, hence there exists a unique (up to a constant) invertible linear map $X : K \otimes \overline{K} \to \overline{K} \otimes K$ intertwining them:

$$(X \otimes \mathrm{id})(u \oplus \bar{u}) = (\bar{u} \oplus u)(X \otimes \mathrm{id}). \tag{24}$$

It is easy to check (cf. formula (6) of [1], also formula (14) above), that such an intertwiner has to satisfy the following 'reality' condition:

$$\sim \overline{X} \sim = \tau X, \tag{25}$$

where τ is a complex number (necessarily of modulus one). The following proposition states compatibility requirements analogous to conditions (50), (51) of [1].

We shall denote by $K^{(l)}$ (resp. $u^{(l)}$) the l-th tensor power of K (resp. u), $K^{(0)} = \mathbf{C}$, $u^{(0)} = I$ (the unit of A). We set $X^{(0)} = \mathrm{id}$ and, for natural l,

$$X^{(l)} = (X \otimes \mathrm{id} \otimes \ldots \otimes \mathrm{id})(\mathrm{id} \otimes X \otimes \ldots \otimes \mathrm{id}) \ldots (\mathrm{id} \otimes \mathrm{id} \otimes \ldots \otimes X) \qquad (l \text{ times}).$$

Then $X^{(l)} : K^{(l)} \otimes \overline{K} \to \overline{K} \otimes K^{(l)}$ is a linear map intertwining $u^{(l)} \oplus \bar{u}$ with $\bar{u} \oplus u^{(l)}$.

Proposition 3.1 *Let (H, u) be a complex matrix quantum group with irreducible u as above. Let $G = H^{\mathrm{R}(s)}$ be a realification of H satisfying (23). Then the invertible intertwining map X in (24) satisfies the following condition:*

$$X^{(n)^{-1}}(\mathrm{id}_{\overline{K}} \otimes \mathrm{Mor}\,(u^{(m)}, u^{(n)}))X^{(m)} = \mathrm{Mor}\,(u^{(m)}, u^{(n)}) \otimes \mathrm{id}_{\overline{K}} \tag{26}$$

for any pair of non-negative integers m, n.

Proof: By (22), (21), the left hand side in (26) equals

$$X^{(n)^{-1}} \text{Mor}\,(\overline{u} \oplus u^{(m)}, \overline{u} \oplus u^{(n)}) X^{(m)} = \text{Mor}\,(u^{(m)} \oplus \overline{u}, u^{(n)} \oplus \overline{u}) = \text{Mor}\,(u^{(m)}, u^{(n)}) \otimes \text{id}_{\overline{K}}$$

Q.E.D.

Remark 3.2 The set $C(u) = \coprod_{m,n} \text{Mor}\,(u^{(m)}, u^{(n)})$ has an obvious structure of a category (objects being parametrized by non-negative integers). By the above proposition, for each $E \in \text{Mor}\,(u^{(m)}, u^{(n)})$ there is exactly one $E^X \in \text{Mor}\,(u^{(m)}, u^{(n)})$ such that

$$X^{(n)^{-1}}(\text{id} \otimes E) X^{(m)} = E^X \otimes \text{id}. \tag{27}$$

It is easy to see that the correspondence $E \mapsto E^X$ defines a bijective functor in $C(u)$ (acting trivially on objects).

Clearly, the map X contains all information about s. From formula (24) it follows that

$$\overline{u} \oplus u = (X \otimes \text{id})(u \oplus \overline{u})(X^{-1} \otimes \text{id}),$$

or

$$\overline{u}_m^k u_n^l = \sum_{ijab} X_{ij}^{kl} u_a^i \overline{u}_b^j (X^{-1})_{mn}^{ab}$$

using the components in a basis, hence

$$s(\overline{u}_m^k \otimes u_n^l) = \sum_{ijab} X_{ij}^{kl}(u_a^i \otimes \overline{u}_b^j)(X^{-1})_{mn}^{ab},$$

or

$$s_{24}(\overline{u} \otimes u) = X_{13}(u \otimes \overline{u})X_{13}^{-1}, \tag{28}$$

using the leg-numbering notation. Moreover, using (24) sufficiently many times, we obtain

$$(X_{(m)}^{(n)} \otimes \text{id})(u^{(n)} \oplus \overline{u}^{(m)}) = (\overline{u}^{(m)} \oplus u^{(n)})(X_{(m)}^{(n)} \otimes \text{id}),$$

where $X_{(0)}^{(n)} = \text{id}$ and, for natural m,

$$X_{(m)}^{(n)} = (\text{id} \otimes \text{id} \otimes \ldots \otimes X^{(n)}) \ldots (\text{id} \otimes X^{(n)} \otimes \ldots \otimes \text{id})(X^{(n)} \otimes \text{id} \otimes \ldots \otimes \text{id}) \quad (m \text{ times}).$$

We have therefore the following formula for the action of s on arbitrary monomials:

$$s_{m+1,m+n+2}(\overline{u}^{(m)} \otimes u^{(n)}) = (X_{(m)}^{(n)})_{1,\ldots,n,n+2,\ldots,n+m+1}(u^{(n)} \otimes \overline{u}^{(m)})(Y_{(m)}^{(n)})_{1,\ldots,m,m+2,\ldots,m+n+1},$$

using the leg-numbering notation, where

$$Y_{(m)}^{(n)} = [X_{(m)}^{(n)}]^{-1}.$$

It is convenient to use here a simplified notation: we simply drop the leg indices and write

$$s(\overline{u}^{(m)} \otimes u^{(n)}) = X_{(m)}^{(n)}(u^{(n)} \otimes \overline{u}^{(m)})Y_{(m)}^{(n)}. \tag{29}$$

If needed, the leg indices can be easily recovered, remembering that s acts always on $\overline{A} \otimes A$ and $X_{(m)}^{(n)}$ acts on $K^{(n)} \otimes \overline{K}^{(m)}$. The simplified notation will be used in the sequel. In particular, we have

$$s(\overline{u} \otimes u) = X(u \otimes \overline{u})X^{-1}. \tag{30}$$

From the above considerations, it is clear that formula (29) can be used in order to construct s for a given X. More precisely, given a complex matrix quantum group (H, u), $Pol(H) = (A, \Delta)$, and a map X satisfying (25), (26), we would like to construct a realification of H. However, in order to show that s is well defined by (29), we need A to be defined in terms of some explicit relations on the generators — the matrix elements of u. The following proposition gives a standard construction of this type (cf. [8,4]).

Proposition 3.3 Let K be a finite-dimensional complex vector space and $E_i: K^{(m_i)} \to K^{(n_i)}$, $i = 1, \ldots, r$, be a finite set of linear maps. Let A be the universal algebra generated by the matrix elements of $u \in \mathrm{End}(K) \otimes A$ satisfying

$$E_i u^{(m_i)} = u^{(n_i)} E_i, \qquad i = 1, \ldots, r \tag{31}$$

(the simplified notation!). Then there is a unique morphism of algebras $\Delta: A \to A \otimes A$ such that $(\mathrm{id} \otimes \Delta)u = u \textcircled{T} u$ (Δ defines then the structure of a coalgebra on the vector space A). If, additionally, $r \geq 2$, $m_1 = 0 = n_2$, $m_2, n_1 \geq 2$, and

$$E_1: \mathbf{C} \to K^{(n_1)}, \qquad E_2: K^{(m_2)} \to \mathbf{C}$$

are non-degenerate in the sense of ([9,4]), then (A, Δ) is a Hopf algebra with bijective antipode.

The complex matrix quantum group (H, u), $Pol(H) = (A, \Delta)$, defined by $E_1, \ldots, E_r$ as in the above proposition will be denoted by $(H, u)_{E_1, \ldots, E_r}$. Let $C_{E_1, \ldots, E_r}(u)$ be the smallest subcategory in $C(u)$ (cf. Remark 3.2) containing $\mathrm{id}_K, E_1, \ldots, E_r$ and closed under linear combinations and tensor products (the subcategory consists of linear combinations of 'monomials' in $E_1, \ldots, E_r$, cf. [9,4]). We set

$$\mathrm{Mor}_{E_1, \ldots, E_r}(u^{(m)}, u^{(n)}) = \mathrm{Mor}(u^{(m)}, u^{(n)}) \cap C_{E_1, \ldots, E_r}(u).$$

Theorem 3.4 Let $(H, u) = (H, u)_{E_1, \ldots, E_r}$ be a complex matrix quantum group as above. Let $X: K \otimes \overline{K} \to \overline{K} \otimes K$ be an invertible linear map satisfying (25) and one of two following conditions:

$$X^{(n)^{-1}}(\mathrm{id}_{\overline{K}} \otimes \mathrm{Mor}(u^{(m)}, u^{(n)}))X^{(m)} = \mathrm{Mor}(u^{(m)}, u^{(n)}) \otimes \mathrm{id}_{\overline{K}} \tag{32}$$

for $(m, n) = (m_i, n_i)$, $i = 1, \ldots, r$, or

$$X^{(n)^{-1}}(\mathrm{id}_{\overline{K}} \otimes \mathrm{Mor}_{E_1, \ldots, E_r}(u^{(m)}, u^{(n)}))X^{(m)} = \mathrm{Mor}_{E_1, \ldots, E_r}(u^{(m)}, u^{(n)}) \otimes \mathrm{id}_{\overline{K}} \tag{33}$$

for $(m, n) = (m_i, n_i)$, $i = 1, \ldots, r$. Then there is exactly one twisted realification $H^{R(s)}$ of H such that s satisfies (30).

Proof: The uniqueness follows from (29). To show the existence, consider the map

$$\tilde{s}: \overline{\tilde{A}} \otimes \tilde{A} \to \tilde{A} \otimes \overline{\tilde{A}}$$

defined by formula (29), where $\tilde{A}$ is the free algebra generated by the matrix elements of u (so that $A = \tilde{A}/J$ where J is the ideal in $\tilde{A}$ generated by the matrix elements of $E_i u^{(m_i)} - u^{(n_i)} E_i$, $i = 1, \ldots, r$). We have

$$(\mathrm{id} \otimes \tilde{s})(\tilde{s}(\overline{u}^{(i)} \otimes u^{(j)}) \otimes u^{(k)}) = (\mathrm{id} \otimes \tilde{s})(X_{(i)}^{(j)}(u^{(j)} \otimes \overline{u}^{(i)}) Y_{(i)}^{(j)} \otimes u^{(k)}) =$$

$$= (X_{(i)}^{(j)} \otimes \mathrm{id})(\mathrm{id} \otimes X_{(i)}^{(k)})(u^{(j)} \otimes u^{(k)} \otimes \overline{u}^{(i)})(\mathrm{id} \otimes Y_{(i)}^{(k)})(Y_{(i)}^{(j)} \otimes \mathrm{id}).$$

Since

$$(X_{(i)}^{(j)} \otimes \mathrm{id})(\mathrm{id} \otimes X_{(i)}^{(k)}) = X_{(i)}^{(j+k)},$$

we obtain

$$(\widetilde{m} \otimes \mathrm{id})(\mathrm{id} \otimes \tilde{s})(\tilde{s} \otimes \mathrm{id})(\overline{u}^{(i)} \otimes u^{(j)} \otimes u^{(k)}) = X_{(i)}^{(j+k)}(u^{(j)} u^{(k)} \otimes \overline{u}^{(i)}) Y_{(i)}^{(j+k)} =$$

$$= \tilde{s}(\overline{u}^{(i)} \otimes u^{(j)} u^{(k)}) = \tilde{s}(\mathrm{id} \otimes \widetilde{m})(\overline{u}^{(i)} \otimes u^{(j)} \otimes u(k)),$$

where $\widetilde{m}$ is the multiplication in $\tilde{A}$. It follows that $\tilde{s}$ satisfies (16) (with $\widetilde{m}$ playing the role of m). Formula (17) is proved similarly. Trivially, $\tilde{s}$ satisfies also (15). By taking the complex conjugation of these formulas we can check that $\sim \overline{\tilde{s}} \sim$ has the same properties. Since $\tilde{s}$ coincides with $\sim \overline{\tilde{s}} \sim$ on $\overline{u} \otimes u$ (by (25)), it follows that $\sim \overline{\tilde{s}} \sim = \tilde{s}$. It is easy to check that condition (12) is also satisfied.

In order to show that $\tilde{s}$ induces a map s from

$$\overline{\tilde{A}/J} \otimes \tilde{A}/J = (\overline{\tilde{A}} \otimes \tilde{A})/(\overline{J} \otimes \tilde{A} + \overline{\tilde{A}} \otimes J)$$

to the same product with reversed order, it is sufficient to show that

$$\tilde{s}(\overline{J} \otimes \tilde{A}) \subset \tilde{A} \otimes \overline{J}, \qquad \tilde{s}(\overline{\tilde{A}} \otimes J) \subset J \otimes \overline{\tilde{A}}. \qquad (34)$$

We shall show the second inclusion. Since $\tilde{s}$ satisfies (16), it follows that

$$\{a \in \tilde{A} : \tilde{s}(\overline{\tilde{A}} \otimes a) \subset J \otimes \overline{\tilde{A}}\}$$

is an ideal in $\tilde{A}$. It is therefore sufficient to show that the matrix elements of

$$\tilde{s}(\overline{u}^{(k)} \otimes (E_i u^{(m_i)} - u^{(n_i)} E_i))$$

belong to $J \otimes \overline{\tilde{A}}$ for $i = 1, \ldots, r$. We fix i and set for simplicity $E_i = E$, $m_i = m$, $n_i = n$. We have

$$\tilde{s}(\overline{u}^{(k)} \otimes (E u^{(m)} - u^{(n)} E)) = (\mathrm{id} \otimes E)\tilde{s}(\overline{u}^{(k)} \otimes u^{(m)}) - \tilde{s}(\overline{u}^{(k)} \otimes u^{(n)})(\mathrm{id} \otimes E)$$

$$= (\mathrm{id} \otimes E)X_{(k)}^{(m)}(u^{(m)} \otimes \overline{u}^{(k)}) Y_{(k)}^{(m)} +$$

$$\qquad - X_{(k)}^{(n)}(u^{(n)} \otimes \overline{u}^{(k)}) Y_{(k)}^{(n)}(\mathrm{id} \otimes E)$$

$$= X_{(k)}^{(n)}[Y_{(k)}^{(n)}(\mathrm{id} \otimes E)X_{(k)}^{(m)}(u^{(m)} \otimes \overline{u}^{(k)}) +$$

$$\qquad -(u^{(n)} \otimes \overline{u}^{(k)}) Y_{(k)}^{(n)}(\mathrm{id} \otimes E)X_{(k)}^{(m)}] Y_{(k)}^{(m)}.$$

Due to our assumptions on X,

$$
\begin{aligned}
Y_{(k)}^{(n)}(\mathrm{id}_{(k)} \otimes E)X_{(k)}^{(m)} &= (Y_{(k-1)}^{(n)} \otimes \mathrm{id}_{(1)})(\mathrm{id}_{(k-1)} \otimes E^X \otimes \mathrm{id}_{(1)})(X_{(k-1)}^{(m)} \otimes \mathrm{id}_{(1)}) \\
&= (Y_{(k-2)}^{(n)} \otimes \mathrm{id}_{(2)})(\mathrm{id}_{(k-2)} \otimes (E^X)^X \otimes \mathrm{id}_{(2)})(X_{(k-2)}^{(m)} \otimes \mathrm{id}_{(2)}) \\
&= \ldots = \tilde{E} \otimes \mathrm{id}_{(k)},
\end{aligned}
$$

where $\tilde{E} \in \mathrm{Mor}\,(u^{(m)}, u^{(n)})$ (in both considered cases of assumptions). (We have denoted the identity of $\overline{K}^{(j)}$ explicitly by $\mathrm{id}_{(j)}$.) This proves the second inclusion in (34).

Taking the complex conjugation of the second inclusion in (34) and using $\sim$ $\bar{\tilde{s}} \sim= \tilde{s}$, we obtain also the first inclusion.

Now, let us note that $\tilde{s}$ is a bijection with an inverse map satisfying

$$
\tilde{s}^{-1}(u^{(n)} \otimes \overline{u}^{(m)}) = Y_{(m)}^{(n)}(\overline{u}^{(m)} \otimes u^{(n)})X_{(m)}^{(n)}.
$$

Using this formula and similar arguments as before one can show that $\tilde{s}^{-1}(J \otimes \overline{\tilde{A}}) \subset (\overline{\tilde{A}} \otimes J)$. To prove it, we have this time to simplify expressions like $X_{(k)}^{(n)}(E \otimes \mathrm{id}_{(k)})Y_{(k)}^{(m)}$, so, instead of the map $E \mapsto E^X$ we use its inverse. We conclude that s is a well defined bijective map.

The required properties of s follow easily from the corresponding properties of $\tilde{s}$.

Q.E.D.

Remark 3.5 In case when u is irreducible, condition (32) is also necessary for the existence of the corresponding realification. The condition can be also formulated as follows:

- For $i = 1, \ldots, r$ and $E \in \mathrm{Mor}\,(u^{(m)}, u^{(n)})$, where $(m, n) = (m_i, n_i)$, there exist $E', E'' \in \mathrm{Mor}\,(u^{(m)}, u^{(n)})$ such that

$$
Y^{(n)}(\mathrm{id} \otimes E)X^{(m)} = E' \otimes \mathrm{id}, \qquad X^{(n)}(E \otimes \mathrm{id})Y^{(m)} = \mathrm{id} \otimes E''. \tag{35}
$$

Condition (33) is equivalent to the following one:

- There exist $E_i', E_i'' \in \mathrm{Mor}\,_{E_1, \ldots, E_r}(u^{(m_i)}, u^{(n_i)})$, $i = 1, \ldots, r$, such that

$$
Y^{(n_i)}(\mathrm{id} \otimes E_i)X^{(m_i)} = E_i' \otimes \mathrm{id}, \qquad X^{(n_i)}(E_i \otimes \mathrm{id})Y^{(m_i)} = \mathrm{id} \otimes E_i''. \tag{36}
$$

Along with (or instead of) the map X, it is convenient to consider $Q \in \mathrm{End}\,(K) \otimes \mathrm{End}\,(\overline{K})$ defined by

$$
Q =\sim X
$$

(cf. [1]). In terms of Q, condition (25) reads as follows

$$
\sim \overline{Q} \sim= \tau Q \qquad (|\tau| = 1) \tag{37}
$$

and equalities in (35) are equivalent to

$$EQ^{(m)} = Q^{(n)} E', \qquad Q^{(n)} E = E'' Q^{(m)}, \tag{38}$$

where $Q^{(j)} = Q \oplus \ldots \oplus Q$ (j times) is an element of $\mathrm{End}\,(K^{(j)}) \otimes \mathrm{End}\,(\overline{K})$. Here we use symbol $\oplus$ in the same way as in the case of $u \in \mathrm{End}\,(K) \otimes A$, the role of the algebra A being played by $\mathrm{End}\,(\overline{K})$.

In case when $\mathrm{Mor}\,(u^{(m)}, u^{(n)})$ is one-dimensional, the existence of E', E'' in (38) is equivalent to the existence of a number $\lambda \in \mathbf{C}$ such that

$$EQ^{(m)} = \lambda Q^{(n)} E. \tag{39}$$

4   Examples

We consider complex matrix quantum groups (H, u) which are deformations of classical groups of the series A_n, B_n, C_n, D_n. Those deformations which are known are all of the form

$$(H, u) = (H, u)_{E_1, E_2, R},$$

where $E_1 : \mathbf{C} \to K^{(n_1)}$, $E_2 : K^{(m_2)} \to \mathbf{C}$, $R : K^{(2)} \to K^{(2)}$, and

1. for the series A_n
$$n_1 = m_2 = n + 1 = \dim K = N$$

2.
$$n_1 = m_2 = 2, \qquad \dim K = N = \begin{cases} 2n + 1 & \text{for} \quad B_n \\ 2n & \text{for} \quad C_n, D_n \end{cases}$$

(see [4] for a concise description of the standard 1-parameter deformation; for the A_n series R can be discarded [9,4]; see [3] for multiparametric deformations, also below for A_n).

Problem: Does any realification of a group from the above list has to satisfy (23)?

In the sequel we consider only realifications satisfying (23). In particular, condition (26) has to be satisfied. They are described by suitable matrices Q. The following proposition stating sufficient conditions on Q, follows immediately from Theorem 3.4.

Proposition 4.1 *Let $E_1, E_2, E_3 = R$, where $R : K^{(2)} \to K^{(2)}$, satisfy the assumptions of Prop.3.3 and let $(H, u) = (H, u)_{E_1, E_2, R}$ be the corresponding complex matrix quantum group. If an invertible $Q \in \mathrm{End}\,(K \otimes \overline{K})$ satisfies (37) and*

$$E_1 Q^{(0)} = Q^{(n_1)} E_1, \qquad E_2 Q^{(m_2)} = Q^{(0)} E_2, \qquad R Q^{(2)} = Q^{(2)} R, \tag{40}$$

then there exists exactly one realification $H^{R(s)}$ of H satisfying (30).

Before passing to concrete examples, we shall show to what extent conditions in (40) are necessary in the case of A_n, B_n, C_n, D_n. The discussion depends on a knowledge of some basic features of representations of those complex matrix quantum groups. Since those elementary features do not seem to be clearly worked out in the literature, we shall state them as assumptions rather than facts (we hope to return to this problem elswhere).

<u>Assumption</u> 0: $\mathrm{Mor}\,(u, u)$ is one-dimensional (i.e. u is irreducible). This assumption makes (35) necessary.

<u>Assumption</u> 1: $\mathrm{Mor}\,(u^{(0)}, u^{(n_1)})$ and $\mathrm{Mor}\,(u^{(m_2)}, u^{(0)})$ are one-dimensional.

In this case it follows from (39) that $E_1 = \lambda_1 Q^{(n_1)} E_1$, $E_2 Q^{(m_2)} = \lambda_2 E_2$. Remembering that $n_1 = m_2$, we have $E_2 Q^{(m_2)} E_1 = \lambda_1 \lambda_2 E_2 Q^{(n_1)} E_1$, hence $\lambda_1 \lambda_2 = 1$. By rescaling $Q_{\mathrm{old}} \mapsto Q_{\mathrm{new}} = \mu Q_{\mathrm{old}}$ with $\mu^{m_1} = \lambda_1$ we obtain first two equalities in (40). For the A_n series, the third equality follows automatically (it is in fact redundant). For B_n, C_n, D_n we use one more assumption.

<u>Assumption</u> 2: In $\mathrm{Mor}\,(u^{(2)}, u^{(2)})$, the only solutions of the Yang-Baxter equation are multiples of R, R^{-1} and of the identity.

By (38), there exist $R', R'' \in \mathrm{Mor}\,(u^{(2)}, u^{(2)})$ such that

$$RQ^{(2)} = Q^{(2)} R', \qquad Q^{(2)} R = R'' Q^{(2)}$$

By Remark 3.2, R' and R'' have to satisfy Yang-Baxter equation. Assumption 2 implies then two possibilities:

1. $R' = \lambda R = \lambda^2 R''$,

2. $R' = \lambda R^{-1} = R''$.

In the first case $\lambda = 1$: using $RE_1 = \xi E_1$ (Assumpt. 1), we have

$$\lambda \xi E_1 = \lambda \xi Q^{(2)} E_1 = \lambda Q^{(2)} R E_1 = R Q^{(2)} E_1 = R E_1 = \xi E_1,$$

and we obtain the third equality in (40). In the second case, a similar calculus yields $\lambda = \xi^2$.

It follows that for the B_n, C_n, D_n series, the considered matrices Q should satisfy (40), or, possibly, (40) with third equation replaced by $RQ^{(2)} = \lambda Q^{(2)} R^{-1}$ ($\lambda = \xi^2$).

Consider now equations (40). They are equivalent to the statement that the matrix elements $Q_j^i \in \mathrm{End}\,(\overline{K})$ of Q (Q is treated as an element of $\mathrm{End}\,(K)$ with values in some algebra) defined in any basis $e_1, e_2, \ldots, e_N$ of K by $Q = \sum_{ij} e_i^j \otimes Q_j^i$ ($e_i^j = e_i \otimes e^j$; $e^1, \ldots, e^N$ is the dual basis), satisfy the defining relations (31) of the algebra A imposed on the elements u_j^i of the matrix u of the fundamental representation. In other words, $u_j^i \mapsto Q_j^i$ realizes a (non-degenerate) representation of A in $\mathrm{End}\,(\overline{K})$.

We shall describe now a particular (the easiest) method for finding many-parameter solutions of (40) and (37) for the series A_n (suggested by the results of [1]).

Example 1. *Method of the diagonal Q for $SL_q(N, \mathbf{C})$.*

We look for solutions such that $Q^i_j = 0$ for $i \neq j$. From the well known relations of $SL_q(N, \mathbf{C})$, we obtain that all Q^i_i commute and

$$Q^1_1 Q^2_2 \cdot \ldots \cdot Q^N_N = I. \tag{41}$$

The 'reality' condition (37) for $Q = \sum_i e^i_i \otimes Q^i_i$ implies that Q^i_i has to be diagonal (as a combination of e^i_i), hence

$$Q = \sum_{ij} \xi_{ij} e^i_i \otimes e^j_j \qquad \text{and} \qquad \overline{\xi_{ij}} = \tau \xi_{ji}. \tag{42}$$

By (41), we obtain

$$\prod_{i=1}^{N} \xi_{ij} = 1. \tag{43}$$

We have

$$\tau^{N^2} = \prod_{ij} \tau \xi_{ji} = \prod_{ij} \overline{\xi_{ij}} = 1,$$

hence τ is not a continuous deformation parameter and we can set $\tau = 1$ (its value for the commutative case $Q = I$). Since ξ's have to be real on the diagonal and the first row of ξ's can be computed from other ξ's, the number of independent real parameters in the matrix Q is equal

$$(N-1) + 2(1 + 2 + \ldots + (N-2)) = (N-1)^2.$$

Example 2. *Method of the diagonal Q for multiparameter quantum $SL(N, \mathbf{C})$.*

We describe here multiparameter deformation of the complex $SL(N, \mathbf{C})$, introduced in [3]. Let q be a non-zero complex number. Let $h = (h^{kl})_{k,l=1,\ldots,N}$ be a matrix of complex numbers such that

$$h^{kl} = \frac{1}{h^{lk}}, \qquad \prod_{k=1}^{N} h^{kl} = 1.$$

We assume $h^{kk} = 1$ (we reject $h^{kk} = -1$: the commutative limit will be $h^{kl} = \delta^{kl}$). We introduce also the matrix $\tilde{h} = (h_{kl})_{k,l=1,\ldots,N}$ such that $h_{kl} = h^{lk}$. Now, for any permutation σ of the set $\{1, \ldots, N\}$ we define the deformed 'signum':

$$\operatorname{sgn}^h_q(\sigma) = \prod_{\substack{k < l \\ \sigma_k > \sigma_l}} (-q h^{\sigma_k, \sigma_l})$$

$$\text{sgn}_{q,h}(\sigma) = \text{sgn}_q^{\tilde{h}}(\sigma) = \prod_{\substack{k < l \\ \sigma_k > \sigma_l}} (-qh_{\sigma_k,\sigma_l}) = \prod_{\substack{k < l \\ \sigma_k > \sigma_l}} (-qh^{\sigma_l,\sigma_k}).$$

The fundamental intertwiners E_1, E_2 are introduced as follows

$$E_1 = \sum_{\sigma \in S_N} \text{sgn}_q^h(\sigma) e_{\sigma_1} \otimes \dots \otimes e_{\sigma_N}$$

$$E_2 = \sum_{\sigma \in S_N} \text{sgn}_{q,h}(\sigma) e^{\sigma_1} \otimes \dots \otimes e^{\sigma_N},$$

where S_N is the set of all permutations of N elements. It is not difficult to see that the intertwiner

$$(E_2 \otimes \text{id} \otimes \text{id})(\text{id} \otimes \text{id} \otimes E_1) \in \text{Mor}\,(u^{(2)}, u^{(2)})$$

is proportional to an operator S, given by

$$S(e_a \otimes e_b) = \begin{cases} e_a \otimes e_b - qh^{ba}e_b \otimes e_a & a < b \\ q^2 e_a \otimes e_b - qh^{ba}e_b \otimes e_a & a > b \end{cases},$$

and $S(e_a \otimes e_a) = 0$. Using this intertwiner we find the following commutation relations for the matrix elements of u:

$$u_c^a u_d^b = q\frac{h^{ba}}{h^{dc}}u_d^b u_c^a \qquad \text{if} \qquad (a-b)(c-d) = 0, (a,c) \neq (b,d)$$

$$u_c^a u_d^b = \frac{h^{ba}}{h^{dc}}u_d^b u_c^a \qquad \text{if} \qquad (a-b)(c-d) < 0$$

$$h_{cd}u_c^a u_d^b - h^{ba}u_d^b u_c^a = (q - \frac{1}{q})u_d^a u_c^b \qquad \text{if} \qquad (a-b)(c-d) > 0.$$

One can easily check that the corresponding solution for diagonal matrices Q is exactly the same as in the preceding example. We remind that the number of real parameters involved in q and h is equal (cf. [3])

$$2\{1 + \binom{N-1}{2}\} = N^2 - 3N + 4.$$

Acknowledgments

I would like to thank to Prof. S. L. Woronowicz for valuable discussions and encouragment. I am also grateful to Prof. H. D. Doebner for his interest and kind hospitality in TU Clausthal. Recent article by P. Podleś ([4]) influenced much the present paper.

References

[1] S. L. Woronowicz and S. Zakrzewski : "Quantum deformations of Lorentz group. Hopf *-algebra level", preprint.

[2] L. D. Faddeev, N. Yu. Reshethikin and L. A. Takhtajan : "Quantization of Lie groups and Lie algebras", *Algebra i analiz* **1** (1989), 178–206 (in Russian).

[3] A. Schirrmacher : "Multiparameter R-matrices and their quantum groups", preprint MPI-Ph/91-24 to appear in J. Phys. A.; "The multiparameter deformation of $GL(n)$ and the covariant differential calculus on the quantum vector space", Z. Phys. C **50** (1991), 321–327.

[4] P. Podleś : "Complex quantum groups and their real representations", preprint, 1991.

[5] P. Podleś and S. L. Woronowicz : "Quantum deformation of Lorentz group", *Commun. Math. Phys.* **130** (1990), 381–431.

[6] S. L. Woronowicz : "Compact matrix pseudogroups", *Commun. Mat. Phys.* **111** (1987), 613–665.

[7] S. Zakrzewski : "Quantum and classical pseudogroups", Part I. "Union pseudogroups and their quantization", *Commun. Math. Phys.* **134** (1990), 347–370.

[8] S. L. Woronowicz : "Standard construction of a Hopf algebra" (unpublished notes)

[9] S. L. Woronowicz : "Tannaka-Krein duality for compact matrix pseudogroups. Twisted $SU(N)$ groups", Inv. Math. **93** (1988), 35–76.

[10] T. H. Koornwinder : "Comments on P.Podles and S.L.Woronowicz *Quantum deformation of Lorentz group*", unpublished notes, September 1990.

SECTION II
NON COMMUTATIVE DIFFERENTIAL GEOMETRY

ON MULTIGRADED DIFFERENTIAL CALCULUS*

ANDRZEJ BOROWIEC, WŁADYSŁAW MARCINEK
and ZBIGNIEW OZIEWICZ
Institute of Theoretical Physics, University of Wrocław,
plac Maxa Borna, 50-204 Wrocław, Poland
uwift13@plwrtu11, Fax: 48-71-201467

Abstract

We outline the differential calculus for the quadratic algebras in the braided monoidal categories. In particular we consider the colour braidings, i.e. the most general multigraded commutative algebras.

Foreword

The important chapter of the noncommutative calculus, the calculus on the (non-commutative) Hopf algebras ($\equiv$ on the quantum groups) has been elaborated by Woronowicz since 1979 (see (Woronowicz 1989) and in this volume). Important contributions to the calculus on associative rings, which do not assume the Hopf algebra structure, have been made by Alain Connes in a series of papers and is described in a book by Connes (1990), by Max Karoubi (1983, 1987) and Michel Dubois-Violette (1988). The calculus on the supercommutative rings, so called *fermionic* supercalculus has been presented completely by Jadczyk and Kastler (1987), Coquereaux and Kastler (1989), Coquereaux, Jadczyk and Kastler (1991), and generalized further by Marcinek (1989-1992) and recently by Matthes (1990-1991 and in this volume). Wess and Zumino (1990) developed a calculus implicitly for the quadratic algebras defined by means of the Hecke algebras and this theory has been enlarged by Hlavathý in 1991. We are not able to summarize in the present short contribution all of these achievements.

For arbitrary associative ring $\mathcal{A}$ there is the universal construction of the bimodule of the differential forms, (see Kunz 1986), which is the extension of the Kähler construction from 1953. The universal bimodule of the Kähler differentials need not to be a free bimodule even for the finitely generated rings.

*The work sponsored by Polish Committee for Scientific Research (KBN) under Grant PB # 2 2419 92 03

R. Gielerak et al. (eds.), Groups and Related Topics, 103–114.
© 1992 *Kluwer Academic Publishers*.

The reason is that the Lie algebra $Der\mathcal{A}$ of the appropriately generalized derivations of the considered algebra $\mathcal{A}$ cannot in general be extended to Lie $\mathcal{A}$-module[1]. The exception seems to be the Zamolodchikov algebras[2], i.e. algebras defined by means of the solutions of the quantum Yang-Baxter braid (or triangle) equation ($\equiv QYBE$ in the sequel). In particular, we are interested here in the *algebras with a braiding*[3], i.e. in the algebras defined by means of the constant (parameter independent) solutions of the QYBE. The solution of the QYBE is called the Yang-Baxter operator or the braiding operator and denoted here by $S \equiv \{S_{V,W}\}$. The braiding operator S, in the contrast to the *symmetry* S, need not to be unital, $S_{V,W} \circ S_{W,V} \neq id_{W,V}$ in general.

As was pointed out by Lyubashenko (1987) if $\mathcal{A}$ is the algebra with the unital Yang-Baxter symmetry S, then the Gurevich's Lie algebra[4] $Der\mathcal{A}$ of the S-dependent derivations of the considered algebra $\mathcal{A}$ is a Lie $\mathcal{A}$-module. We ask whether this holds for the arbitrary braiding.

In fact we like to know the largest class of algebras for which one can define $Der\mathcal{A}$ (in the S-dependent meaning as indicated by Gurevich and by Lyubashenko) *as a Lie $\mathcal{A}$-module*. This seems to be the case for the Zamolodchikov algebra determined by means of the solutions of the QYBE.

The Zamolodchikov algebras seems to be the largest class of algebras for which one can construct the canonical differential calculus in the complete analogy with the commutative case, along the lines developed by Coquereaux, Jadczyk and Kastler for fermionic supercalculus and generalized by Marcinek and recently by Matthes. The ultimate goal, which is outside of the present preliminary note, is the explicit formulation of the differential calculus for such algebras.

In particular in the present note we are illustrating the calculus for the specific Yang-Baxter operators, so called *the colour[5] braidings*. The algebras with the colour braidings (or colour quasi-symmetries) are the most general multigraded commutative algebras although the grading group is not encoded explicitly into the colour braiding. It is known that the arbitrary abelian grading group[6] determine the colour symmetry (see for example (Marcinek 1989 and 1990)). The colour braidings has been considered recently by Majid (1991) on the example of the anyonic braiding. Another example is the quotient algebra of the tensor algebra generated by the given vector space E, with respect to the ideal generated by Hecke operator

---

[1] What we call here after Nelson (1967) *the Lie module* has the rich terminology. Let us remind some of the other names still in use: *Lie differential ring* (Palais in 1961), *Lie-Cartan pair* (Kastler and Stora in 1985, Jadczyk in 1987, Marcinek in 1989), *differential Lie algebra* (Kosmann-Schwarzbach and Magri in 1990).

[2] also called Yang-Baxter algebras or S-*braided* (rather than S-symmetric) algebras

[3] also called *algebra with a quasi-symmetry*

[4] S-dependent Lie algebras wrt the unital Yang-Baxter and Hecke symmetries S have been introduced by Gurevich (1983, 1987 and 1988).

[5] The name invented by Trostel in 1984 who considered colour symmetries (unital braidings).

[6] In general the grading object is not a group but a Hopf algebra constructed by Lyubashenko (1987).

$S \in End(E \otimes E)$ ($\equiv$ a Hecke *symmetry*). The Wess-Zumino (1990) calculus is implicitly the calculus *for* the Hecke symmetry. The notion of the algebra with the *symmetry* operator S, generalizing symmetric and skew-symmetric algebras, has been invented by Lyubashenko (1986-1987) and by Gurevich (1987-1988). The particular, so called *colour* or *multigraded symmetries* has been considered implicitly firstly by Lukierski, Rittenberg and Wyler in 1978, by Scheunert (1979), Mosolova (1981), Trostel (1984) and explicitly invented later on independently by Marcinek (1989 and 1990). The algebra $\mathcal{A}$ for Yang-Baxter *colour* operators appears to be the most general *multigraded commutative* algebra for E.

In particular, here we are stressing that "coloured" algebra $\mathcal{A}$ with the *colour anyonic braiding* S such that $S^n = id$, is described through *the pair* of the bilinear forms (one symmetric and other skew-symmetric) on the grading group. We point out that if the symmetric form vanish and if moreover the skewsymmetric has a particular form, then we arrive at the example described by Matthes (1990-1991). From the other side if the skewsymmetric form vanish (which is always the case for n=odd), then our coloured braiding generalize to the multigraded case the *anyonic braiding* introduced by Majid (1991).

Braided monoidal category

Monoidal categories were introduced by Mac Lane in 1963 under the name "categories with multiplication" (Mac Lane 1971). Shortly, a *monoidal* category $\{\mathcal{C}, \otimes\}$ is a category $\mathcal{C}$ with a bifunctor[7]

$$\otimes : \mathcal{C} \times \mathcal{C} \longrightarrow \mathcal{C},$$

which has a two-sided identity and is associative. The identity object in the monoidal category as well as the associativity of the product $\otimes$ are up to the *natural* isomorphisms. For example we wish that the pair of the ternary functors

$$\otimes \circ (\otimes \times id) \quad \text{and} \quad \otimes \circ (id \times \otimes)$$

are *naturally* isomorphic. The bifunctor $\otimes$ assigns also to each pair of arrows (morphisms) f and g an arrow $f \otimes g$.

Mac Lane (1971) introduced a *symmetric* monoidal categories which are equipped with the *natural* isomorphism (a *symmetry* [8]) $S \equiv \{S_{V,W}\}$,

$$S_{V,W} : \quad V \otimes W \longrightarrow W \otimes V,$$

such that

$$S_{V,W} \circ S_{W,V} = id_{W \otimes V}. \tag{1}$$

---

[7]We use convention to write the product as $\otimes$, however this could be "$\times$" or "$\oplus$" or anything else.

[8]In terminology adopted by Lyubashenko (1986-1987) the family of morphisms $\{S_{V,W}\}$ is called a *vectorsymmetry data*.

Here V and W are the objects in the monoidal category [9]. In this case one says that the category satisfies "permutation symmetry". The compatibility (the coherence) of commutativity (the symmetry S) and associativity (this tell us that S is a morphism), expressed by means of hexagon identity by Mac Lane (1963, 1971), imply that S must solve QYBE. The QYBE for S assure that S is a morphism. Then the *coherence theorem* by Mac Lane asserts that "every diagram commute".

The symmetric monoidal category is known also under the names: *tensor category* (Deligne and Milne in 1982), *vectorsymmetry* category (Lyubashenko 1986-1987), Yang-Baxter category (Manin 1987).

When the unital condition (1) is dropped, then the compatibility of commutativity S with associativity need to be expressed by means of the *pair* of hexagon identities, which imply QYBE. This is known since 1986 and has been shown independently by Majid in 1990. Omitting the associativity, the hexagons take the form

$$S_{V\otimes W,U} = (S_{V,U} \otimes id_W) \circ (id_V \otimes S_{W,U}), \qquad (2)$$
$$S_{V,W\otimes U} = (id_W \otimes S_{V,U}) \circ (\dot{S}_{V,W} \otimes id_U).$$

Moreover

$$(S_{W,U} \otimes id_V) \circ S_{V,W\otimes U} = (id_U \otimes S_{V,W}) \circ S_{V\otimes W,U},$$

which is a QYBE. The corresponding category is called the *braided monoidal category* or the *quasi-tensor category* or *quasi-symmetric* (by Rieffel). The resulting coherence theorem for the braided categories has been proved by Joyal and Street in 1986 (see Joyal and Street 1991). One says that the category satisfies "braid symmetry".

The category is *rigid* if every object has a dual object. In the present paper we deal with the rigid braided (monoidal) category. Evidently the symmetric or tensor category is a particular case of the braided category.

In the categories whose objects are *finite*-dimensional vector spaces or modules the notion of the minimal polynomials of the braiding $S \equiv S_{V,W}$ is meaningfull. Let $T \equiv \{T_{V,W}\}$ denote the natural *canonical* isomorphism (the transposition), which commutes with the projections. Consider the compositions

$$S \circ T \equiv \{S_{V,W} \circ T_{W,V} \in End(W \otimes V)\}, T \circ S = \{T_{V,W} \circ S_{W,V} \in End(W \otimes V)\}.$$

By minimal polynomial of the braiding S in the monoidal category $\mathcal{C}$ we mean the minimal polynomials of $S \circ T$ and $T \circ S$. They do not need to be the same. Generaly the minimal polynomial of $S \circ T$ (and similar for $T \circ S$) has the form

$$(S \circ T - q_1)^{r_1}(S \circ T - q_2)^{r_2} \cdots (S \circ T - q_m)^{r_m} = 0, \qquad (3)$$

where $\{r_i\}$ are the multiplicities and $\sum r_i \leq dim(W \otimes V)$.

---

[9]We are considering categories whose objects are *finite-* dimensional vector spaces or modules.

Example: algebras with Hecke symmetries

Let R be commutative (ground) ring with unit. The tensor product $\otimes$ means $\otimes_R$. The tensor algebra of tensor powers of free R-module E is denoted by $E^\otimes$.

Let $B_n, H_n(q)$ and $S_n = H_n(1)$ denote the braid group, the Hecke algebra and the permutation (symmetric) group. By ρ we denote the braid representation (of the braid group in $End(E^{\otimes n})$) which factors through the Hecke algebra

$$B_n \longrightarrow H_n(q) \xrightarrow{\ \rho\ } End(E^{\otimes n+1}). \tag{4}$$

This representation is generated by the Hecke q-*symmetry* operator $S \equiv \{S_{E,E}\}$ (S is a solution of the constant QYBE and represent the elementary braid) with the minimal polynomial:

$$(S - id_{E\otimes E}) \circ (S + q \cdot id_{E\otimes E}) = 0. \tag{5}$$

Let I be an ideal in $E^\otimes$, stable wrt the Hecke representation ρ (1), which will be indicated as $\rho I \subset I$. This means that the Yang-Baxter representation ρ (as well as the Hecke operator S itself) in $E^\otimes$ both project to the Hecke representation in the factor algebra $\mathcal{A} \equiv (E^\otimes)/I$. We are interested in the *quadratic* algebras, where the ideal I is generated by the image of $(S - id_{E\otimes E})$ or $(S + q \cdot id_{E\otimes E})$. This factor algebra $\mathcal{A}$ is called *the algebra with Hecke symmetry* S. The axiomatic definition of *the associative algebra with the symmetry* is given by Gurevich (1987).

The more restrictive way to introduce the exterior algebra $\mathcal{A}$ use projection in $E^\otimes$, so that $\mathcal{A}$ is embeded into $E^\otimes$, $\mathcal{A} \hookrightarrow E^\otimes$. In particular consider the representation ρ (4) of the Hecke algebra. Then the linear space of $\mathbb{Z}$-homogeneous elements in $\mathcal{A}$ is generated by the image of the q-dependent idempotent S-alternator (or S-symmetrizer) [10]

$$Alt_n(Sym_n) \equiv \frac{1}{[n]_q!} \sum_{\pi\ in S_n} (sgn_q\pi) \cdot \rho_\pi \in End(E^{\otimes n}). \tag{6}$$

Here $[n]! \equiv [n][n-1]\cdots[1]$ and $[n] \equiv \frac{q^n - q^{-n}}{q - q^{-1}}$. $Alt_n(Sym_n)$ are an idempotents. Note that $Sym + Alt \neq id$ for $n > 2$. The $sgn_q\pi$ for the Hecke symmetry has been calculated by Jimbo (1986). Gurevich (1988) gave the recurssion formula.

Colour braidings

DEFINITION. The braiding $S \equiv \{S_{V,W}\}$ is called a *colour braiding* if $S = \eta \cdot T$, where $T \equiv \{T_{V,W}\}$ denote the transposition and $\eta \equiv \{\eta_{V,W}\}$ is a scalar-valued commutation factor, $\eta_{V,W} : V \otimes W \longrightarrow R$.

---

[10]Unfortunately we have *two* names for one object.

We like to explain how an abelian grading group Γ determine the colour braiding. Let ρ be a representation of the braid group generated by the *trivial* transposition T. Let E be a Γ-graded free R-module and E_Γ denote the set of Γ-homogeneous vectors in E. We say that $t \in E^{\otimes n}$ is a χ-vector if

$$\rho_\pi t = \chi(\pi, grade\, t) \cdot t, \tag{7}$$

for *every* $\pi \in B_n$ and *grade* $t \in \overset{n}{\oplus} \Gamma$. The multiplier χ generalize to the graded situation the sign of the braid. This is the map

$$\chi : \qquad B_n \times (\overset{n}{\oplus} \Gamma) \longrightarrow R, \tag{8}$$

such that

$$\chi(id, \dots) = 1 \in R, \tag{9}$$
$$\chi(a \circ b, \gamma) = \chi(a, b\gamma) \cdot \chi(b, \gamma). \tag{10}$$

Note that the particular example of the super χ-multipliers for the permutation group S_n ("groupoid characters") has been introduced by Jadczyk and Kastler (1987, Appendix B). If t is a *simple* ($\equiv$ a *decomposable*) Γ-homogeneous tensor in $E^{\otimes n}$ then the grade of t is denoted by $grade_\Gamma t \equiv |t| \in \overset{n}{\oplus} \Gamma$. The above multiplier conditions follows from $\rho_{\alpha \circ \beta} = \rho_\alpha \circ \rho_\beta$.

Instead of (6) we have the *coloured* alternator- symmetrizer in $E^{\otimes n}$

$$Alt_n(Sym_n) \equiv \frac{1}{|S_n|} \sum_{\pi \in S_n} \{\chi(\pi, \dots)\}^{-1} \cdot \rho_\pi. \tag{11}$$

The exponent (-1) is taken for convenience of the next formula (12). The multiplier condition (10) is equivalent to idempotency of Sym_n.

One can check that

$$Alt_n(Sym_n) \circ (\rho_\pi - \chi(\pi, |\dots|) \cdot id) = 0 \in End(E^{\otimes n}). \tag{12}$$

This means that image of $(\rho_\pi - \chi(\pi, \dots))$ is in the kernel of $Alt - Sym$. To show this one need the relation $\{\chi(\pi, \gamma)\}^{-1} = \chi(\pi^{-1}, \pi\gamma)$, which follows from the definition (10).

¿From (12) it follows that the range of the coloured symmetrizers (11) generate the χ-commutative algebra

$$A \ni \qquad f \cdot g = \chi(\pi, |f \otimes g|)(g \cdot f) \tag{13}$$

where $\rho_\pi(a \otimes b) = b \otimes a$. In (13) and in the sequel, by abuse of notation, the image of f and $g \in E^\otimes$ in A under $Sym(Alt)$ are denoted by the same symbols: $m(f \otimes g) \equiv f \cdot g \equiv f \otimes g \mod I$.

We denote

$$\epsilon(|x|, |y|) \equiv n(x \otimes y) \equiv \chi(elementary\ braid, |t|). \tag{14}$$

Because every permutation can be presented as the composition of the transpositions (such presentation is not unique but this is irrelevant) therefore it is not surprising that the χ-multiplier can be presented as the product of the commutation factors (14)

$$\chi(\pi, |t|) = \prod_{i<j,\ i\circ\pi>j\circ\pi} \epsilon(|t_i|, |t_j|). \tag{15}$$

¿From above considerations it is clear that the χ-algebra $\mathcal{A}$ is quadratic algebra and can be presented as the quotient of the tensor algebra $E^\otimes$ through the ideal generated by the image of $(S \pm id_{E\otimes E})$ where S is a colour braiding

$$S(x \otimes y) \equiv \eta(x \otimes y) \cdot (y \otimes x). \tag{16}$$

Let ρ be a representation of B_n generated by the above colour braiding, then χ-algebra is generated by eigenmultivectors $f \in E^{\otimes n}$, such that $\rho_\pi f = f$ for every π in B_n. Comparing symmetrizers (6) and (7) we see that

$$\rho_\pi(\text{for colour braiding}) = \{\chi(\pi, \ldots)\}^{-1} \cdot \rho_\pi(\text{for transposition}).$$

Note that Lyubashenko proposed more sophisticated way to relate the grading group (or the grading Hopf algebra) to braid symmetries. Inserting the colour Ansatz $S_{V,W} = \epsilon_{V,W} \cdot T$ into the "hexagons" (2) we get

$$\epsilon(\alpha + \beta, \gamma) = \epsilon(\alpha, \gamma) \cdot \epsilon(\beta, \gamma), \tag{17}$$
$$\epsilon(\alpha, \beta + \gamma) = \epsilon(\alpha, \beta) \cdot \epsilon(\alpha, \gamma).$$

These last conditions are exactly related to QYBE for colour braiding and they assure the Lie structure of the set of the generalized derivations (see below).

Anyonic braiding (Majid 1991)

DEFINITION. The colour braiding is said to be *anyonic* if the minimal polynomial (3) of $S \equiv S_{E,E}$ take the form $S^n = id$.

In this case one should separately consider an even and odd n. For commutation factor we get:

$$\epsilon(\alpha, \beta) \cdot \epsilon(\beta, \alpha) \cdots \epsilon(\alpha, \beta) = 1. \tag{18}$$

Here there is n factors and the last term is different for even and odd n. The commutation factors on the grading abelian group, has been introduced by Scheunert in 1979 for unital (n=2) case without of any relation to QYBE.

The general solution for the condition (18) alone is

$$\epsilon = e^{\frac{2\pi i}{n} \cdot} z^\omega, \tag{19}$$

where z is in the ground ring[11], s is a symmetric map and ω is a skewsymmetric map on on $\Gamma \times \Gamma$ (cf. Oziewicz 1990). The set of values of ω depends on the grading group Γ. Evidently if n is odd then $\omega \equiv 0$. If $\omega \equiv 0$ then the colour braiding is exactly the anyonic one considered by Majid (1991).

Inserting (19) into the "hexagon" conditions (17) force the maps s and ω to be *linear* on $\Gamma \otimes_{\mathbf{z}} \Gamma$ (Oziewicz 1990).

If $\epsilon = const$ ($\Rightarrow \epsilon^n = 1$) then $\chi(\pi, \ldots) = \epsilon^{length\ of\ \pi}$.

Note that the colour braiding is not compatible with the Hecke symmetry because for $q \neq 1 \Rightarrow \epsilon(\alpha, \beta) = 0$ *for* $\alpha \neq \beta$.

Derivations

Let m denote the multiplication morphism,

$$m: \qquad A \otimes A \longrightarrow A.$$

For $f, g \in A$ let

$$l_f g \equiv r_g f \equiv m(f \otimes g) \equiv (f \otimes g\ mod\ I) \in A.$$

Here l and r denote the left and right multiplication in A. Associativity of m could be expressed in several equivalent ways:

$$
\begin{aligned}
m \circ (id \otimes m) &= m \circ (m \otimes id), & (20)\\
l_f \circ m &= m \circ (l_f \otimes id),\\
r_f \circ m &= m \circ (id \otimes r_f),\\
l_f \circ r_g &= r_g \circ l_f.
\end{aligned}
$$

The algebra $\{A, m\}$ is said to be S-commutative, if $m = m \circ S$.

Let T denote the *natural* transposition. The module endomorphism $D \in End_R A$ is said to be a T-derivation of a T-commutative algebra A if

$$D \circ m = m \circ (D \otimes id + id \otimes D).$$

We have $id \otimes D = T \circ (D \otimes id) \circ T$. Therefore the above can be rewritten as (identical formula holds for right adjoint r)

$$D \circ l_f = l_{Df} + (m \circ T)(D \otimes l_f). \qquad (21)$$

For A and B in $EndA$ we put $m(A \otimes B) \equiv A \circ B$. The S-commutator is well defined for any braiding S,

$$[A, B] \equiv \{m \circ (id - S)\}(A \otimes B). \qquad (22)$$

---

[11] Matthes (this volume) use notation q for our z and m for our ω. We denote by (-q) the *root* of the minimal polynomial of the braiding S.

Then the formulae (21) take the compact form

$$[D, l_f] = l_{Df} \qquad and \qquad [D, r_f] = r_{Df}. \tag{23}$$

$End\mathcal{A}$ is a S-Lie algebra wrt the S-commutator (Gurevich and Lyubashenko, compare also with a general approach by Karasev and Maslov (1990)). The S-commutator is an "S-inner derivation" which implies that the commutator of derivations is again the derivation. This is equivalent to S-Jacobi identity. Note that D^2 is a derivation iff $S(D \otimes D) = const \cdot (D \otimes D)$ with $const \neq 1$.

Evidently one can adopt the formula (23) as the definition of the S-Lie algebra $Der\mathcal{A}$ of S-derivations for any braiding S. However with such definition is not evident whether $Der\mathcal{A}$ is or is not the $\mathcal{A}$-module? Let us try to assure the $\mathcal{A}$-module structure. Consider again the T-commutative case when the T-derivation should be compatible with the identity $m = m \circ T$. In order that $D \circ m = D \circ m \circ T$ we need to define D as $D \circ m = m \circ (D \otimes id) + m \circ (D \otimes id) \circ T$. Such set $Der\mathcal{A}$ has by the construction the $\mathcal{A}$-module structure as it follows from (20). This can be generalized for any braiding S,

$$D \circ m = m \circ (D \otimes id_\mathcal{A}) \circ (id_{\mathcal{A} \otimes \mathcal{A}} + S), \tag{24}$$

The formulas (23) and (24) are compatible thanks of the identities in $\mathcal{A}$. The choice of the ideal play here the crucial role. ¿From the above construction is evident that $Der\mathcal{A}$ is a sub-$\mathcal{A}$-module of $End\mathcal{A}$.

PROPOSITION. The space $Der\mathcal{A}$ of S-dependent derivations is a S-Lie-sub-$\mathcal{A}$-module.

Illustration 1. Consider the Hecke symmetries (4). Let's define the derivations in the S-commutative factor algebra $\mathcal{A} \equiv E^\otimes / I$, where I is the ideal in $E^\otimes$ generated by the image of the operator $(S - id_{E \otimes E})$ and $S \in End(E \otimes E)$ is a Hecke q-symmetry (4).

Consider the map from $Hom(E, E^\otimes)$ into $Hom(E \otimes E, E^\otimes)$ given by formula

$$D_1 \longmapsto D_2 \equiv (D_1 \otimes id_E) \circ (q \cdot id_{E \otimes E} + S). \tag{25}$$

By construction $D_2 \circ (S - id) = 0$. This sugest to define $D_n \equiv D_1 \circ Alt_n$. This S-derivation D_2 is evidently compatible with the S-commutativity $m = m \circ S$.

Illustration 2. Let S be the colour symmetry (16), then according to (13):

$$\begin{aligned} D(f \cdot g) &= Df \cdot g + \chi(\alpha, |f \otimes g|) Dg \cdot f \\ &= Df \cdot g + \chi(\alpha, |f \otimes g|) \cdot \chi(\beta, |Dg \otimes f|) f \cdot Dg. \end{aligned} \tag{26}$$

One can show (see Oziewicz 1990), using (15), that the above product of χ-multipliers is g-independent iff commutation factor ϵ (14) satisfy the hexagon conditions (18) and (17) for n=2.

Differential one-forms

We will consider the following braiding and notation

$$A \otimes Der A \xrightarrow{\ s\ } Der A \otimes A \xrightarrow{\ j\ } A.$$

DEFINITION. Let $f \in A$ and $D \in Der A$.
  The S-differential $d \equiv d_S$ is introduced as follows

$$(d_S f)D \equiv (j \circ S)(f \otimes D). \tag{27}$$

For the colour braiding S,

$$S(f \otimes D) \equiv \epsilon(|f|, |D|) \cdot (D \otimes f), \tag{28}$$

(cf with (16)). Those definitions are forced by (24) in order to get the Leibniz rule

$$d(f \cdot g) = df \cdot g + f \cdot dg.$$

Matthes (this volume) adopt the formula (28) for the colour *symmetry*, $S^2 = id$.
  The space of one-forms $\Lambda^1 \equiv Hom_A(Der A, A)$ is a A-bimodule with

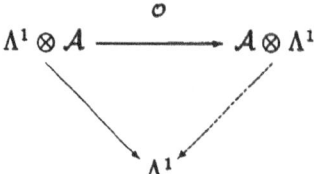

The braiding $\mathcal{O}$ define the calculus and must satisfy consistency conditions derived by Wess and Zumino (1990).

Acknowledgments

It is a pleasure to thank Rainer Matthes for conversation at the First Max Born Symposium and for sending to us the papers by Gurevich and Lyubashenko. Discussions with Jerzy Różański are gratefully acknowledged.

One of us (WM) is greatly thankful to Ph. D. students Wojciech Hann, Wojciech Mulak, Zbigniew Lipiński, Cezary Juszczak and many other for time-consumming help in everyday transportation from home to the University and back.

References

Coquereaux Robert and Daniel Kastler (1989), *Remarks on the differential envelopes of associative algebras.* Pacific Journal of Mathematics **137** (2) 245-263

Coquereaux Robert, Arkadiusz Jadczyk and Daniel Kastler (1991), *Differential and integral geometry of Grassmann algebras*, Rev. Math. Phys. **3** 63-99

Connes Alain (1990), *Geometrie non commutative*, Inter Editions, Paris

Dubois-Violette Michel, (1988), *Dèrivations et calcul diffèrentiel non commutatif*, C.R. Acad. Sci. Paris, **307** (Sèrie I) 403-408

Gurevich D.I. (1983), *Generalized shift operators on Lie groups*, Izviestia Armenian Acad. Sci., **18** (4) 305-317

Gurevich D.I. (1987), *Quantum Yang-Baxter equation and a generalization of the formal Lie theory*, in *Seminar on Supermanifolds No 15*,
edited by D. Leites. Matematiska Institutionen, Stockholms Universitet, Report 1987-No 19, 33-123.

Gurevich D.I. (1988), *Hecke symmetry and quantum determinants*, Doklady Akademi Nauk SSSR **303** (3) 542-546, Soviet Math. Dokl. **38** (1989) (3) 555-559.

Hlavatý Ladislav (1991), *Yang-Baxter matrices and differenial calculi on quantum hyperplanes*,
Czechoslovak Academy of Sciences Report, Prague, PRA-HEP-91/1

Jadczyk Arkadiusz and Daniel Kastler (1987), *Graded Lie-Cartan pairs. Part II: The fermionic differential calculus*, Annals of Physics **179** 169-200

Jimbo Michio (1986), *A q-analoque of $U(gl_{N+1})$, Hecke algebra and Yang-Baxter equation*, Lett. Math. Physics **11** 247-252

Jimbo Michio (1989), *Introduction to the Yang-Baxter equation*,
International Journal of Modern Physics **A4** (15) 3759-3777

Joyal A. and R. Street (1991), *The geometry of tensor calculus I*, Advances in Mathematics **88** 55-112

Kähler Erich (1953), *Algebra und Differentialrechnung*,
Bericht über die Mathematikertagung in Berlin.

Karasev M.V. and Maslov V.P. (1990), *Non-Lie commutations relations*, Soviet Mathematical Sciences Uspekhi (Moscow) **45** 41-79 (in Russian)

Karoubi Max (1983), *Homologie cyclique des groupes et des algèbres*,
C.R. Acad. Sc. Paris, **297** (Sèrie I) 381-384.

Karoubi Max (1987), *Homologie cyclique et K-thèorie*,
Asterisque 149, Socièté Mathèmatique de France.

Kunz E. (1986), *Kähler Differentials*, Viehweg, Braunschwig - Wiesbaden.

Lukierski Jerzy and V. Rittenberg (1978), Physical Review **D18** 385

Lyubashenko V.V. (1986), *Berezinian in some monoidal categories*,
Ukrainian Mathematical Zhurnal **38** (5) 588-592.

Lyubashenko V.V. (1986), *Hopf algebras and vectorsymmetries*,
Soviet Mathematical Sciences Uspekhi (Moscow) **41** (5) 185-186

Lyubashenko V.V. (1987), *Vectorsymmetries*, in *Seminar on Supermanifolds No 14*, edited by D. Leites. Matematiska Institutionen, Stockholms Universitet, Report 1987-No 19, 1-77.

Manin Yu. I. (1987), *Some remarks on Koszul algebras and quantum groups*, Annales de L'Institut Fourier, Grenoble, **37** (4) 191-205

Mac Lane Saunders (1971), *Categories for Working Mathematician*, Graduate Texts in Mathematics # 5, Springer-Verlag

Majid Shahn (1990), *Quasitriangular Hopf algebras and Yang-Baxter equations*, International Journal of Modern Physics A, **5** (1) 1-91

Majid Shahn (1991), *Anyonic quantum group*, University of Cambridge (U.K.) preprint DAMTP/91-16

Marcinek Władysław (1989), *Generalized Lie-Cartan pairs*, Reports on Math. Phys. **27** 385-400

Marcinek Władysław (1989), *Algebras based on Yang-Baxter operators*, Wrocław University Report ITP UWr 731/89, Reports on Math. Phys. in press

Marcinek Władysław (1992), *Graded algebras and geometry based on Yang-Baxter operators*, J. Math. Phys. March 1992. Wrocław University Report ITP UWr 745 (1990).

Matthes Rainer (1990), *Vector fields and differential forms on* $\mathbf{C}_q^n$, unpublished manuscript

Matthes Rainer (1991), *An example of a differential calculus on the quantum complex n-space*, Seminar Sophus Lie (Darmstadt) **1** 23-30

Matthes Rainer (1991), *A covariant differential calculus on the "quantum group"* C_q^n, Proceedings of the Wigner Symposium, Goslar.

Matthes Rainer (1992), *"Quantum group" structure and "covariant" differential calculus on symmetric algebras corresponding to commutation factors on* $\mathbb{Z}^n$, this volume

Oziewicz Zbigniew (1990), *Lie algebras for arbitrary grading group*. In *Differential Geometry and its Applications*, edited by Josef Janyška and Demeter Krupka, World Scientific, Singapore, 148-154

Scheunert M. (1979), *Generalized Lie algebras*, J. Math. Phys. **20** 712

Trostel Robert (1984), *Colour analysis, variational self-adjointness and Poisson (super)algebras*, J. Math. Phys. **25** (11) 3183-3189

Wess Julius and Bruno Zumino (1990), *Covariant differential calculus on the quantum hyperplane*, Nuclear Physics **18B** 303-312, Proceedings, suplements. Volume in honor of R.Stora.

Woronowicz Stanisław L. (1989), *Differential calculus on compact matrix pseudogroups* (quantum groups), Comm. Math. Phys. **122** 125-170 and this volume.

Yang Mills fields and symmetry breaking:
From Lie super-algebras to non commutative geometry

R. Coquereaux

Centre de Physique Théorique
CNRS Luminy - Case 907
F 13288 - Marseille Cedex 9 (France)

Abstract

We describe a formalism using both ideas of non commutative geometry and of Lie super-algebras to describe the geometry of Yang Mills fields and symmetry breaking

Keywords : Standard Model, Higgs fields, Lie Superalgebras, non commutative geometry.

This work was partly supported by the Swiss National Foundation.

R. Gielerak et al. (eds.), Groups and Related Topics, 115–127.

Yang Mills fields and symmetry breaking:
From Lie super-algebras to non commutative geometry

Introduction

The purpose of this contribution is to present a new mathematical description of the phenomenon of symmetry breaking and Higgs mechanism in Physics. The traditional description uses differential geometry and involves the reduction of a principal bundle with structure group G to a sub-bundle with structure group $H \in G$. The Higgs field responsible for the reduction is related to a global section of a fiber bundle with fibers diffeomorphic with the homogeneous space G/H. This is not the road that we are going to follow: we shall describe the Higgs field as part of a "generalized connection". Of course, the very definition of what is meant by connection, in this wider sense, should be precised in the sequel. Conceptually, this description can be related to the study of symmetries of Yang-Mills on spaces that can be written locally as $M \times F$. When M and F are smooth manifolds, tools of usual differential geometry suffice. When this is not the case, for instance when F is chosen as a discrete set, tools of non- commutative geometry become relevant. One important ingredient is then the replacement of the commutative algebra of functions on Space(-Time) by a non-commutative algebra which, in the simple example that we are going to study, is just a matrix algebra of functions. Another conceptual change involves the replacement of the Dirac operator coupled to a Yang-Mills field by the Yukawa operator -interpreted as a mass operator- coupled to a Higgs field. It is rather surprising to notice that the "old" Yukawa operator provides the simplest non trivial example of a Dirac operator in non commutative geometry.

The formalism that we are going to present was, at the beginning, strongly influenced by A. Connes ideas and in particular by the article [1]. However, evolution follows different paths. The relation between the approach that will be described below and, for example, the ideas found in the book [2] are still to be clarified. A simple presentation of our ideas can be found in [3] and many comments and extensions are discussed in [4] (see also [5]). Because the mathematical tools are new we will refrain of adopting a dogmatic style. This means that we shall only describe what we do and what we get in particular examples and leave to others (and to to the future) the work of inventing general mathematical definitions that would allow our calculations to fall in a precisely described category. Our conviction is that a given branch of mathematics (and the modelisation of physical phenomena) evolves generally from concrete examples to the definition of an abstract framework, not the converse.

A new mathematical description of the phenomenon of symmetry breaking leads, *ipso facto*, to an alternative description of the standard model of electroweak interactions. Since it is a new description, not a new theory, physical consequences, numerical constraints *etc.* are the same as usual. The first advantage is conceptual. Indeed, many "theoretical inputs" of the standard model, that usually appear as quite artificial (for instance the mere existence of Higgs fields and the description of their self interaction via a fourth degree polynomial) can be "deduced" from the new mathematical structure. Another advantage is that a new formalism often leads to the expression of new (and useful) physical hypothesis. It is indeed obvious that a relation between physical quantities may look quite natural in one formalism and quite unnatural in another one. The full description of the Standard Model involves 18 parameters (and the minimal -and very natural- extension of the model incorporating right neutrinos involves 24 parameters). These parameters are independent and may be chosen at will. One should not forget that these parameters have also to be renormalized since one should not stop at the classical level but study the corresponding fully interacting quantum field theory. Any numerical constraint (even totally unrealistic from the physical point of view) between these independent parameters can be implemented at the quantum level and lead to a decent quantum field theory. Therfore there is no hope of deducing these values from perturbative quantum field theory alone. The value of these parameters are therefore taken from experiment (in those cases where experiment is precise enough .... This is of course rather sad. One possible belief is that a more fundamental theory (like superstrings) will allow one, someday to compute them. Another possible belief is that the collection of these (renormalized) parameters can be somehow guessed, in the sense that this collection of numbers would correspond to a precise "geometry" of Space-Time at small scale. Which kind of geometry to postulate is of course totally unknown, yet, but our hope is that the present description of symmetry breaking and the corresponding algebraico-geometrical interpretation of the Yukawa operator will be useful in this quest for uncovering the veil under which what we call "Nature" hides herself.

The $U(1) \times U(1)$ example
The Dirac-Yukawa operator

In the simplest non-trivial case, the Dirac-Yukawa operator coupled to a generalized connection and acting on the field Ψ will be written as

$$\mathcal{P} = \begin{pmatrix} \partial\!\!\!/ & \mu \\ \mu & \partial\!\!\!/ \end{pmatrix} + g \begin{pmatrix} i\gamma^\mu L_\mu & \phi/\sqrt{2} \\ \bar\phi/\sqrt{2} & i\gamma^\mu R_\mu \end{pmatrix}$$

$$\Psi = \begin{pmatrix} \psi_L \\ \psi_R \end{pmatrix} = \begin{pmatrix} (\frac{1-\gamma_5}{2})\psi \\ (\frac{1+\gamma_5}{2})\psi \end{pmatrix}$$

(1)

Here $\partial\!\!\!/ = \gamma^\mu \partial/\partial x^\mu$ denotes the free (and massless) Dirac operator defined in flat Minkowski space, μ is an arbitrary positive real number (to be interpreted as a mass), g is an arbitrary real number (to be interpreted as a coupling constant), ϕ is a complex scalar field, L_μ and R_μ are the components of two 1-forms (to be interpreted in terms of $U(1)$-connections fields). Also, γ_5 denotes the chirality operator of the complexified Clifford algebra (the dimension of the underlying manifold is indeed even), ψ_L and ψ_R are half spinors and $\psi = \psi_L + \psi_R$ is a Dirac spinor.

The fermionic lagrangian may be defined as

$$\mathcal{L}_{Fermion} = \overline{\Psi}\slashed{\mathcal{D}}\,\Psi$$

$$= \overline{\psi}\gamma^\mu\partial_\mu\psi + i\frac{g}{\sqrt{2}}\overline{\psi}P\psi - i\frac{g}{\sqrt{2}}\overline{\psi}\gamma_5 Z\psi + m\overline{\psi}\psi + \frac{g}{\sqrt{2}}(\overline{\psi}_L\phi\psi_R + \overline{\psi}_R\overline{\phi}\psi_L)$$

$$(2)$$

This describes one massive Dirac fermion of mass $m = \mu$ with two bosonic fields $P = \frac{(L+R)}{\sqrt{2}}$ and $Z = \frac{(L-R)}{\sqrt{2}}$. The mixing angle that comes out naturally is equal to $\pi/4$. If one sets $\phi' = \mu + \phi$ and express $\mathcal{L}_{Fermion}$ in terms of ϕ', the massive term $m\overline{\psi}\psi$ disappears. By rescaling μ and ϕ, i.e. , by replacing them by $k\mu$ and $k\phi$, with $k \in R^+$, we would get a mass term with $m = k\mu$ and a coefficient in front of the Yukawa interaction term equal to $\frac{gk}{\sqrt{2}} = \frac{gm}{\mu\sqrt{2}}$ (so that it appears as proportionnal to the fermionic mass). We could have also introduced two coupling constants g_1 and g_2 (one for L_μ and one for R_μ). In this case, the mixing angle between the two fields would have been arbitrary. Notice that the constant μ in the expression of the Dirac- Yukawa operator appears as a discrete analogue of $\slashed{\partial}$ and that the scalar field ϕ appears as a discrete analogue of the gauge fields L_μ and R_μ. Therefore, the second term in the expression of the Dirac-Yukawa operator appears as expressiong the coupling of Ψ to a generalized connection incorporating scalar fields (0- forms) and gauge fields (1-forms). Let us consider the associative algebra $\mathcal{C}$ of 2×2 matrices generated by $\Omega_+ = \begin{pmatrix} 0 & 1 \\ 0 & 0 \end{pmatrix}$ and $\Omega_- = \begin{pmatrix} 0 & 0 \\ 1 & 0 \end{pmatrix}$. We call $I = \Omega_+\Omega_- = \begin{pmatrix} 1 & 0 \\ 0 & 0 \end{pmatrix}$ and $Y = \Omega_-\Omega_+ = \begin{pmatrix} 0 & 0 \\ 0 & 1 \end{pmatrix}$. This associative algebra appears as a kind of discrete analogue of the Clifford algebra since the second term of the Dirac-Yukawa operator can be considered as an element of the tensor product of two associative algebras: the usual Clifford algebra and $\mathcal{C}$. Matrices Ω_+ and Ω_- are then an analogue of the γ^μ. This associative algebra $\mathcal{C}$ is also Z_2 graded (the Ω's being odd by definition). Of course, it is enough to know the odd part to know the whole algebra. Whenever one has a Z_2-graded associative algebra, one may construct the corresponding Lie super-algebra by using graded commutators. In the present case, the situation is slightly too simple to exhibit generic features but it is clear that we get in this way is the Lie super-algebra $SU(1|1)$. This emergence of a Lie super-algebra comes only from the fact that we have an associative Z_2-graded algebra to start with and has nothing to do with supersymmetries (and we shall never try to gauge a Lie super- algebra).

The generalized connection

Rather than working with the Clifford algebra, we define the generalized connection in terms of 0-forms and 1-form as the antihermitian matrix

$$\mathcal{A} = \begin{pmatrix} L & i\mu^{-1}\phi \\ i\mu^{-1}\overline{\phi} & R \end{pmatrix}$$

$$(3)$$

Here L and R are dimensionless one-forms so that L_μ and R_μ defined by $L = L_\mu dx^\mu$ and $R = R_\mu dx^\mu$ have dimension of a mass. $\mathcal{A}$ is an element of a Z_2-graded associative and differential algebra that is constructed as the graded tensor product of two graded associative and differential algebras. The first is the algebra of 2×2 complex matrices. Its Z_2-grading is defined as follows: even elements are diagonal matrices. The differential of a matrix a is defined as $da = i[\eta, a]_S$, where $[.,.]_S$ denotes the graded commutator and where $\eta = \cos \gamma \, \tau_1 + \sin \gamma \, \tau_2$. Here γ denotes an arbitrary phase factor and $\tau_{1,2}$ are Pauli matrices. For instance, chosing $\eta = \tau_1$ leads to

$$a = \begin{pmatrix} a_{11} & a_{12} \\ a_{21} & a_{22} \end{pmatrix}, da = i \begin{pmatrix} a_{21} + a_{12} & a_{22} - a_{11} \\ a_{11} - a_{22} & a_{21} + a_{12} \end{pmatrix} \tag{4}$$

It is easy to check that d is a graded derivation. The second graded differential associative algebra (that is even graded commutative in this case) is the algebra of differential forms. When x is a homogeneous element in a Z_2-graded algebra –i.e. when its grading is well defined– we call ∂x its Z_2- grading. The graded tensor product is defined as

$$(a \otimes B) \odot (a' \otimes B') = (-1)^{\partial B \partial a'} (a \cdot a') \otimes (B \wedge B') \tag{5}$$

and the differential as

$$d(a \otimes B) = da \otimes B + (-1)^{\partial a} a \otimes dB \tag{6}$$

For arbitrary matrices X and X' of differerential forms,

$$X = \begin{pmatrix} A & C \\ D & B \end{pmatrix}, X' = \begin{pmatrix} A' & C' \\ D' & B' \end{pmatrix} \tag{7}$$

one gets

$$X \odot X' =$$
$$\begin{pmatrix} A \wedge A' + (-1)^{\partial C} C \wedge D' & C \wedge B' + (-1)^{\partial A} A \wedge C' \\ D \wedge A' + (-1)^{\partial B} B \wedge D' & B \wedge B' + (-1)^{\partial D} D \wedge C' \end{pmatrix} \tag{8}$$

and

$$dX = \begin{pmatrix} dA + i(e^{i\gamma}C + e^{-i\gamma}D) & -dC - ie^{-i\gamma}(A - B) \\ -dD + ie^{i\gamma}(A - B) & dB + i(e^{i\gamma}C + e^{-i\gamma}D) \end{pmatrix} \tag{9}$$

It is easy to check that d is a graded derivation (for the total Z_2 grading). For more details, cf [3].

Generalized curvature and bosonic lagrangian

From the expression of the generalized connection $\mathcal{A}$ given previously and from the covariant derivative $\nabla = d + \mathcal{A}$, one gets $\mathcal{F} = \nabla^2 = d\mathcal{A} + \mathcal{A} \odot \mathcal{A}$. Explicitely, one finds

$$\begin{aligned} \mathcal{F}_{11} &= F^L - \mu^{-2}(\mu(e^{i\gamma}\phi + \overline{e^{i\gamma}\phi}) + \phi\overline{\phi}) \\ \mathcal{F}_{12} &= -i\mu^{-1}(\nabla\phi + \mu e^{-i\gamma}(L - R)) \\ \mathcal{F}_{21} &= -i\mu^{-1}(\nabla\overline{\phi} - \mu e^{i\gamma}(L - R)) \\ \mathcal{F}_{22} &= F^R - \mu^{-2}(\mu(e^{i\gamma}\phi + \overline{e^{i\gamma}\phi}) + \overline{\phi}\phi) \end{aligned} \tag{10}$$

with

$$\nabla\phi = d\phi + L\phi - \phi R$$
$$\nabla\overline{\phi} = \overline{\nabla\phi} = d\overline{\phi} - \overline{\phi}L + R\overline{\phi} \qquad (11)$$

The Yang-Mills action itself is defined as

$$\mathcal{L} = \|\mathcal{F}\|^2 \doteq Tr < \overline{\mathcal{F}}, \mathcal{F} >= \|\mathcal{F}_{11}\|^2 + \|\mathcal{F}_{12}\|^2 + \|\mathcal{F}_{21}\|^2 + \|\mathcal{F}_{22}\|^2 \qquad (12)$$

where $\overline{\mathcal{F}}$ denotes the hermitian conjugate of $\mathcal{F}$. The symbol $< .,. >$ refers to a global scalar product in the exterior algebra. Clearly p-forms and q-forms are orthogonal whenever $p \neq q$ and the scalar product in teh space of 1-forms is directly defined via the space-time metric. However one is free to introduce unrelated scaling factors r_0, r_1 and r_2 in the definition of scalar products of 0-forms, 1-forms, and 2-forms. The factor r_2 can be reabsorbed in a global rescaling of the bosonic lagrangian and one gets

$$\mathcal{L} = -\frac{1}{4}((F_{\mu\nu}^L)^2 + (F_{\mu\nu}^R)^2) + 2r_1^2\overline{D_\nu\phi}D^\nu\phi + 2r_0^2(\mu(e^{i\gamma}\phi + \overline{e^{i\gamma}\phi}) + \phi\overline{\phi})^2 \qquad (13)$$

with

$$D_\nu\phi = \nabla_\nu\phi + \mu e^{-i\gamma}(L_\nu - R_\nu)$$
$$D_\nu\overline{\phi} = \overline{D_\nu\phi} = \nabla_\nu\overline{\phi} - \mu e^{i\gamma}(L_\nu - R_\nu) \qquad (14)$$

It is convenient to introduce coupling constant $\mathcal{L} = 1/g^2 \|\mathcal{F}\|^2$ and to rescale the Yang-Mills fields by setting $iL = L^{old}/g$ and $iR = R^{old}/g$. We introduce a factor i so that L and R are hermitian and also rescale the scalar field in order to get a conventional kinetic energy term in the Lagrangian. We therefore obtain a $U(1) \times U(1)$ Yang-Mills action with a symmetry breaking Higgs potential. The potential is already shifted onto an absolute minimum, no further shift is necessary. As it is clear from the above expressions, the freedom of choice for γ in the definition of the derivation d on the algebra of 2×2 matrices amounts to choose the position of the vacuum (the origin) on the circle of minima of the potential. The gauge field $L - R$ becomes massive since a term $(L - R)^2$ appears in the Lagrangian. To get a kinetic term that is diagonal in the dynamical variables, one redefines the fields as follows

$$Z = (R - L)/\sqrt{2} \; P = (L + R)/\sqrt{2} \qquad (15)$$

Because of the freedom in the definition of the scalar product in the space of differential forms (constants r_0, r_1), the mass of the Higgs particle associated with the field ϕ and the mass of the Z are not related (and are not related with the mass of the lepton). However, one can play with the idea of imposing a "natural" scalar product ... Also, the back-shifted lagrangian is unvariant under $U(1) \times U(1)$ and gauge freedom alone allows one to rescale differently the gauge fields, $i.e.$ to introduce two independent coupling constants, thus destroying the value $\theta = \pi/4$ found for the mixing angle. This is however quite unnatural, considering the way we obtained the lagrangian.

Generalizations

The approach followed in the previous section can be carried out as soon as we have a Z_2-graded differential associative algebra to start with. One can then, in turn, build the corresponding Lie super-algebra generated by graded commutators. The previous construction rests on the structure given by an associative algebra and the corresponding Lie super-algebra is only a by-product. However, for model building purposes, it is convenient to start from the data given by a Lie super-algebra and one of its representations. The reason is that finite dimensional Lie super-algebras have been classified and their representations are known. Our building receipe can therefore be described as follows. 1) Choose a Lie super-algebra 2) Choose a representation (reducible or not). 3) Consider the odd generators in this representation (call them Ω) and build the *associative* algebra generated by these Ω's. 4) This algebra is Z_2 graded and one can give it the structure of a graded differential algebra by mimicking the construction explained previously. If its dimension is even ($2N \times 2N$ matrices) one can can consider it has a 2×2 matrix whose elements are blocks. If its dimension is odd, one can embedd it in even dimension by adding a line and a raw of zeros. The definition of the Z_2-grading and of the differential then exactly follows the description given in the last section (provided we think in terms of 2×2 block- matrices). If the dimension is even, one defines the covariant differential as $\nabla = d + \mathcal{A}$ and gets, as before, the curvature $\mathcal{F} = d\mathcal{A} + \mathcal{A} \odot \mathcal{A}$. If the dimension is odd, the problem is that the d of a matrix that has a last line and last column filled with zeros does not have the same structure. One has then to define the covariant derivative as $\nabla = p\,d + \mathcal{A}$ where p is the projector that projects back onto the space of matrices of interest (we suppose that $\mathcal{A} = p\,\mathcal{A}\,p$. For instance, in the case of 3×3 matrices embedded into 4×4 matrices, one takes $p = diag(1,1,1,0)$. Then $\mathcal{F} = \nabla^2$ is not given by the previous equation but by $\mathcal{F} = d\mathcal{A} + \mathcal{A} \odot \mathcal{A} + p\,dp\,dp$. This is similar to what happens when one performs differential geometric calculations in a smooth manifold while using a description that is not intrinsic but that uses explicitly some kind of embedding. In all cases anyway, one gets as structure group the group whose Lie algebra corresponds to the even part of the Lie super-algebra we started with. The difference between the choice of one representation or another is reflected in the pattern of quantum numbers and mixing angles that emerges.

The Lagrangian describing electroweak interactions (the standard model) can be recovered by following the previous method. One may start with the Lie super-algebra $SU(2|1)$. Its use in weak interactions was advocated long ago (*cf.* [6-8]) but its meaning was not correctly recognized since many physicists tried (without succes) to gauge it. Needless to say, this is of course not what we intend to do. In all cases, (*i.e.* for different choices of –graded– representations), the emerging bosonic lagrangian will be the same and is the Lagrangian of the Standard Model (*cf.* [4] for details). The only gauge invariance of the theory is described by the group $SU(2) \times U(1)$. In all cases the graded sum of weak hypercharges vanishes (this means physically that the average electric charge of left handed particles is equal to the average electric charge of right handed particles). Let us consider different choices for the algebra C. In all cases this algebra is generated by four elements that we call $\Omega_\pm$ and $\Omega'_\pm$. We set

$$\{\Omega_\pm, \Omega_\pm\} = \{\Omega_\pm, \Omega_\mp\} = \{\Omega'_\pm, \Omega'_\pm\} = \{\Omega'_\pm, \Omega'_\mp\} = 0 \tag{16}$$

and call

$$
\begin{aligned}
Y &= \{\Omega_+, \Omega'_-\} + \{\Omega_-, \Omega'_+\} \\
2I_3 &= \{\Omega_+, \Omega'_-\} - \{\Omega_-, \Omega'_+\} \\
\sqrt{2} I_\pm &= \{\Omega_\pm, \Omega'_\pm\}
\end{aligned}
\tag{17}
$$

We also set $Q = \{\Omega_+, \Omega'_-\} = I_3 + Y/2$. The generator Y is called weak hypercharge and Q is the electic charge. In all the cases that we consider next (except in the last paragraph) Lie super-algebra generated by $\Omega_\pm$, $\Omega'_\pm$, $I_\pm$, I_3 and Y is isomorphic with $SU(2|1)$. The associative algebra generated by the Ω's is always bigger. It may be of interest to notice that the Ω's satisfy two non trivial polynomial relations: the first is of degree 4 and the other of degree 6. These two relations are related to the existence of Casimir operators [9] of degree 2 and 3 in $SU(2|1)$. Being of degree 2 or 3 in the generators of the Lie super-algebra means indeed being of degree 4 or 6 in terms of the matrices Ω since even generators of $SU(2|1)$ are themselves expressed as expressions of degree two in terms of the odd ones.

Elementary leptons.

The Ω matrices are described by

$$
P \begin{pmatrix} \mathbf{0}_{2\times 2} & \begin{pmatrix} \Omega'_- \\ -\Omega'_+ \end{pmatrix} \\ (\Omega_+ \quad \Omega_-) & 0 \end{pmatrix} P^{-1} = \begin{pmatrix} \mathbf{0}_{2\times 2} & \begin{pmatrix} \Omega'_-/\alpha \\ -\Omega'_+/\alpha \end{pmatrix} \\ (-\alpha\Omega_+ \quad \alpha\Omega_-) & 0 \end{pmatrix}
\tag{18}
$$

Here $P = diag(1, 1, \alpha)$ where α is an arbitrary constant. The notation means that, in order to get the expression of the generator Ω_+, for example, one replace the symbol Ω_+ by 1 in the previous matrix and the others by 0. In the present case, one gets $Y = diag(-1, -1, -2)$ and $Q = diag(-1, 0, -1)$. This corresponds to the (one of the two) fundamental representations $[\ell]$ of $SU(2|1)$ which is of dimension 3. It is non-typical (the number of right- handed fields is not equal to the number of and left-handed fields). Under $SU(2) \times U(1)$ we have the branching rule

$$[\ell] \longrightarrow (I = \frac{1}{2})_{y=-1} \oplus (I = 0)_{y=-2} \tag{19}$$

This describes therefore a left doublet (e_L, ν_L) and a right singlet (e_R). The other fundamental representation gives hypercharges $Y = diag(1, 1, 2)$ and therefore describes the corresponding antiparticles. Actually, it is natural to explicitely add both contributions to get the leptonic Lagrangian [4]. The generalized connection reads

$$\mathcal{A} = \begin{pmatrix} \mathbf{L} & i\mu^{-1}\Phi \\ i\mu^{-1}\overline{\Phi} & \mathbf{R} \end{pmatrix}, \quad \Phi = \begin{pmatrix} \phi_0 & 0 \\ \phi_+ & 0 \end{pmatrix} \tag{20}$$

where $\mathbf{L}$ and $\mathbf{R}$ anti-hermitian 2×2 matrices. Therefore

$$\mathcal{A} = \begin{pmatrix} \frac{i}{\sqrt{2}} \vec{\tau} \vec{W} + \frac{i}{\sqrt{6}} \mathbb{1} W_8 & \frac{i}{\mu} \begin{pmatrix} \phi_0 & 0 \\ \phi_+ & 0 \end{pmatrix} \\ \frac{i}{\mu} \begin{pmatrix} \overline{\phi}_0 & \overline{\phi}_- \\ 0 & 0 \end{pmatrix} & \begin{pmatrix} +\frac{2i}{\sqrt{6}} W_8 & 0 \\ 0 & 0 \end{pmatrix} \end{pmatrix} \tag{21}$$

We denoted by W_8 the abelian gauge field associated with the $U(1)$ gauge field because of the formal analogy with Gell-Mann matrices. Notice that $\mathcal{A}$ contains a line and a raw of zeros, but it is convenient to keep them in order to use the results of the previous section. One then computes the curvature $\mathcal{F}$ as it was explained previously and obtains the expression of the bosonic lagrangian of the Standard Model. Massive gauge fields are, here again, described by a term $(L - R)^2$, i.e. , one gets a term proportionnal to $(W_1^2 + W_2^2 + (W_3 - \frac{1}{\sqrt{3}} W_8)^2)$. To diagonalize the kinetic term for the gauge fields, one sets

$$Z = -\cos \theta \, W_3 + \sin \theta \, W_8$$
$$P = \sin \theta \, W_3 + \cos \theta \, W_8 \tag{22}$$
$$\theta = \frac{\pi}{6}$$

One can also define as usual $W_\pm = \frac{1}{\sqrt{2}}(W_1 \pm iW_2)$. Notice finally that the mass term for leptons comes only from the neutral Higgs field so that the mass matrix is $\mu(\Omega_+ + \Omega'_-)$.

Quarks.

The calculations are very similar. Ω matrices are given by

$$\mu P \cdot \begin{pmatrix} 0_{2 \times 2} & \begin{pmatrix} \frac{2}{3} \Omega_+ & \frac{1}{3} \Omega'_+ \\ \frac{2}{3} \Omega_- & -\frac{1}{3} \Omega'_- \end{pmatrix} \\ \begin{pmatrix} \Omega'_- & \Omega'_+ \\ -\Omega_- & \Omega_+ \end{pmatrix} & 0_{2 \times 2} \end{pmatrix} \cdot P^{-1} \tag{23}$$

Here P is the matrix $diag(1, 1, \alpha, \beta)$ and the two arbitrary constants that enter its expression come from the arbitrariness in the normalization of scalar products between the representation spaces $I = 0$ and $I = 1/2$ of $SU(2)$. Two representations differing by the action of P are equivalent but not unitarily equivalent. The above matrices Ω generate an algebra $\mathcal{C}$ of 4×4 matrices acting on a graded vector space $[q]$ of dimension 4. One gets here $Y = diag(1/3, 1/3, 4/3, -2/3)$ and $Q = diag(2/3, -1/3, 2/3, -1/3)$ which is the right pattern for quarks. Again, notice that $\Sigma_L Y = \Sigma_R Y$. This corresponds to the (one of the two) fundamental representations of $SU(2|1)$ which is of dimension 4. It is typical (the number of right-handed fields is equal to the number of and left-handed fields, here 2). $\mathcal{C}$ contains $SU(2|1)$ and in particular an homomorphic image of the Lie algebra of $SU(2) \times U(1)$. Under this Lie algebra we have the branching rule (with $y = 1/3$)

$$[q] \longrightarrow (I = \frac{1}{2})_y \oplus (I = 0)_{y-1} \oplus (I = 0)_{y+1} \tag{24}$$

This therefore describes a doublet of left-handed quarks along with their right-handed partners. The calculation to be carried out in the bosonic sector is analoguous to the one done previously. The only difference is that the natural mixing angle turns out to be defined by $tan^2\theta = 9/11$. More generally it is easy to show that, with the gauge group $SU(2) \times U(1)$ and for given I_3 and Y generators, one gets $tan^2\theta = TrI_3^2/Tr(Y/2)^2$. There is also a corresponding representation describing antiquarks and it is natural to explicitly add both contributions to get the Lagrangian for quarks [4].

A more general example.

We now discuss a more general example incorporating color, right neutrinos and family mixing in both quark and leptonic sectors. This corresponds to an algebra of 48 × 48 matrices acting on a graded vector space of dimension 48 = 24 + 24 with an equal number of "left" and "right" dimensions. We shall not give here the full matrix structure for the matrices Ω but just mention that calculation of Y and Q and of the branching rules show that it describes three families of quarks $[q_1]$, $[q_2]$, $[q_3]$ coming in three colors with mixing between generations (we use the notation $[q_1] \in [q_2] \in [q_3]$ to indicate that it should correspond to a reducible (but indecomposable) representation of $SU(2|1)$ (cf. [10]). The restriction of the Ω matrices to the corresponding subspace have the following structure (each block refers to a 4 × 4 matrix) :

$$P \cdot \begin{pmatrix} A & B & B \\ 0 & A & B \\ 0 & 0 & A \end{pmatrix} \cdot P^{-1} \tag{25}$$

where P is an invertible matrix, involving constants α, β, etc. and blocks A and B are given by

$$A = \begin{pmatrix} 0_{2\times2} & \begin{pmatrix} \frac{2}{3}\Omega_+ & \frac{1}{3}\Omega'_+ \\ \frac{2}{3}\Omega_- & -\frac{1}{3}\Omega'_- \end{pmatrix} \\ \begin{pmatrix} \Omega'_- & \Omega'_+ \\ -\Omega_- & \Omega_+ \end{pmatrix} & 0_{2\times2} \end{pmatrix} \quad ; \quad B = \begin{pmatrix} 0_{2\times2} & \begin{pmatrix} \Omega_+ & -\Omega'_+ \\ \Omega_- & \Omega'_- \end{pmatrix} \\ 0_{2\times2} & 0_{2\times2} \end{pmatrix} \tag{26}$$

We also want to describe three extended families of leptons, each family being denoted by $([\ell] \in [1])$ where [1] refers to the trivial representation and descibes a right-handed neutrino (with zero hypercharge and isospin) which has interactions with the other leptons via Yukawa couplings as well as a direct coupling to its left-handed partner. The restriction of the Ω matrices to the corresponding subspace reads

$$\begin{pmatrix} 0_{2\times2} & \begin{pmatrix} \Omega'_-/\alpha \\ -\Omega'_+/\alpha \end{pmatrix} & \begin{pmatrix} \epsilon\Omega_- \\ \epsilon\Omega_+ \end{pmatrix} \\ \begin{pmatrix} (-\alpha\Omega_+ & \alpha\Omega_-) \\ (\quad 0 & \quad 0\ ) \end{pmatrix} & 0_{2\times2} \end{pmatrix} \tag{27}$$

Moreover the several extended leptonic families may interact via these neutrinos. What we get at the end is therefore a description of 24 elementary particles schematized as

$$3([q_1] \Subset [q_2] \Subset [q_3]) \oplus ([\ell_1] \Subset [1]) \Subset ([\ell_2] \Subset [1]) \Subset ([\ell_3] \Subset [1]) \qquad (28)$$

Here again, the symbol $\Subset$ refers to the existence of reducible but indecomposable (*i.e.* not fully reducible) representations. Here again, the definition of the matrices Ω incorporate a number of (arbitrary) parameters that can be interpreted later in terms of masses and mixing matrices for quarks and leptons. Although we have no room for this discussion here, we want to point out that such indecomposable representations imply (in the case of quarks) the existence of a non diagonal Y generator. The corresponding (unwanted) contributions may be cancelled by taking into account the contribution of a representation describing the anti-fermions. The number of parameters of the model is now 24 (another 24!) since, on the top of the usual parameters of the Standard Model, we have the three masses of the (Dirac) neutrinos and the four parameters entering the Kobayashi-Maskawa like matrix for leptons. One may notice that for a family of particles incorporating leptons and anti-quarks, one gets $\Sigma_L Y^3 = \Sigma_R Y^3$ as it should (cancellation of the anomalies). The value found for the Weinberg mixing angle is now $tan^2\theta = 3/5$ –use the general formula given previously– *i.e.* $sin^2\theta = 3/8$. The existence or not of right neutrinos do not modify this calculation. As already mentionned in the introduction, this value 3/8 is "natural" in the present approach but cannot be justified on the grounds of gauge invariance alone that would allow for arbitrary rescalings of the different components of the non-simple Lie group $SU(2) \times U(1)$. Notice also that the $12 + 4 + 4 = 20$ parameters describing masses and mixing parameters are encoded in the matrix $\Omega_+ + \Omega'_-$. From the present point of view, this matrix (directly related to the matrix of Yukawa coupling constants in the usual approach) appears as part of a kind of discrete Dirac operator whose modulus is related to particle masses and whose phase is related to the Kobayashi-Maskawa-Cabibbo mixing coefficients. Since the value of these parameters lies anyway beyond quantum field theory (their renormalized values may be postulated at will without harming the corresponding quantum field theory), it is tempting to hope that these physical values (that are still unknown, because of experimental incertainties and because it is not obvious how to reconstruct the previous operator from the usually tabulated quantities) characterizing the structure of Space-Time at very small distances should correspond to a mathematically natural (meaning here aesthetic) discrete Dirac operator describing an exceptionnally simple "non-commutative" -or discrete- geometry. This hope motivates the search for educated ansatz concerning the structure of this operator. For instance, in the case of only two generations of quarks, postulating a very natural ansatz leads to a calculation [4] of the Cabibbo angle in good agreement with experiment. One finds

$$|\theta_c| = \frac{1}{2} \left[\arcsin \frac{2\sqrt{m_d m_s}}{m_d + m_s} - \arcsin \frac{2\sqrt{m_u m_c}}{m_u + m_c} \right] \approx \sqrt{\frac{m_d}{m_s}} - \sqrt{\frac{m_u}{m_c}} \qquad (29)$$

This may be taken as a good indication that something simple has still to be discovered in the physical and mathematical structures of the Standard Model.

A last example

Incorporation of color degrees of freedom is done by considering a separated $SU(3)$ gauge group (or a separated associative algebra of 3×3 matrices). Some new ideas in this direction recently appeared ([2]). We shall not discuss them here. However, in order to prevent possibly wrong extrapolations and also for the sake of illustrating again some general features of our construction, we consider the following example (that will turn out to have nothing to do with the Standard Model). Let us start with the Lie super-algebra $SU(3|2)$ or better, consider its odd generators along with the associative algebra C they generate. It acts on a vector space of dimension $5 = 3+2$ (this could describe for instance a fermionic multiplet containing three left-handed fermions and two right-handed ones). The emerging gauge group is $G = SU(3) \times SU(2) \times U(1)$ with gauge fields respectly denoted by W_α, with $\alpha \in \{1,2,\ldots,8\}$, A_i, with $i \in \{1,2,3\}$ and a field U_{15}. Again, by adding a line and a column of zeros, one can consider the corresponding matrices as 2×2 matrices, each element being a 3×3 block. Spontaneous breaking of the symmetry leads from $SU(3) \times SU(2) \times U(1)$ to a $SU(2) \times U(1)$ subgroup whose embedding is specified below. Indeed, a subset of the gauge fields becomes massive (there is a $(L - R)^2$ term in the lagrangian). It is easy to see that one gets a mass term proportionnal to

$$(W_1 - A_1)^2 + (W_2 - A_2)^2 + (W_3 - A_3)^2 + \Sigma W^7_{\alpha=4} W^2_\alpha + (W_8 - W_{15}/\sqrt{15})^2 \quad (30)$$

This gives four –unmixed– massive gauge fields, three massive gauge fields –with a mixing angle of $\pi/4$– and one massive gauge field –with a mixing angle defined by $tan\theta = \frac{1}{\sqrt{5}}$. Four gauge fields –corresponding to an $SU(2) \times U(1)$ subgroup – stay massless, namely $\frac{W_i + A_i}{\sqrt{2}}$, $i \in \{1,2,3\}$ and the abelian field $\frac{1}{\sqrt{6}} W_8 + \frac{5}{\sqrt{6}} W_{15}$. Needless to say –and despite of the emergence of a familiar structure group– the pattern of symmetry breaking exhibited in this last example has nothing to do with the one usually associated with the description of strong and electrowek interactions!

Remarks

The fact that it is possible to write the whole lagrangian (bosonic as fermionic) describing gauge theories with symmetry breaking in terms of very simple structures using non-commutative geometry suggests that an analoguous reformulation could be usefull at the quantum field theory level (quantum effective action). This is clearly to be investigated. As already mentionned in the introduction, the tools developed in [2][11] are more general than those presented here and also extend the field of possible models that could be used to described new physical systems. We hope that the present exposition will trigger new ideas and new research in these "non-commutative directions".

References

[1] A. Connes, J. Lott, *Particle models and Non-commutative geometry*, in "Recent Advances in Field Theory", Nucl. Phys. (Proc. Suppl.) **B18**, eds P. Binetruy *et al.* North Holland (1990) 29-47.

[2] A. Connes, *Geometrie Non Commutative*, InterEditions, Paris, (1990)

[3] R. Coquereaux, G. Esposito-Farèse, G. Vaillant, *Higgs fields as Yang-Mills fields and discrete symmetries*, Nucl. Phys. **B353** (1991) 689-706.

[4] R. Coquereaux, G. Esposito-Farese and F. Scheck *The theory of electroweak interactions described by $SU(2|1)$ algebraic superconnections.* CPT-91 /P.E. 2464

[5] R. Coquereaux, R. Häußling, N.A. Papadopoulos, F. Scheck *Generalized gauge transformations and hidden symmetry in the Standard Model*, Int. Journ. of Mod.Phys.A (in print)

[6] Y. Ne'eman, *Irreducible gauge theory of a consolidated Salam- Weinberg model*, Phys. Lett. **B81** (1979) 190-194

[7] Y.Ne'eman, J. Thiery-Mieg, *Exterior gauging of an internal supersymmetry and $SU(2)$ quantum asthenodynamics*, Proc. Nat. Acad. Sci. USA 79 (1982) 7068-7072

[8] P.H. Dondi, P.D. Jarvis, *A supersymmetric Weinberg-Salam model*, Phys. Lett. **B84** (1979) 75- 78

[9] M. Scheunert, W. Nahm, V. Rittenberg, *Irreducible representations of the $osp(2,1)$ and $spl(2|1)$ graded Lie algebras*, J. Math. Phys. **18** (1977) 155- 162

[10] M. Marcu, *The representations of $spl(2|1)$*, J. Math. Phys. **21** (1980) 1277-1283

[11] A. Connes, *Non-Commutative Differential Geometry*, Publ. Math. IHES 62 (1985) 41.

Differential and Integral Calculus on the Quantum C-Plane

J. Rembieliński
University of Łódź, Dep. of Theoretical Physics,
ul. Pomorska 149/153, 90-236 Łódź, POLAND

1 Quantum space

Quantum space is an associative coordinate algebra Q equipped with a set $\mathcal{F} = \{x^i\}$ of generators x^i, $i = 1, 2, \ldots n$ [1]. The reordering rule for generators is postulated in the so called Bethe Ansatz form [2]:

$$(x \times x) = B(x \times x) \tag{1}$$

where B is a $\mathbb{C}$-valued $n^2 \times n^2$ matrix, $\times$ denotes the direct product and x is the column matrix built from the generators x^i

$$x = \begin{bmatrix} x^1 \\ x^2 \\ \vdots \\ x^n \end{bmatrix} \tag{2}$$

Now, let us introduce two matrices B_{12} and B_{23} acting in the n^3-dimensional space $x \times x \times x$: namely $B_{12} = B \times I_n$ and $B_{23} = I_n \times B$. Here I_n is the $n \times n$ identity matrix. By means of the Bethe Ansatz and the assumed associativity of Q we have

$$\underset{1}{x} \times \underset{2}{x} \times \underset{3}{x} = (\underset{1}{x} \times \underset{2}{x}) \times \underset{3}{x} = B_{12} \, \underset{2}{x} \times (\underset{1}{x} \times \underset{3}{x}) = B_{12} B_{23} (\underset{2}{x} \times \underset{3}{x}) \times \underset{1}{x} = B_{12} B_{23} B_{12} \, \underset{3}{x} \times \underset{2}{x} \times \underset{1}{x}$$

and

$$\underset{1}{x} \times \underset{2}{x} \times \underset{3}{x} = \underset{1}{x} \times (\underset{2}{x} \times \underset{3}{x}) = B_{23} (\underset{1}{x} \times \underset{3}{x}) \times \underset{2}{x} = B_{23} B_{12} \, \underset{3}{x} \times (\underset{1}{x} \times \underset{2}{x}) = B_{23} B_{12} B_{23} \, \underset{3}{x} \times \underset{2}{x} \times \underset{1}{x}$$

Therefore the sufficient condition for associativity of Q reads

$$B_{12} B_{23} B_{12} = B_{23} B_{12} B_{23} \tag{3}$$

R. Gielerak et al. (eds.), Groups and Related Topics, 129–139.
© 1992 Kluwer Academic Publishers.

and it is known as the quantum Yang-Baxter equation.

Notice, that the *Bethe* matrix B is not fixed uniquely: the same reordering rules are obtained for

$$B' = I_{n^2} - M(I_{n^2} - B) \tag{4}$$

where M is an arbitrary nonsingular $n^2 \times n^2$ complex matrix. It is expected that B' may not fulfill the Yang-Baxter equations.

2   Quantum external differential algebra

Quantum external differential algebra is obtained by the affiliation to $\mathcal{F}$ of a set of differentials dx^k i.e. t o the extension of the generator set to $\mathcal{F} \cup d\mathcal{F}$. Next the re-ordering rules are extended to this set under consistency with the graded Leibnitz rule as well as with linearity and nilpotency of the external differential operator. Sometimes it is necessary to use a set of differential forms ω^k rather than differentials to guarantee a consistency. Finally we define a substitute of the external differential algebra as the sum of the $Q(\mathcal{F})$-modules Λ^k spanned by the k-forms $dx^{i^1} \ldots dx^{i^k}$ (or $\omega^{i^1} \ldots \omega^{i^k}$). According to the above prescription let us introduce the reordering rules in the form of the Bethe-like Ansatz [2]:

$$(x \times dx) = C(dx \times x) \tag{5}$$

where C is a $\mathbb{C}$-valued $n^2 \times n^2$ matrix. The Leibnitz rule holds if the sufficient condition

$$(I_{n^2} - B_{12})(I_{n^2} + C_{12}) = 0 \tag{6}$$

is satisfied. Associativity is guaranteed by the following Yang-Baxter-like equations

$$B_{12}C_{23}C_{12} = C_{23}C_{12}B_{23} \tag{7}$$

$$C_{12}C_{23}C_{12} = C_{23}C_{12}C_{23} \tag{8}$$

Here C_{12} and C_{23} is defined in the same way as B_{12} and B_{23}. Now, applying the d operation to the both sides of the Bethe-like Ansatz (5) we obtain by means of the graded Leibnitz rule the reordering prescription for differentials:

$$(dx \times dx) = -C(dx \times dx) \tag{9}$$

With help of the above three reordering rules (1,5,9) we can construct $Q(\mathcal{F})$-modules Λ^k spanned by the k-forms $dx^{i^1} \ldots dx^{i^k}$. The direct sum $\Lambda = \bigoplus_k \Lambda^k$ is an analogon of the external differential algebra.

3 Algebraic geometry of Q

Algebraic geometry aspects of Q are developed by the affiliation of the corresponding "*transformation groups*" (**quantum groups**) (Drinfeld [3]). Fadeev [4], Woronowicz [5]). **Quantum Group** $\Re$ is a quantum space equipped with the Hopf algebra structure. We follow the definition of $\Re$ by Woronowicz [5]. Let us $g = [g_{ik}]$, $i, k = 1, 2, \ldots n$, be the set of generators of the associative algebra $\Re$ ordered in the matrix form g. The Hopf structure is defined by:

1. The homomorphism $\Re \xrightarrow{\Delta} \Re \otimes \Re$:

$$\Delta(g) = g \otimes g \qquad \text{-co-product;} \tag{10}$$

here $\Delta(g)_{ik} = \Delta(g_{ik}) = \sum_j g_{ij} \otimes g_{jk}$.

2. The homomorphism $\Re \xrightarrow{\epsilon} \mathbb{C}$:

$$\epsilon(g) = I_n \qquad \text{-co-unity.} \tag{11}$$

3. The anti-homomorphism $\Re \xrightarrow{S} \Re$:

$$S(g) = g^{-1}, \qquad S(g^\tau) = g^{\tau^{-1}} \qquad \text{-antipode.} \tag{12}$$

An appropriate differential calculus for quantum groups can be also introduced (Woronowicz [6]).

Let us call the ith-partial derivative of a function $f(x^1, \ldots, x^n)$, the function $\partial_i f(x^1, \ldots, x^n)$ such that

$$df(x^1, \ldots, x^n) = \sum_{i=1}^n dx^i \partial_i f(x^1, \ldots, x^n). \tag{13}$$

It is easy to verify, that:

$$\partial_i x^j = \delta_i^j$$

if $\{x^i\}_{i=1}^n$ is a set of lineary independent generators.

The m-th derivative in i^{th}-direction is defined by iterations:

$$\partial_i^{(m)} f\left(x^1, \ldots, x^n\right) = \partial_i \left(\partial_i^{(m-1)} f\left(x^1, \ldots, x^n\right)\right). \tag{14}$$

Now the scale invariance condition under the mapping $x^k \to \alpha^k x^k$, $\alpha^k \in \mathbb{C}$, reads: if function f has the form:

$$f\left(x^1, \ldots, x^n\right) = \left(x^i\right)^m g\left(x^1, \ldots, \hat{x}^i, \ldots, x^n\right)$$

then

$$\partial_i^{(m+1)} f = 0. \tag{15}$$

A classification of differential calculi satisfying the above conditions was given in [8].

4   Quantum Complex Plane.

Now, we restrict ourselves to the two-dimensional case and according to the eqs.
(1.5) we define matrices B and C ($B, C \in End(\mathbb{C}^2 \otimes \mathbb{C}^2)$) such that ($x^1 = x, x^2 = y$):

$$\vec{x} \otimes \vec{x} = B\vec{x} \otimes \vec{x} \tag{16}$$

$$\vec{x} \otimes d\vec{x} = C d\vec{x} \otimes \vec{x}, \tag{17}$$

where $\otimes$ denotes usual direct product.

Consistency of a differential calculus with commutation relations (16) means
that the algebra $\mathcal{U} = \Lambda^0 \oplus \Lambda^1 \oplus \Lambda^2$ is an associative graded algebra generated by
$x, y \in \Lambda^0$, $dx, dy \in \Lambda^1$ and $dxdy \in \Lambda^2$, where Λ^1 and Λ^2 are modules over Λ^0. The
module Λ^2 is one-dimensional, hence it is comfortable to work in basis such, that
$dxdx = 0$ and $dydy = 0$. In fact we have to demand that such basis exists.

Quantum (Manin's) plane $\mathbb{M}_q^{2|0}$ [1] is an algebra over $\mathbb{C}$ generated by two ele-
ments x, y, obeying defining relation

$$xy = qyx, \tag{18}$$

where q is a parameter. It has a structure of bialgebra, given by the relations:
$\Delta(x) = x \otimes x$, $\Delta(y) = y \otimes 1 + x \otimes y$, $\epsilon(x) = 1$, $\epsilon(y) = 0$, where Δ is coproduct and
ϵ is counit. The corresponding B matrix is of the form

$$B = \begin{bmatrix} 1 & 0 & 0 & 0 \\ 0 & 1-s^{-1} & qs^{-1} & 0 \\ 0 & q^{-1} & 0 & 0 \\ 0 & 0 & 0 & 1 \end{bmatrix} \tag{19}$$

The $*$-algebra structure.
An immediate possibility of introducing a complex structure in the Manin's plane
lies in observation that $\mathbb{M}_q^{2|0}$ admits $*$-algebra structure ($*$ is an antilinear involu-
tion) if q is real and

$$x^* = y. \tag{20}$$

This observation suggests to consider quantum complex variable [8]

$$\xi = x. \tag{21}$$

and its complex (hermitean) conjugation $\xi^* = y$. Thus

$$\xi\xi^* = q\xi^*\xi \tag{22}$$

Note, that this $*$-structure is not compatible with the coproduct Δ, defined above.
In fact this space, denoted below as $\mathbb{C}_q$ ($\mathbb{C}_q$ is also called an algebra of polynomials
on the Euclidean quantum plane [7]), is a quantum space and does not demand

bialgebra structure. Consider quantum group of motions on the $\mathbb{C}_q$ plane, namely the $ISO(2)_q$, generated by the matrix:

$$\begin{bmatrix} a & 0 & u \\ 0 & a^* & u^* \\ 0 & 0 & I \end{bmatrix} \tag{23}$$

with defining relations:

$$\begin{aligned} aa^* &= a^*a = 1 \\ uu^* &= qu^*u \\ au &= qua \end{aligned} \tag{24}$$

and a natural bialgebra structure

$$\begin{aligned} \Delta(a) &= a \otimes a, & \Delta(u) &= a \otimes u + u \otimes I \\ \epsilon(a) &= 1, & \epsilon(u) &= 0 \\ S(a) &= a^*, & S(u) &= -a^*u \end{aligned} \tag{25}$$

If we define the following comodule action $\delta : \mathbb{C}_q \longrightarrow ISO(2)_q \otimes \mathbb{C}_q$:

$$\delta\left(\begin{bmatrix} \xi \\ \xi^* \\ I \end{bmatrix}\right) = \begin{bmatrix} a & 0 & u \\ 0 & a^* & u^* \\ 0 & 0 & I \end{bmatrix} \otimes \begin{bmatrix} \xi \\ \xi^* \\ I \end{bmatrix} \tag{26}$$

then we can easily check the invariance of the reordering rule (22). The problem of finding possible differential calculi on the quantum plane is equivalent to the searching for a matrix C, which will satisfy conditions (16, 17). The most general matrix C satisfying all required conditions (16, 17) under the relations (20 - 22) have the form [8]:

$$C = \begin{bmatrix} p & 0 & 0 & 0 \\ 0 & 0 & q & 0 \\ 0 & q^{-1} & 0 & 0 \\ 0 & 0 & 0 & p^{-1} \end{bmatrix};$$

where p is a real parameter. Therefore we obtain a two-parametric family of the differential calculi of the form:

$$\begin{aligned} \xi\xi^* &= q\xi^*\xi \\ \xi d\xi &= pd\xi\xi, & \xi d\xi^* &= qd\xi^*\xi, & d\xi d\xi^* &= -qd\xi^*d\xi \\ \partial_\xi\xi &= 1 + p\xi\partial_\xi, & \partial_\xi\xi^* &= q^{-1}\xi^*\partial_\xi, & \partial_\xi\partial_{\xi^*} &= q\partial_{\xi^*}\partial_\xi \\ \partial_\xi d\xi &= p^{-1}d\xi\partial_\xi, & \partial_\xi d\xi^* &= q^{-1}d\xi^*\partial_\xi \\ (d\xi)^* &= d\xi^*, & (\partial_\xi)^* &= -p^{-1}\partial_{\xi^*}. \end{aligned} \tag{27}$$

Notice that the invariance under the $ISO(2)_q$ group implies that $p = q^{-1}$. It is interesting that the set of all holomorphic functions $B(\mathbb{C}_q)$ forms an algebra and every holomorphic function can be written as a formal power series of one variable ξ. This last statement follows immediately from the definition of a holomorphic function, (a function $h(\xi,\xi^*) \in \mathbb{C}_q$ is holomorphic if $dh(\xi,\xi^*) = d\xi \cdot h(\xi,\xi^*)$ i.e. $\partial_{\xi^*}h(\xi,\xi^*) = 0$).

5   Quantum Integral

From the algebraic point of view integration connected with the invariant measure (Haar measure) is the most interesting one. However, on the Hopf algebra level the existence of the invariant measure is proved only for commutative (co-semi-simple) and for finite-dimensional Hopf algebras which are useless in construction of quantum groups. On the C^*-algebra level Woronowicz [6] proved the existence of Haar measure for compact quantum groups.

In this section we define an integral on quantum complex plane $\mathbb{C}_q$. We show that our integral is continuous with respect to changing of deformation parameters. We also show that classical Riemann's integral and Grassmanian Berezin's integral are two limit cases of our deformation.

In the following we use the abbreviation: $f(\xi) = \sum f_n \xi^n$, $f(\xi^*) = \sum f_n \xi^{*n}$, where $f_n \in \mathbb{C}$.

Let us consider first a possibility of a definition of the inverse operation for the derivative ∂_ξ, i.e. $\mathbb{C}$-linear operation $\overset{pq}{\int} d\xi : \mathbb{C}_q^* \longrightarrow \mathbb{C}_q^*$, such that $\partial_\xi \left\{ \overset{pq}{\int} d\xi f(\xi) \right\} = f(\xi)$ (analogically we can look for $\mathbb{C}$-linear operation $\overset{pq}{\int} d\xi^* : \mathbb{C}_q^* \longrightarrow \mathbb{C}_q^*$ such that $\partial_{\xi^*} \left\{ \overset{pq}{\int} d\xi^* f(\xi^*) \right\} = f(\xi^*)$). It is enough to restrict ourselves to the monomials ξ^n, (ξ^{*n}) $n \in \mathbb{N}$, and then we find immediately, that:

$$
\begin{aligned}
\overset{pq}{\int} d\xi \xi^n &= \frac{\xi^{n+1}}{[n+1]_p} + const., \\
\overset{pq}{\int} d\xi^* \xi^{*n} &= p^n \frac{\xi^{*n+1}}{[n+1]_p} + const.
\end{aligned}
\tag{28}
$$

where, $[n]_p \equiv (1 - p^n)/(1 - p)$.

Now we define an analogue of the definite integral, i.e. a linear, continuous functional over elements of our involutive algebra. We propose the following definition:

$$
\overset{pq}{\underset{[\alpha,\beta]}{\int}} d\xi \xi^n = \left(\beta^{n+1} - \alpha^{n+1} \right) \frac{a(n,0;p,q)}{[n+1]_p}
\tag{29}
$$

$$
\overset{pq}{\underset{[\alpha,\beta]}{\int}} d\xi^* \xi^{*n} = \left(\beta^{n+1} - \alpha^{n+1} \right) \frac{b(0,n;p,q)}{[n+1]_{1/p}}
\tag{30}
$$

where $\alpha, \beta \in \mathbb{C}$ and a, b are coefficients taking the value 1 for $p = q = 1$. Universality of our definition demands a symmetry under the replacement ξ and ξ^* and simultaneously p, q and p^{-1}, q^{-1} respectively. Consequently

$$
b(0,n;p,q) = a(n,0;p^{-1},q^{-1})
\tag{31}
$$

Note that our integral, except linearity, satisfy:

$$\int\limits_{[\alpha,\beta]}^{pq} [\cdot] + \int\limits_{[\beta,\gamma]}^{pq} [\cdot] = \int\limits_{[\alpha,\gamma]}^{pq} [\cdot] \quad \text{and} \quad \int\limits_{[\alpha,\alpha]}^{pq} [\cdot] = 0 \tag{32}$$

for any $\alpha, \beta, \gamma \in \mathbb{C}$, so

$$\int\limits_{[\alpha,\beta]}^{pq} [\cdot] = - \int\limits_{[\beta,\alpha]}^{pq} [\cdot].$$

It is easy to check that $*$-operation is in the agreement with the definition (29) and (30), i.e.

$$\left[\int\limits_{[\alpha,\beta]}^{pq} d\xi \xi^n \right]^* = \int\limits_{[\alpha,\beta]}^{pq} \xi^{*n} \, d\xi^* \tag{33}$$

iff

$$\bar{b}(0,n;p,q) = a(n,0;p,q) \tag{34}$$

With these preliminary definition we can start to construct q-like integral along the path. For now on we will reserve letter z for the complex variable, and letter ξ for the q-complex variable. The classical complex plane $\mathbb{C}$ plays the role of the "shadow manifold". Let Γ be any path in $\mathbb{C}$. We propose the following definition of the q-integral "along the path Γ":

$$\int\limits_{\Gamma}^{pq} d\xi \xi^n \xi^{*n} = a(n,m;p,q) \frac{n+1}{[n+1]_p} \cdot \int dz \, z^n \cdot \bar{z}^m \tag{35}$$

$$\int\limits_{\Gamma}^{pq} \xi^n \xi^{*m} d\xi^* = b(n,m;p,q) \frac{m+1}{[m+1]_p} \cdot \int d\bar{z} \cdot \bar{z}^m z^n \tag{36}$$

Let now D be a domain in $\mathbb{C}$. We define:

$$\int\limits_{D}^{pq} d\xi \xi^n \wedge d\xi^* \xi^{*m} = $$
$$= c(n,m;p,q) \cdot \frac{n+1}{[n+1]_p} \cdot \frac{m+1}{[m+1]_{1/p}} \cdot \int\limits_{D} dz \wedge d\bar{z} \cdot z^n \cdot \bar{z}^m \tag{37}$$

In the definitions (35-37) the coefficients a, b, c are restricted by the assumed $*$ invariance and by the symmetry under the replacement $\xi \to \xi^*$ and $p, q \to p^{-1}, q^{-1}$, namely

$$\bar{b}(n,m;p,q) = a(m,n;p,q) \tag{38}$$

$$\bar{c}(n,m;p,q) = c(m,n;p,q) \tag{39}$$

$$b(n,m;p,q) = q^{n(m+1)} a(m,n;p^{-1},q^{-1}) \tag{40}$$

$$c(n,m;p,q) = q^{(n+1)(m+1)}c(m,n;p^{-1},q^{-1}) \tag{41}$$

The existence of the classical limit of our integral implies

$$a(n,m;1,1) = b(n,m;1,1) = c(n,m;1,1) = 1 \tag{42}$$

Finally we will demand that integration defined above obeys *Stokes'* theorem, i.e. if $D \subset \mathbb{C}$ is a domain, ∂D is a boundary of D and $\omega(\xi,\xi^*)$ is a one-form, then:

$$\int\limits_{\partial D}^{pq} \omega\,(\xi,\xi^*) = \int\limits_{D}^{pq} d\omega\,(\xi,\xi^*) \tag{43}$$

This last assumption implies the following relations

$$a(n,m;p,q) = c(n,m-1,;p,q) \tag{44}$$

$$b(n,m;p,q) = c(n-1,m;p,q) \tag{45}$$

Therefore, putting

$$c(n,m;p,q) = q^{(n+1)(m+1)/2}\gamma(n,m;p,q) \tag{46}$$

we obtain finally that

$$\int\limits_{\Gamma}^{pq} d\xi\,\xi^n\xi^{*m} = \gamma(n,m-1;p,q)\,q^{(n+1)m/2}\frac{(n+1)}{[n+1]_p}\cdot\int dz\,z^n\cdot\bar{z}^m \tag{47}$$

$$\int\limits_{\Gamma}^{pq} \xi^n\xi^{*m}d\xi^* = \gamma(n-1,m;p,q)\,q^{n(m+1)/2}\frac{(m+1)}{[m+1]_p}\cdot\int d\bar{z}\cdot\bar{z}^m z^n \tag{48}$$

$$\int\limits_{D}^{pq} d\xi\,\xi^n\wedge d\xi^*\xi^{*m} = \gamma(n,m;p,q)\cdot q^{(n+1)(m+1)/2}\frac{(n+1)}{[n+1]_p}\cdot\frac{(m+1)}{[m+1]_{1/p}}\cdot$$
$$\cdot\int\limits_{D} dz\wedge d\bar{z}\cdot z^n\cdot\bar{z}^m \tag{49}$$

where

$$\gamma(n,m;p,q) = \gamma(m,n;p^{-1},q^{-1}) \tag{50}$$

$$\bar{\gamma}(n,m;p,q) = (q/|q|)^{(n+1)(m+1)}\gamma(m,n;p,q) \tag{51}$$

and

$$\gamma(n,m;1,1) = 1 \tag{52}$$

We have not a definitive criterion to fix the coefficient γ. We can choose $\gamma(n,m;p,q) = 1$ to obtain an integral quite analogical to the standard one. However, in that a case does not exist the Berezin limit ($p = -1$). A simplest choice guarantying the proper Berezin limit is $\gamma(n,m;p,q) = [(1+p)(1+p^{-1})]^{(n+m+2)/2}$. Note that the factors $q^{(\cdots)}$ in the front of the integrals (47-49) are essentially connected with the noncommutativity of the quantum complex numbers.

Quantum Cauchy's theorems:

Let C_a be a circle $z\bar{z} = a$, then:

$$\int\limits_{C_a}^{pq} d\xi\, \xi^n \xi^{*m} = \begin{cases} 2\pi i\, q^{(n+1)^2/2}\, \dfrac{n+1}{[n+1]_p}\, a^{n+1}\gamma(n,n;p,q) & \text{for } m = n+1 \\ 0 & \text{for } m \neq n+1 \end{cases} \tag{53}$$

And for special case $n+1 = m = 1$ and $a = 1$ these equations reduce to:

$$\int\limits_{C_1}^{pq} d\xi\, \xi^* = 4\pi i\, (1/p+1)\, q^2\gamma\,(1,1;p,q)$$

The above equations give the following *Cauchy's q-integral formula*:

$$q^{1/2}\, F(a) = \frac{1}{2\pi i r^2} \int\limits_{\Gamma}^{pq} d\xi\, \xi^* f(\xi) \tag{54}$$

where Γ is a circle $|z - a| = r$ and $f(\xi) = \sum f_n\xi^n$ is a quantum holomorphic function while the function $F(a)$ is given by

$$F(a) = \sum f_n \left(p^{-1/2}a\right)^n (n+1)\bigg/[n+1]\cdot\gamma(n,0;p,q) \tag{55}$$

where $[n]$ means *symmetric* quantum n, namely $[n] = [n]_p\cdot p^{(1-n)/2}$. It is remarkable that for

$$\gamma(n,m;p,q) = [n+1][m+1]\bigg/(n+1)(m+1)$$

we obtain eq. (54) in a very suggestive form

$$q^{1/2}f\left(p^{-1/2}a\right) = \frac{1}{2\pi i r^2} \int\limits_{\Gamma}^{pq} d\xi\, \xi^* f(\xi) \tag{56}$$

especially for $p = q^{-1}$.

Now it is possible to give the quantized version of the (local) *Cauchy's theorem*. Let $f(\xi)$ be any quantum holomorphic function. $f(\xi) \sum\limits_{n=0} f_n\xi^n$ and Ω be a domain in $\mathbb{C}$ such that $F(z)$ (given by eq. (54)) is continuous in Ω and the series (54) is absolutely convergent in Ω. Then

$$\int\limits_{\Gamma}^{pq} d\xi\, f(\xi) = 0. \tag{57}$$

for any closed road $\Gamma \subset \Omega$.

Using the prescription for q-integration one can easy give a quantum version of *Cauchy's global theorem*, i.e. if $f(\xi)$ is a quantum holomorphic function and $\Omega \subset \mathbb{C}$ is a simply connected domain such that the function $F(z)$ is analytic in $cl(\Omega)$, then

$$\int\limits_{\Gamma}^{pq} d\xi f(\xi) = 0, \tag{58}$$

for any closed road $\Gamma \subset \Omega$.

Finally we can define quantum analog of the square measure of a (flat) domain. Let $D \subset \mathbb{C}$ be a domain. The number:

$$|D|_{pq} = \frac{1}{2} \cdot \left| \int\limits_{D}^{pq} d\xi \wedge d\xi^* \right| \tag{47}$$

is called a *quantum square measure* of D.

It is easy to verify, that $|D|_{pq} = |q\gamma(0,0;p,q)| \cdot |D|$, where $|D|$ denotes a classical square measure of D.

This paper is supported by the KBN Grant 2 0218 91 01.

References

[1] Yu. I. Manin, Quantum groups and noncommutative geometry, Preprint Montreal University, CRM-1561 (1988)
    Yu. I. Manin, Commun. Math. Phys. **123**, 163(1989)

[2] J. Wess, B. Zumino, Covariant differential calculus on the quantum hyperplane, Preprint CERN-TH-5697/90. LAPP-TH-284/90

[3] V.G. Drinfeld, Quantum groups, Proceedings of the International Congress of Mathematicians, Berkeley, CA, USA (1986)

[4] L.D. Fadeev, N.Yu. Reshetikhin and L.A. Takhtajan, Quantization of Lie-groups and Lie algebras. Algebraic Analysis, Vol. 1, Academic Press 1988

[5] S.L. Woronowicz, Compact matrix pseudogroups, Commun. Math. Phys. **111**, 613(1987)

[6] S.L. Woronowicz, Differential calculus on compact matrix pseudogroups (quantum groups). Commun. Math. Phys. **122**, 125(1989)

[7] S.L. Woronowicz, *Unbounded Elements Affiliated withC^*-Algebras and Non-Compact Quantum Groups.* Commun. Math. Phys. **136**, 399-432(1991)

[8] T. Brzeziński, H. Dabrowski and J. Rembieliński, *On the quantum differential calculus and the quantum holomorphicity*, (1991) Preprint IFUL 1(46), January 1991, to appear in J. Math. Phys.

J. Rembieliński, Lecture given in University *La Sapienza*, Rome (1991)

SECTION III
INTEGRABLE SYSTEMS

Rigorous Approach to Abelian Chern-Simons Theory

by

S. Albeverio and J. Schäfer
Department of Mathematics Ruhr-Universität Bochum (FRG)

Abstract:

We introduce a rigorous mathematical model of abelian quantized *Chern-Simons theory* (C.S. theory) based on the theory of infinite dimensional oscillatory integrals developed by Albeverio and Høegh-Krohn. We construct a gauge-fixed C.S. path integral as a Fresnel integral in a certain Hilbert space. *Wilson loop* variables are defined as Fresnel integrable functions and it is shown in this context that the expectation value of products of Wilson loops w.r.t. the C.S. path integral is a *topological invariant* which can be computed in terms of pairwise *linking numbers* of the loops, as conjectured by Witten. We also propose a lattice C.S. action which converges to the continuum limit.

1. Introduction

In recent years there has been an increasing interest in the relationship between quantum field theory and topology. In his paper "Quantum Field Theory and the Jones Polynomial" [1] E. Witten conjectured that there should be a remarkable connection between special quantum field theories which are based on the so called C.S. action and the famous Jones polynomial – a link invariant. The abelian version of his theory seemed to be connected with the linking number that is also a topological invariant of a link. Our aim with the present work (based on [2]) is to give at least in the abelian case a *rigorous*, i.e. *mathematical* model of the theory. We also construct a lattice C.S. action in the framework of [3], which may be suited for a lattice C.S. theory (perhaps also in the non-abelian case). Let us briefly recall the setting of [1].

Let M be a smooth compact n-dimensional oriented manifold without boundary. Let $\mathcal{G}$ be a (compact) Lie group – the "gauge group" – and denote by $P(M, \mathcal{G}, \pi)$ a principal fiber bundle which for simplicity is assumed to be trivial. In this case a connection A on P may be identified with a Lie algebra g-valued one-form. We take in the three dimensional abelian case, i.e. with $\mathcal{G} = U(1) \cong S^1$, an action of the following special form

$$S_{C.S.}(A) = \frac{k}{4\pi} \int_M A \wedge dA, \qquad k \in \mathbf{R} \qquad (1.1)$$

R. Gielerak et al. (eds.), Groups and Related Topics, 143–152.

– the C.S. action. One should remark that (1.1) does *not* depend on any given Riemannian structure of M. This fact has given rise to the conjecture, see [1], that (1.1) should be the action for a topological field theory. Now let us introduce the metric independent and gauge invariant observables of the theory – the Wilson loops – which are defined as follows. Let c be an oriented closed curve in M and compute the holonomy of A around c. In the abelian case this determines a unique element of $\mathcal{G}$, otherwise it is unique up to conjugacy and one takes the trace w.r.t some fixed representation $\mathcal{R}$ of $\mathcal{G}$, which defines the Wilson loop variable $W_{\mathcal{R}}(c)$ depending on A. One can easily compute that in the abelian case

$$W_{\mathcal{R}}(c) = \exp\left(i\alpha_{\mathcal{R}} \int_c A_i \, dx^i \right) \tag{1.2}$$

where $\alpha_{\mathcal{R}} \in \mathbb{Z}$ corresponds to a representation of $U(1)$. The normal heuristic quantization of the theory is formulated in a Feynman path integral formalism, i.e. one tries to compute formally integrals of the type $Z_{C.S.}^{-1} \int DA \, F(A) e^{iS_{C.S.}(A)}$, with the so called partition function

$$Z_{C.S.} := \int DA \, e^{iS_{C.S.}(A)}. \tag{1.3}$$

DA represents a heuristic "flat measure" over all gauge orbits. Loosely speaking (1.3) is an integral over all equivalence classes of connections modulo gauge orbits. Following the standard heuristic Faddeev-Popov construction we choose the familiar Lorentz gauge condition $\delta A = 0$ with δ the formal adjoint to d w.r.t some fixed metric on M and we introduce an auxiliary field φ which is a three-form over M to get heuristically

$$Z_{C.S.} = \int DA \, D\varphi \, e^{iS_{C.S.}(A) + i\xi \int (\delta A)\varphi}, \qquad \xi \in \mathbb{R}. \tag{1.4}$$

The following heuristic integral

$$\left\langle \prod_{l=1}^{m} W_{\mathcal{R}_l}(c_l) \right\rangle := Z_{C.S.}^{-1} \int DA \, D\varphi \, e^{iS_{C.S.}(A) + i\xi \int (\delta A)\varphi} \prod_{l=1}^{m} W_{\mathcal{R}_l}(c_l)(A) \tag{1.5}$$

where $\{(c_l)\}$, $l = 1, \ldots m$ is a link and $\mathcal{R}_l$ is a representation of $\mathcal{G}$ for each l should then be connected with linking numbers. We have two objectives. First we want to give a precise mathematical definition of the quantities in (1.5). Moreover we want to show rigorously that there exists indeed a connection of such expectations (1.5) with the linking number. To make sense out of (1.4) we want to interpret it as a Fresnel integral [4]. So let us briefly recall the definition of Fresnel integrals. For motivation, details and applications we refer to this reference.

Definition 1.1   *Let $\mathcal{H}$ be a real separable Hilbert space. Let $C_b(\mathcal{H})$ be the space of continuous bounded complex-valued functions on $\mathcal{H}$. We denote by $M(\mathcal{H})$ the Banach-algebra of bounded complex Borel-measures on $\mathcal{H}$. Then we call*

$$\mathcal{F}(\mathcal{H}) := \{\, f \in C_b(\mathcal{H}) \mid \exists \mu_f \in M(\mathcal{H}) : f(x) = \int e^{i(x,y)} \mu_f(dy) \,\} \qquad (1.6)$$

the space of Fresnel integrable functions on $\mathcal{H}$.

Remark: The relation $\phi : M(\mathcal{H}) \to \mathcal{F}(\mathcal{H})$, $\mu_f \overset{\phi}{\longrightarrow} f$ is linear, 1-1 and continuous (with suitable natural topologies on the spaces).

Definition 1.2   *Let B a densely defined symmetric operator on $\mathcal{H}$. Then $(x,y) \mapsto (x, By)$ $\forall x, y \in D(B)$ is a symmetric bilinear form on the domain $D(B)$ of B. Assume further that B is invertible with $D(B^{-1}) = \mathcal{H}$ and $B^{-1} \in B(\mathcal{H})$, i.e. that B^{-1} is continuous (which is a slightly stronger assumption than in [4]). Then $\forall f \in \mathcal{F}(\mathcal{H})$ we call $I_B(f)$, defined by*

$$I_B(f) := \int\limits_{\mathcal{H}} e^{\frac{i}{2}(x, Bx)} f(x)\, dx := \int\limits_{\mathcal{H}} e^{-\frac{i}{2}(x, B^{-1}x)} \mu_f(dx), \qquad (1.7)$$

the Fresnel integral of f with respect to B.

Of course the 2nd term is just a mnemonic notation defined by the 3rd one. This notation and the name *integral* is justified by several properties of the above defined linear functional, that it has in common with "normal" integrals, see [4].

2. The Model

Let M be a smooth compact n-dimensional manifold without boundary. Let $T_p^* M$ be the cotangent space of M in p and denote by $\Lambda^k(T_p^* M)$ the space of alternating k-linear functions on $T_p M$. Then $\Lambda^k := \bigcup_{p \in M} \Lambda^k(T_p^* M)$ is the bundle of k-forms, a smooth section of it is called a smooth k-form over M and we denote the space of such forms by $\Omega^k(M)$ or for notational convenience simply Ω^k. A Riemannian metric g on M defines fiber metrics on $\Lambda^k(T_p^* M)$ in the standard way. Assume henceforth that M is oriented and denote by $dvol$ the oriented volume element. The Hodge $*$ operator $* : \Lambda^k(T_p^* M) \mapsto \Lambda^{n-k}(T_p^* M)$ is defined by the identity:

$$\omega \wedge *\omega = (\omega, \omega) dvol \quad \forall \omega \in \Omega^k \qquad (2.1)$$

where $(\cdot, \cdot)$ denotes the inner product defined by the metric.

Now we restrict ourself to the three dimensional case. Ω^k are pre-Hilbert spaces with scalar products defined by

$$(\omega_1, \omega_2) \equiv \int_M \omega_1 \wedge *\omega_2 \quad \forall \omega_1, \omega_2 \in \Omega^k. \qquad (2.2)$$

Therefore $\Omega := \Omega^1 \oplus \Omega^3$ is in an obvious way also a pre-Hilbert space and we will denote an element ω of this space by $\omega = (A, \varphi)$ where $A \in \Omega^1$ and $\varphi \in \Omega^3$. Let $H := \mathcal{L}^2(\Omega)$ be the closure of Ω with respect to the norm induced by the scalar product and denote by $H^s(\Omega)$ the Sobolev space of order s defined by the use of a partition of unity that yields a local trivialization of the Λ^k. These spaces are (M being assumed to be compact) independent of the metric g. We define the following operators (with δ the formal adjoint to d w.r.t. $(\cdot, \cdot)$):

$$J(A, \varphi) := \begin{pmatrix} 1 & 0 \\ 0 & -1 \end{pmatrix} \begin{pmatrix} A \\ \varphi \end{pmatrix}$$

$$(d + \delta)_{odd} : \Omega^1 \oplus \Omega^3 \mapsto \Omega^0 \oplus \Omega^2$$

$$(d + \delta)_{odd} := (d + \delta)|_{odd \ forms}$$

$$* : \bigoplus_{k=0}^{3} (\Omega) \mapsto \bigoplus_{k=0}^{3} (\Omega) \quad (even \leftrightarrow odd)$$

$$B := *(d + \delta)_{odd} J \quad i.e.$$

$$B(A, \varphi) = \begin{pmatrix} *d & d* \\ -d* & 0 \end{pmatrix} \begin{pmatrix} A \\ \varphi \end{pmatrix} \tag{2.3}$$

The following lemmata are proven in [5]:

Lemma 2.1 B *is a first order elliptic differential operator. B is closable and extends to $\overline{B} : H^1(\Omega) \mapsto \mathcal{L}^2(\Omega)$ which for the sake of notational convenience is also denoted by B. B is self-adjoint, $B^2 = \Delta = (d + \delta)^2$ and the kernel of B consists of the space of harmonic forms on M.*

Lemma 2.2   *Define $T := B + H$ where H is the projector on the space of harmonic forms. Then $T : \mathcal{L}^2(\Omega) \supset H^1(\Omega) \mapsto \mathcal{L}^2(\Omega)$ is a self-adjoint unbounded operator which is 1-1 and onto. The inverse $T^{-1} : \mathcal{L}^2(\Omega) \mapsto H^1(\Omega) \subset \mathcal{L}^2(\Omega)$ is a compact self-adjoint operator.*

Remark: One can also give an "explicit" formula for the operator T^{-1}. If we denote by G the "Greens" operator of the familiar Laplace-Beltrami operator Δ, i.e. the partial inverse to Δ on $N(\Delta)^{\perp}$, then T^{-1} is given by $T^{-1} = BG + H$.

Now we can give the definition of the abelian C.S. functional-integral, because T satisfies all conditions stated in the definition 1.2 of Fresnel integrals.

Definition 2.3   *Let $H = \mathcal{L}^2(\Omega)$ be the real separable Hilbert space of $\mathcal{L}^2$-forms as defined above. Let T be the above-mentioned operator and $k \in \mathbf{R}$ arbitrary. Then*

$$B_{C.S.}^k := \frac{k}{2\pi} T \tag{2.4}$$

defines a functional-integral through

$$Z^k_{C.S.}(f) := I_{B^k_{C.S.}}(f) = \int_{\mathcal{H}} e^{\frac{i}{2}(x, B^k_{C.S.}sx)} f(x)\, dx, \qquad (2.5)$$

defined for all $f \in \mathcal{F}(\mathcal{H})$, which we call Chern-Simons functional-integral (in Lorentz-gauge). k is the coupling constant.

Remark: 1.) In general, functions that are members of $\mathcal{F}(\mathcal{H})$ depend on *both* variables A *and* φ. If we want to interpret $Z^k_{C.S.}$ as a functional or "path integral" on the physically relevant space of connections or gauge-fields $A \in \Omega^1$ we have to restrict our functional to the space $\mathcal{F}(\overline{\Omega}^1)$, where $\overline{\Omega}^1$ is the $\mathcal{L}^2$-closure of Ω^1. This is possible, because there exists a natural imbedding of $\mathcal{F}(\overline{\Omega}^1)$ into $\mathcal{F}(\mathcal{H})$ and we get explicitly:

Theorem 2.4   *Let $f \in \mathcal{F}(\overline{\Omega}^1)$ and denote by μ_f the unique measure associated with f on $\overline{\Omega}^1$. Then*

$$Z^k_{C.S.}(f) = \int_{\overline{\Omega}^1} e^{-\frac{ix}{k}\left(A,(*dG+H)A\right)} \mu_f(dA) \qquad (2.6)$$

Remark: 2.) Before proceeding we would like to give an answer to the question, in which sense the construction above yields a realization of the physical model discussed by Witten [1] and others (see e.g. references in [2]). Hence let us compute the action $S_{C.S.}$. If we choose for simplicity a manifold M with $\mathcal{H}^1_{dR}(M) = 0$, i.e. with vanishing first de Rham-cohomology, then we get

$$S_{C.S.} := \frac{1}{2}((A,\varphi), B^k_{C.S.}(A,\varphi)) = \frac{k}{4\pi}\left(\int_M A \wedge dA - 2(\varphi, d * A) + (\varphi, H\varphi)\right)$$

$$= \frac{k}{4\pi}\left(\int_M (A \wedge dA + 2(\delta A)\varphi) + (\varphi, H\varphi)\right). \qquad (2.7)$$

The first term represents the gauge-invariant C.S. functional while the second term corresponds to the gauge fixing term with $\xi = \frac{k}{2\pi}$. The third term is due to the "zero-modes", i.e. zero eigenvalues of the Laplace-Beltrami operator and is independent of A. If we had not chosen a manifold M with $\mathcal{H}^1_{dR}(M) = 0$ we would have got a fourth term depending on the projection of A into the kernel of the Laplace- Beltrami operator. This corresponds to the fact that in general, depending on the topology of M, the Lorentz-gauge does *not* completely fix the gauge.

So we have constructed a model that shares the same formal properties as the heuristic model in [1], which immediately leads to the second question whether or not there really exists a relationship between expectation values of Wilson loops w.r.t. the abelian C.S. functional-integral and the linking number.

3. Wilson Loops and Linking Number

In order to define Wilson loops *regularization* is necessary, because the holonomy concept is formulated in the category of *smooth* objects and in our context Fresnel integrable functions depend upon $\mathcal{L}^2$-objects. That is to say that an expression of the form (1.2) is meaningless since A is only in $\mathcal{L}^2(\Omega)$ and cannot be integrated over a one dimensional submanifold. For our purposes the theory of de Rham currents is suited, for details see [6]. A current is a linear functional on the space of C^∞-forms on a (compact) manifold that is continuous in the sense of distributions. Let c be an oriented closed curve in M and interpret for an arbitrary smooth form ϕ

$$\phi \mapsto c[\phi] := \int_c \phi \tag{3.1}$$

as a linear functional. Then c is continuous in the sense of distributions and therefore defines a de Rham current. Now a theorem for regularizations of currents, see [6], guarantees the existence of linear operators R_ϵ and A_ϵ $\epsilon \in \mathbf{R}^+$ acting on the space of de Rham currents such that: for every current T $R_\epsilon T$ and $A_\epsilon T$ are also currents the support of which are contained in an arbitrary neighbourhood of the support of T if ϵ is only small enough and

$$R_\epsilon T - T = \partial A_\epsilon T + A_\epsilon \partial T \tag{3.2}$$

holds, where ∂T is defined by duality $\partial T[\phi] := T[d\phi]$ for all smooth forms ϕ. For $\epsilon \to 0$ we have:

$$R_\epsilon T[\phi] \to T[\phi] \qquad A_\epsilon T[\phi] \to 0 \tag{3.3}$$

The main point is that R_ϵ is a *smooth* form called regulator of T. Relation (3.2) ensures that the regularization respects homological properties. If we had chosen an arbitrary regularization then we would have got theorem 3.5 below *only* in the limit $\epsilon \to 0$.

Definition 3.1   *Let c be a loop in M. Then we call*

$$j_\epsilon(c) := *R_\epsilon(c), \tag{3.4}$$

a current associated to c (is the Hodge star operator).*

Definition 3.2   *The Wilson loop-class $\mathcal{W}_c^\alpha$ which is associated to a given loop c and a representation $\mathcal{R}_\alpha$, $\alpha \in \mathbf{Z}$, of $U(1)$ consists of the following functions:*

$$\mathcal{W}_c^\alpha := \{\, f : \overline{\Omega}^1 \to \mathbf{C} \mid f(A) =: \exp i\alpha(j_\epsilon(c), A) : \,\}, \tag{3.5}$$

where : : denotes Wick ordering, i.e. $: \exp h(A) :\equiv \exp h(A)/Z_{Bc.s.}^k (\exp h(A))$.

We then have easily the following

Lemma 3.3   *Each member of the class W_c^α is Fresnel integrable, gauge invariant and independent of the metric.*

$\square$

We call any member of W_c^α a representative of W_c^α.

So we have associated to a given loop a *class* of functions. The main point is now that this class contains the whole homological information of the loop.

Lemma 3.4   *Let $R_\varepsilon(c)$ be a regulator of c. Let T_c be a tubular neighborhood of c. Let $j : T_c \hookrightarrow M$ be the inclusion. If ε is small enough then $j^*(R_\varepsilon(c)) \in \mathcal{H}_c^2(T_c)$ ($\mathcal{H}_c^k$ denotes the compact de Rham cohomology classes) and $[\eta_c] := [j^*(R_\varepsilon(c))]$ is the compact Poincaré dual (see e.g. [7]) to c in T_c.*

For the proof again we refer to [5]. Now we are able to state the main theorem:

Theorem 3.5   *Let $L = \bigcup_{l=1}^m c_l$ be a link. Each component c_l of L, which we assume to be homologous to zero, carries a representation $\mathcal{R}_{\alpha_l}$ of $U(1)$. Let $W_{\varepsilon_l}^{\alpha_l}(c_l)(A)$, $l = 1,\ldots,m$ be representatives of $W_{c_l}^{\alpha_l}$, $l = 1,\ldots,m$. We assume that the ε_l's are chosen small enough to ensure that the supports of the associated $j_{\varepsilon_l}(c_l)$ are pairwise disjoint, which is posssible for small ε_l. Then*

$$Z_{C.S.}^k\left(\prod_{l=1}^m W_{\varepsilon_l}^{\alpha_l}(c_l)(A)\right) = \exp\left(-\frac{i\pi}{k}\sum_{l\neq l'}\alpha_l\alpha_{l'}LK(c_l,c_{l'})\right), \qquad (3.6)$$

where $LK(c_l,c_{l'})$ denotes the linking number of c_l and $c_{l'}$.

Sketch of the proof of 3.5: (For details we refer to [2] or [9].) The proof is based upon the fact that it is possible to calculate the linking number via differential forms. We compute

$$Z_{C.S.}^k\left(\prod_{l=1}^m W_{\varepsilon_l}^{\alpha_l}(c_l)(A)\right) = \exp\left(-\frac{i\pi}{k}\sum_{l\neq l'}^m\alpha_l\alpha_{l'}\left(j_{\varepsilon_l}(c_l),(*dG + H)j_{\varepsilon_{l'}}(c_{l'})\right)\right). \quad (3.7)$$

If we define $\omega_{c_l} := \delta GR_{\varepsilon_l}(c_l)$ then $d\omega_{c_l} = R_{\varepsilon_l}(c_l)$ holds $\forall l = 1,\ldots,m$, because c_l is homologous to zero hence there exists a two dimensional submanifold D_{c_l} of M such that $\partial D_{c_l} = c_l$ and therefore

$$d\omega_{c_l} = d\delta GR_{\varepsilon_l}(c_l) = R_{\varepsilon_l}(c_l) - HR_{\varepsilon_l}(c_l) = R_{\varepsilon_l}(c_l) - HR_{\varepsilon_l}(\partial D_{c_l}) = R_{\varepsilon_l}(c_l) \quad (3.8)$$

holds since $HR_{\varepsilon_l}(\partial D_{c_l}) = HdR_{\varepsilon_l}(D_{c_l}) = 0$. Now an easy computation shows

$$\left(j_{\varepsilon_l}(c_l),(*dG + H)j_{\varepsilon_{l'}}(c_{l'})\right) = \int_M R_{\varepsilon_l}(c_l) \wedge \omega_{c_{l'}}. \qquad (3.9)$$

Since $R_{\varepsilon_l}(c_l)$ is a representative of the compact Poincaré dual and (3.8) holds this shows that the right hand side of (3.9) is the linking number, due to a theorem of [7], p. 230 pp..

$\square$

So we have shown that the expectation value of products of Wilson loops depends *indeed* only upon a topological invariant, as conjectured in [1].

Remark: 1.) Let us conclude with a remark on the "framing" procedure as mentioned in various papers, see for instance [1]. It has turned out that framing is *not* necessary in order to remove *singularities* but to get non trivial *topological* invariants from *single* loops. In our definition the Wick ordering normalizes the expectation values of single loops, i.e. links with only one component to the trivial value one. Without this normalizing procedure we would not have got a topological invariant due to the problem of "self linking". Nevertheless it is possible to introduce the framing in our model and to get in this way topological invariants for framed links which are even in the one component case non trivial (but frame dependent).

2.) One can also verify that in the case of $M = S^3$ the identification of an open set in S^3 that contains the link with $\mathbf{R}^3$ leads directly to the famous Gauß formula for the linking number. Also one can straightforwardly *prove* certain formal properties like charge-parity-invariance and factorization property which are connected with the axioms of topological field theory as proposed by Atiyah [8], see [9].

4. The Simplicial Action

Unfortunately a rigorous quantized C.S. theory as a lattice quantum field theory does not yet exist, at least up to our knowledge. Recently there have been the approaches of Turaev, Viro and Schrader, see also Schrader's lecture in this proceeding, which lean heavily on the possibility to triangulate an arbitrary topological 3-manifold and which use Hopf algebras. But it is not clear how to interpret them as lattice quantum field theories. On the other side triangulations were used by Albeverio and Zegarlinski in [3] to construct *simplicial* quantum field theories which converge under certain prerequisites to their continuum counterparts. The first step towards a C.S. simplicial quantum field theory is the construction of the appropriate lattice action. We propose one possible action that we can prove to converge to the continuum C.S. action and which may be suited perhaps in connection with the theory of quantum groups for a lattice C. S. theory. For reader's convenience we briefly describe the setting, however, because of the lack of space, we have to refer to [3] for details.

Let M be a smooth compact differentiable manifold of dimension d and h a triangulation of M, i.e. a homeomorphism $h : |K| \longrightarrow M$, and K be a simplicial complex in a euclidean space with underlying topological space $|K|$. We assume h to be smooth, i.e. there exists a smooth extension of h to the hyperplanes in which each simplex is imbedded. A q-simplex σ_i^q, $q = 1, \ldots, d$ is denoted by

$$\sigma_i^q \equiv [p_{0_i}, \ldots, p_{q_i}], \tag{4.1}$$

with p_{q_i} the vertices. Denote by $C_q(K)$ the vector space of formal linear combinations of simplexes. We have a derivation $d^c \colon C_q(K) \xrightarrow{d^c} C_{q+1}(K)$, defined through

$[d^c(\phi)](\sigma^{q+1}) := c[\partial\sigma^{q+1}]$ with ∂ denoting the boundary operator on simplexes. Recall that we have a map from $C_q(K)$ to the space of $\mathcal{L}^2$ q-forms, denoted by $\Omega^q(M)$ – the Whitney map W:

Definition 6.1   For $[p_i] \in C^0(K)$ let $W[p_i] := b_{p_i}$ and for $\sigma^q = [p_0, \ldots, p_q]$ let

$$W\sigma^q := q! \sum_{i=0}^{q}(-1)^i b_{p_i} \, db_{p_0} \wedge \ldots \wedge d\hat{b}_{p_i} \wedge \ldots \wedge db_{p_q}, \qquad (4.2)$$

where b_{p_i} are barycentric coordinates and $\hat{}$ denotes omission as usual.

We define a scalar product in $C_q(K)$ by $(\sigma, \sigma') := (W\sigma, W\sigma)$. The de Rham map is a map from $\Omega^q(M)$ to $C_q(K)$, defined through $\omega \longrightarrow \sum_{i \in I_K} \left(\int_{\sigma_i^q} \omega\right)\sigma_i^q$. Obviously $RW\sigma = \sigma \; \forall \sigma \in C_q(K)$. We have convergence with respect to refinements of the triangulation, namely:

Theorem 6.2   For all $f \in \Omega^q(M)$, $0 \le q \le d$, $W_n R_n f \longrightarrow f$ holds with $n \to \infty$, where convergence means $\mathcal{L}^2$-convergence and W_n or R_n denotes the Whitney or de Rham map respectively, according to the nth standard subdivision.

For simplicial complexes a cup product $\cup$ is defined through

$$\langle c^p \cup c^q, [v_0, \ldots, v_{p+q}]\rangle := \langle c^p, [v_0, \ldots, v_p]\rangle \cdot \langle c^q[v_p, \ldots, v_{p+q}]\rangle. \qquad (4.3)$$

For our purposes another more symmetrical cup product is better suited:

Definition 6.3   Let $v_0 \le \ldots \le v_n$ be an arbitrary partial order of vertices of K, such that for every vertex $v_0 \le v_1 \le \ldots \le v_{p+q}$ holds. $\dot{\cup}$ is defined through bilinear extension of

$$\dot{\cup} : C_p(K) \times C_q(K) \longrightarrow C_{p+q}(K)$$

$$\langle c^p \dot{\cup} c^q, [v_0, \ldots, v_{p+q}]\rangle := N(p,q) \sum_{\substack{i_0, \ldots, i_{p+q} \\ \in \sigma(0, \ldots, p+q)}} \langle c^p, [v_{i_0}, \ldots, v_{i_p}]\rangle \cdot$$

$$\langle c^q, [v_{i_p}, \ldots, v_{i_{p+q}}]\rangle \, \text{sign} \begin{pmatrix} 0 \ldots p+q \\ i_0 \ldots i_{p+q} \end{pmatrix} \qquad (4.4)$$

with $N(p,q) := ((p+q+1)!)^{-1}$ and σ denoting permutation.

Theorem 6.4 $\dot{\cup}$ is independent of the choice of the partial order and hence welldefined. $\dot{\cup}$ is bilinear and

$$d^c(c^p \cup c^q) = (d^c c^p) \cup c^q + (-1)^p c^p \cup (d^c c^q). \qquad (4.5)$$

$$[c^p] \cup [c^q] = (-1)^{pq}[c^q] \cup [c^p] \qquad \forall[c^p] \in \mathcal{H}^p(K), [c^q] \in \mathcal{H}^q(K). \qquad (4.6)$$

hold.

Remark: $\dot{\cup}$ induces the "normal" cup product in the cohomology classes.

Proof: We refer to [9] or [2].

Definition 6.5

$$S_{C.S.}^c(A^c) := \int_M W(A^c \, \dot\cup \, d^c A^c) \quad A^q \in C_q \tag{4.7}$$

is called the lattice Chern-Simons action.

Remark: $S_{C.S.}^c$ is gauge invariant, i. e. $S_{C.S.}^c(A^c) = S_{C.S.}^c(A^c + d^c\phi^c) \ \forall \phi^c \in C_0(K)$. The above defined action converges in the $\mathcal{L}^2$ sense to $A \wedge dA$:

Theorem 6.6 $\quad W_n(R_n A \, \dot\cup \, d^c R_n A) \xrightarrow[\mathcal{L}^2]{} A \wedge dA. \tag{4.8}$

Proof: see [9] or [2].

On the basis of the above results and those in [2] one can expect to have convergence of averages of relevant observables to the continuum limit. This and the relation with other appoaches, e.g. [10], should be discussed further, see [9].

Acknowledgments: We thank B. Zegarlinski for very stimulating discussions.

References

[1] E. Witten: *"Quantum Field Theory and the Jones Polynomial"*, Commun. Math. Phys. **121**, 353-389 (1989).

[2] J. Schäfer: *Abelsche Chern-Simons Theorie – eine mathematische Konstruktion*, Diplomarbeit, Bochum (1991).

[3] S. Albeverio, B. Zegarlinski: *"Construction of Convergent Simplicial Approximation of Quantum Fields on Riemannian Manifolds"*, Commun. Math. Phys. **132**, 34-71 (1990).

[4] S. Albeverio, R. Høegh-Krohn: *"Mathematical Theory of Feynman Path Integrals"*, Lecture Notes in Mathematics 523, Springer, Berlin (1976).

[5] S. Albeverio, J. Schäfer: *"A Mathematical Model of Abelian Chern-Simons Theory"*, to appear in "Proceedings of the 3rd Conference of Stochastic Processes, Physics and Geometry in Locarno", World Scientific.

[6] G. de Rham: *"Differentiable Manifolds, Forms, Currents, Harmonic Forms"*, A Series of Comprehensive Studies in Mathematics, Springer, Berlin (1984).

[7] R. Bott, L. W. Tu: *"Differential Forms in Algebraic Topology"*, Graduate Texts in Mathematics 82, Springer, Berlin (1982).

[8] M. Atiyah: *"Topological Quantum Field Theories"*, Publications Mathématiques **68**, Institut Des Hautes Études Scientifiques (IHES), 175–186 (1988).

[9] S. Albeverio, J. Schäfer: *"Abelian Chern-Simons theory and linking numbers via oscillatory integrals"*, to appear.

[10] F. Nill: *"A Constructive Approach to Abelian Chern-Simons Theory"*, presented at the "XXIV Symposium on the Theory of Elementary Particles", Berlin-Gosen (1990).

The Conformal Block Structure of
Perturbation Theory in Two Dimensions

Rainald Flume

Physikalisches Institut, Nußallee 12, 5300 Bonn 1

Abstract

We describe some field theoretical aspects of marginal and relevant interactions perturbing two-dimensional conformally invariant field theories. Repercussions of the critical (non-Gaussian) fixed point theory for the structure of the perturbation series and in particular deformations of braid group representations in the perturbed theory are investigated. Some speculations about the importance of those deformations for an adequate understanding of renormalisation group flows in various models are added.

R. Gielerak et al. (eds.), Groups and Related Topics, 153–163.

1   Introduction

It was realised soon after the revival of interest into conformally invariant two-dimensional field theories [?], that the study of perturbations leading off criticality may be important for at least two purposes:

to broaden the range of possible applications to statistical mechanics systems as well as to improve the understanding of various series of critical models themselves. Zamolodchikov [?], Cardy and Ludwig. [?] - [?], following earlier work of Kadanoff [?] and Kadanoff and Brown, [?] initiated a program to investigate consequences of perturbations of critical theories given by

$$S = S_o + \Delta S$$
$$\Delta S = \lambda_o \int d^2x \ \varphi(x, \bar{x}) \qquad (1.1)$$

S_o denotes here a (possibly fictitious) action going along with the critical theory and ΔS a perturbation thereof with $\varphi(x, \bar{x})$ being a primary operator of the critical theory. λ_o is a coupling constant.

It was shown by Zamolodchikov [?] that some of the perturbations (1.1) of rational conformal field theories give rise to integrable massive field theories. Much interesting work has been devoted since the appearance of [?] to derive consequences for mass spectra following from integrability and to construct ensueing scattering matrices [?]. I will not follow this line of approach here but instead pursue the field theoretical track of references [?] - [?].

The concrete models and their perturbations to be dealt with in section III will be

> i) rational conformal field theories with central charge smaller than one perturbed through interactions of relevant (that is, smaller than two) scaling dimensions

> ii) generalized Gross-Neveu models with current times current interactions as perturbation

> iii) hypothetical models with truely marginal perturbations (interactions of dimension two which do not develop a renormalisation group β - function). Supersymmetric σ - models with values in Calabi - Yau spaces may be of this type.

I will start in section II with those properties of the (regularised) perturbation expansion which are shared by all of the examples above. Particular emphasis will be put on those features which might be considered as the vestiges of the respective critical system in the off-critical perturbed theory.

2   a) First order perturbation theory

The material of this subsection is taken from [?]. Let us consider the Gell-Mann Low series of the perturbation (1.1) establishing formally for an arbitrary correlation function the connection between the critical (unperturbed) and the generically off-critical (perturbed) theories:

$$(\varphi_1(y_1 , y_1) \cdots \varphi_e(y_e , \bar{y}_e))_{S_o+\Delta S} = \left(\varphi_1 \cdots \varphi_e \ e^{\lambda \int d x^2 \varphi(x , \bar{x})}\right)_{S_o} \tag{2.1}$$

$\varphi_1 \cdots , \varphi_e$ are here arbitrary operators of the critical theory. $(\ )_{S_o}$ and $(\ )_{S_o+\Delta S}$ refer to vacuum expectation values of the unperturbed and of the perturbed theory respectively. I assume that the model characterised through S_o is quasi - rational (with the meaning that only a finite number of primary operators appear in the fusion product of two arbitrary local operators). One infers therefrom that the integrands of the perturbative series on the r. h. s. of (2.1) can be decomposed into finite sums of products of holomorphic and anti-holomorphic functions - so called conformal blocks (for the axiomatised properties of these blocks cf. [?] - [?]) - as follows:

$$\langle \varphi_1(y_1 , \bar{y}_1) \cdots \varphi_e(y_e , \bar{y}_e) \varphi(x_1 , \bar{x}_1) \cdots \varphi_m(x_m , \bar{x}_m) \rangle$$
$$\equiv X (y , x) =$$
$$\sum_{\sim} \prod_{(\cdots)} C_{\sim(\cdots)} \tilde{f}_\sim(y , x) f_\sim(\bar{y} , \bar{x}) \tag{2.2}$$

The functions $f_\sim(\bar{f}_\sim)$ are supposed to be holomorphic (anti-holomorphic) away from coincidence configurations $x_i = x_j$, or $y_i = y_k$ etc.. They carry genuinely non-trivial monodromy transformations induced through analytic continuation along homotopically non-trivial pathes. Together with it show up - loosely speaking as square root of monodromy - non-trivial representations of the braid group. The constants $C_{\sim(\cdots)}$ in (2.2) denoting coefficients of the operator product algebra are fixed up to an overall normalisation through the requirement that the monodromies of the holomorphic and antiholomorphic functions $f_\sim$ and $\bar{f}_\sim$ cancel each other appropriately (see [?] - [?] for details). We assume for simplicity that $\varphi_1 \cdots \varphi_e$ as well as φ are Lorentz scalar operators. The constant $C_{\sim(\cdots)}$ are then to be chosen such that $X(x,y)$ is a one-valued function with respect to all variables (x,y) varying in the complex plane.

Let us inspect more closely the first order term of Eq. (2.1)

$$\int d^2 x \, (\varphi_1 (y_1 , \bar{y}_1) \cdots \cdots \varphi_e (y_e , \bar{y}_e) \varphi (x , \bar{x}))$$
$$= \int d^2 x \bar{f}_\sim(\bar{y} , \bar{x}) \, f_\sim(y , x) \tag{2.3}$$

where we have suppressed the (for our purposes) irrelevant operator product coefficients. The vectors $f_\sim(\bar{f}_\sim)$ may be considered as analytic (antianalytic) functions of x at fixed $y \cdots y$ in the simply connected domain made up by cutting the complex plane along lines from the points y_i , $i = 1 \cdots e$ to infinity. Let $F(y , x)$ be integral of $f(y , x)$ in the cut plane,

$$\frac{\partial F_\sim}{\partial x}(y , x) = f_\sim(y , x)$$

$$\frac{\partial F_\sim}{\partial \bar{x}}(y , x) = 0$$

We apply Stoke's theorem to rewrite (2.3) as

$$\int d^2x \, \bar{f}_\sim \, f_\sim = \int d^2x \bar{f}_\sim \frac{\partial}{\partial x} F_\sim = \frac{1}{2i} \sum_{i=1}^{e} \int_{\bar{C}_i} d\bar{x} \, f(\bar{x}) \, F_\sim(\bar{x}) = \qquad (2.4.a)$$

$$= \frac{1}{2i} \sum_{i=1}^{e} F_\sim(y)_i \int_{\bar{C}_i} d\bar{x} \, \bar{f}_\sim(\bar{x}) \qquad (2.4.b)$$

$\bar{C}_i$ denotes here a counterclockweise contour around the cut from ∞ to $\bar{y}_i$. The equality of (2.4.a) with (2.4.b) is a consequence of the one-valuedness of the product $\bar{f}_\sim \cdot f_\sim$ (cf. [?]).

Symmetrisation among holomorphic and anti-holomorphic factors leads from (2.4.b) to

$$\int d^2x \, \bar{f}_\sim(x) \, f_\sim(x) =$$

$$= \frac{1}{4i} \left\{ \sum_{i=1}^{e} \int_{\bar{C}_i} d\bar{x} \bar{f}_\sim(\bar{x}) \cdot \frac{1 + P_i}{1 - P_i} \int_{C_i} dx f_\sim(x) \right.$$

$$\left. + \sum_{j,k>1} \varepsilon(j-k) \int_{\bar{C}_j} d\bar{x} \bar{f}_\sim(\bar{x}) \int_{C_k} dx f_\sim(x) \right\} \qquad (2.5)$$

$$\varepsilon(x) = \begin{cases} 1, & x > 0 \\ -1, & x < 0 \end{cases}$$

The matrix P_i acting on the vector $f_\sim$ represents the effect of the monodromy if x is moved on a closed path counterclockwise around the point y_i. There are two possible sources of singularities in the formal expression (2.5).
1) The matrices $(1 - P_i)$ may not be invertible
2) The integrals along C_i may diverge at large distances. The former singularities are of ultra-violet provenience. Both kinds of singularities can regularised through suitable addition of free field vertex operators ([?]). Eq. (2.5) may then be considered as an analytically regularised amplitude in the spirit of Speer [?]. Advantage has in the present situation been drawn from the fact that the starting point is a two-dimensional quasi rational conformal field theory, namely from the factorisation of correlations into finite sums of holomorphic and anti-holomorphic conformal blocks. Note that this factorisation survives the first order of perturbation theory.

b) Recursions

One may of course select depending on the purpose one is pursuing other bases of contours than the one chosen above. The contours C_i in (2.5) are particularly suited for displaying the structure of ultraviolet singularities, which are encoded into the matrix factors $\frac{1+P_j}{1-P_j}$ - the contour integrals $\int_{C_j} dx$, $\int_{\bar{C}_j} d\bar{x}$ being free of singularities (disregarding for the moment possible infrared divergencies). Those matrices can be evaluated in a basis of conformal blocks, in which the monodromy of a closed path around y_i is given in diagonal form, as

$$\left(\frac{1+P_j}{1-P_j}\right)_{kl} = (-i)\delta_{kl}\frac{\cos\pi\Delta_j^e}{\sin\pi\Delta_j^e}$$
$$\Delta_j^e = d(\varphi_e) - d(\varphi_j) - d(\varphi). \tag{2.6}$$

The state space on which this matrix acts is labeled by the (primary) operators φ_e which couple in the operator product expansion to φ_j and φ . $d(x)$ denotes the chiral scaling dimension of an operator x.

Pole terms arise in Eq. (2.6) for integer valued Δ_j^e. A suitable addition of free field operators has the regularising effect to shift the scaling dimensions and therewith the Δ_j^e away from integer values. Renormalisation can be achieved similiarly as in a scheme based on dimensional regularisation through subtraction of pole terms.

It is rather cumbersome to keep track of locality in higher orders of perturbation theory using the basis of contours C_i and their multiple generalisation. More convenient is in this context and therewith for a recursive organisation of the perturbative series a basis introduced by Felder [?], to be called $C_{F,j}(\bar{C}_{F,j})$, which comprise the points $y_1 \cdots y_{j-1}$ and close at y_j. Expressing the contours C_j through $C_{F,j}$ one finds instead of (2.5) (cf. [?], [?]):

$$\int d^2x \, (\varphi_1(y_1, \bar{y}_1) \cdots \varphi_e(y_e, \bar{y}_e) \, \varphi(x, \bar{x}))$$
$$= -\frac{1}{2i} \sum_{k=2}^{e} \int_{\bar{C}_{F,k}} d\bar{x}(\cdots) \left(\frac{1}{1-P_{[k]}} - \frac{1}{P_{[k-1]}}\right) \int_{\bar{C}_{F,k}} dx(\cdots) \tag{2.7}$$

$P_{[e]}$ denotes the monodromy matrix corresponding to a closed path surrounding the points $y_1 \cdots y_e$. The main point of the representation (2.7) is, that the correlations modified through the contours $C_{F,i}$ have the same structure as the unperturbed correlations (in particular what concerns the cancellation of monodromies of chiral and anti-chiral parts). Second order perturbation theory can so be considered as equivalent to a first order perturbation in a system with a modified conformal block structure (modified through addition of Felder's contours). One achieves therewith the completely recursive organisation of the perturbation expansion.

c) Deformations of braid group representaitons

A neat description of the recursive construction of braid group representations and their deformations in perturbation theory has been given by Constantinescu and

Lüdde [?]. I want to sketch here in an informal manner their ideas. To start with I consider, following [?], a group G of homotopy classes of closed contours with a fixed base point, call it x_O. Let $y_1 \cdots y_e$ be distinct points in the plane, which may be assumed for the sake of a simple visualisation to be placed on a straight line not crossing the base point x_O. As generators of G are taken (the homotopy classes of) the contours $\hat{C}_j$ starting and ending at x_O. Two representations of the braid group are induced in G trough interchange of the homotopy generating points $y_1 \cdots y_e$. Let τ_i denote the operation of interchange of the neighbouring points y_i and y_{i+1} and $T_i^{\pm}$ its representation as automorphism of G ($\pm$ refering to the two homotopically inequivalent realisations of the more τ_i in the plane). One obtains

$$T_i^+ \left(\hat{C}_j \right) = \begin{cases} \hat{C}_j^{-1} * \hat{C}_{j+1} * C_j, & \text{for } j = i \\ \hat{C}_{j-1}, & \text{for } j = i + 1 \\ \hat{C}_j, & \text{otherwise} \end{cases} \tag{2.8}$$

$$T_i^- \left(\hat{C}_j \right) = \begin{cases} \hat{C}_j * \hat{C}_{j-1} * C_j^{-1}, & \text{for } i = j - 1 \\ \hat{C}_{j+1}, & \text{for } i = j \\ \hat{C}_j, & \text{otherwise} \end{cases} \tag{2.9}$$

$*$ denotes here the group operation in G (prolongation of contours). The inverse $\hat{C}^{-1}$ is by definition the same contour as $\hat{C}_i$ but with reversed orientation. Representations (2.8) and (2.9) of the braid group through automorphisms of the (free) group G are due to Artin [?]. They are loosely speaking to be folded into the representations carried by the conformal blocks of the correlations $(\varphi_1 \cdots \varphi_e \varphi)$. We need for this purpose some new notations. Let $B_i^{\pm}$ denote the matrices representing on the conformal blocks the move τ_i introduced above. $(B_i^+ B_i^- = 1)$. The $B_i^{\pm}$ act on a tensor product of states built up through chiral vertex operators (cf. [?] - [?]). We double the dimensions of the state spaces by assigning an additional Z_2 - charge to each tensor factor which is defined to be one or zero if a contour is encircling or not encircling respectively the point y_i labeling the tensor factor under consideration. Let $P_0^{(i)}$ and $P_1^{(i)}$ denote projectors onto the eigenstates of the local Z_2 charge and $\sigma^{(i)}$ a flip operator,

$$\left(\sigma^{(i)} \right)^2 = 1, \sigma^{(i)} P_k^{(i)} \sigma^{(i)} = P_{k+1}^{(i)}, k \epsilon \, Z \bmod 2.$$

Denoting with θ_j the monodromy operator associated with a move of the operator φ along the closed path $\hat{C}_j$ we are in the position to quote the braid group representation, call it $\{\hat{B}_i\}$, formed from the original representation $\{B_i\}$ through the addition of contours $\hat{C}$. We content ourselves to write down the representation related to Eq. (2.8) and assume that the conventions are chosen such that (2.8) is compatible with the realisation $\{B_i^+\}$.

$$\hat{B}_j^+ = \left\{ \left(1 - \theta_j^{-1} \theta_{j+1} + \theta_j \, \sigma^{(j+1)} \cdot \sigma^{(j)} \right) P_1^{(j)} + \right. $$
$$\left. \left(P_0^{(j+1)} + \sigma^{(j+1)} P_1^{(j+1)} \cdot \sigma^{(j)} \right) P_0^j \right\} B_j^+ \tag{2.10}$$

Iteration of Eq. (2.10) renders in general more and more complicated (and of course larger) braid group representations. One should note that the braid group representations of all known rational conformal field theories can in the sense of Eq. (2.10) be constructed recursively (the recursion ending after a finite number of steps) with an abelian representation as starting point. This is a transcription of the fact that the known models have free field representations.

3   a) Relevant perturbations of models with central charge smaller than one

The techniques displayed in the previous section can be applied to the perturbation of any rational conformal field theory by a relevant, Lorentz scalar, primary operator. (We want to remain within the framework of relativistic field theory.) One of the most discussed examples [?] - [?], [?] - [?] is the perturbation given in Kac-notation by

$$\Delta = \lambda_O \int d^2x \, \varphi_{13} \, (x, \bar{x}) \tag{3.1}$$

of the unitary conformal models with central charges

$$C_m = 1 - \frac{6}{m(m+1)} \, , \, m \epsilon Z, \, m \geq 3. \tag{3.2}$$

The perturbation is, what concerns the ultraviolet structure particularly simple since two operators couple in the operator product algebra to the vacuum state, to φ_{13} itself and to irrelevant operators. It means that renormalisation is selfcontained with one coupling constant.

One of the unsolved problems coming along with the perturbation (3.1) is the resolution of infrared singularities. Infrared poles appear in $n - th$ order of perturbation theory if n is such that

$$n \cdot y \, \epsilon Z, \, d(\varphi_{13}) = 1 - y$$

where $d(\varphi_{13})$ denotes the chiral scaling dimension of φ_{13}. One may regularise through induction of a shift of y to an irrational value. The meaning of such a procedure is however not clear. Zamolodchikov, [?], has on the other hand devived through a formal use of the equation of motion, connected with (3.1), dynamical conservation laws identifying therewith (3.1) as so called integrable perturbation. Consequences of integrability for the mass spectrum of bound states habe been verified with various calculational techniques on statistical mechanics systems whose scaling limit is supposed to be governed by (3.1), cf. [?] - [?]. The success of these studies can be taken as indication that (regularised)perturbation theory is not meaningless.

A massive integrable field theory as mentioned above is supposed to be reached choosing a positive coupling constant in (3.1). It was suggested in [?], [?] that a negative coupling constant gives rise to a renormalisation group flow from the unperturbed theory with central change C_m towards an effective critical theory at large distances with central charge C_{m-1}. The argument for that is taken from second order perturbation theory which produces through a heuristic oversubtraction scheme a renormalisation group β function hinting to the long distance critical fixed point. It seems to me that a systematic justification of the procedure is of paramount importance for progress in this field. One may ask more concretely whether it is possible to establish a connection between the (deformed) braid group representations in the perturbative (essentially short distance) regime and the representation of the infrared stable fix point theory with smaller central charge.

b Generalized Gross - Neveu models

We want to consider here a perturbation of a Wess-Zumino-Witten (W-Z-W) model [?], i. e., a conformally invariant σ - model taking (as classical theory) values in a group manifold, of the form

$$\Delta S = \lambda_0 \int d^2 \, j_\sim \cdot \bar{j}_\sim \qquad (3.3)$$

with $j_\sim$ denoting the chiral currents corresponding to the Lie-Algebra generators of G and $\bar{j}_\sim$ their anti-chiral counterparts. The global symmetry $G \times G$ of the W - Z - W model is through (3.3) broken down to the diagonal subgroup G. Choosing for G an orthogonal or unitary group and taking the Kac-Moody central charge at unit value, one encounters the chiral Gross-Neveu models [?]. The W - Z - W model has in fact under these circumstances a realisation in terms of free fermions [?]. ΔS becomes a four fermion interaction. The Gross-Neveu model and its generalisations (to arbitrary compact groups and with arbitrary integer values of the Kac-Moody central charge) are - for a positive sign of the coupling constant - asymptotically free field theories. The structure of the terms of the perturbation expansion is governed by the current Ward identities of the unperturbed W - Z - W model which read as

$$\langle j^{1a}(z_1) \, j^{a_2}(z_2) \cdots j^{a_n}(z_n) \varphi_1(y_1, \bar{y}_1) \cdots \varphi_k(y_k, y_k) \rangle$$

$$= -k \sum_j \frac{\delta^{a_1 a_j}}{(z_1 - z_j)^2} \, \langle j^{a_2}(z_2) \cdots \hat{j}^{a_j} \cdots \rangle$$

$$+ \sum_j \frac{f^{a_1 a_j c}}{z_1 - z_j} \, \langle j^{a_2}(z) \cdots j^c(z_j) \cdots \rangle$$

$$+ \sum_e \frac{\delta_a^e}{z_1 - y_e} \, \langle j^{a_2}(z_2) \cdots \varphi_k(y_k y_k) \rangle \qquad (3.4)$$

with k being the central charge of the W - Z - W model, f^{abc} denoting the structure constants of the Lie algebra of $G, \varphi_1 \cdots \varphi_k$ denoting primary fields carrying irreducible representations of $G \times G$ and δ^k indicating the variation of φ_k under infinitesimal chiral G -transformations. Analogous complex conjugate relations hold for the anti-chiral currents $\bar{j}$.

Proceeding as described in general terms in section II one arrives with the use of Eq. (3.4) after subtraction of pole terms corresponding to wave function and coupling constant renormalisation at the following result for the first order term of the perturbative series.

$$\int d^2x \ (j_\sim(x)\bar{j}_\sim(\bar{x}) \ \varphi_1(y_1,\bar{y}_1) \ \cdots \varphi_k(y_k,\bar{y}_k))$$

$$= 2\pi i \sum_a \sum_{e=1}^k \left(\varphi_1(y_1,\bar{y}_1) \ \cdots \ \left(\delta_a \colon \varphi_e \bar{\chi}^a \colon - \bar{\delta}_a \colon \varphi_e \chi^a \colon \right) \cdots \colon \varphi_k(y_k,\bar{y}_k) \right) \qquad (3.5)$$

$\colon$ indicates here normal ordering of the product of the primary field φ_e and the (anti-) chiral current potential $\chi^a(\bar{\chi}^a)$ which is defined through $\partial_z \chi^a = j^a (\partial_{\bar{z}} \bar{\chi}^a = \bar{j}^a)$. The rather compact organisation of the first order perturbative term with help of current potentials can be generalized in higher orders using iterated current potentials. This approach might be helpful for implementing Bernards program [?]who suggests a characterisation of the generalized Gross-Neveu models as integrable systems with dynamical non-local conservation laws of the Lüscher-Pohlmeyer type [?].

d) truely marginal interactions

Interactions are called truely marginal if they are of physical scaling dimensions two and have a vanishing renormalisation group β function to all orders of perturbation theory. The current interactions of the preceding section are of this type if and only if the charges belonging to these currents commute with each other [?]. The explicitly known models with truely marginal interactions are in fact all connected with an abelian current algebra. Truely marginal interactions - i. e. quantum moduli - of a different kind, namely genuine primary fields of dimension two, may occur in some superconformal σ - models.

A criterion for true marginality up to second order perturbation theory has been worked out some time ago by Kadanoff and Brown [?]. It says that a perturbation of scaling dimension two has a vanishing β - function if two perturbing operators do not couple in their operator product algebra to themselves or any other operator of dimension two. The Kadanoff-Brown criterion is easily generalized to higher orders of perturbation theory in view of the recursive structure of the perturbation series: there will be no β - function in $r - th$ order if the conformal block system modified to $(r-1) - th$ order does not provide for couplings to operators of dimension two. A more compact version of this criterion would be desirable.

References

[1] A. A. Belavin, A. M. Polyakov and A. B. Zamolodchikov, *Nucl. Phys.* **B241** (1984), 333.

[2] A. B. Zamolodchikov, *JETP Lett.* **43** (1988) 730.

[3] J. L. Cardy, *Nucl. Phys.* **B270** (1986) 186.

[4] A. Ludwig, *Nucl. Phys.* **B285** (1987) 97.

[5] A. Ludwig and J. L. Cardy, *Nucl. Phys.* **B285** (1987) 687.

[6] L. P. Kadanoff, *J. Phys.* **A11** (1978) 1399.

[7] L. P. Kadanoff and A. C. Brown, *Ann. Phys.* **121** (1979) 318.

[8] A. B. Zamolodchikov, *Int. J. Mod. Phys.* **A4** (1989) 4235.

[9] P. Christe and G. Mussardo, *Int. J. Mod. Phys.* **A5** (1990) 4581 and references therein.

[10] F. Constantinescu and R. Flume, *J. Phys.* **A23** (1990) 2971.

[11] P. Chaselon, F. Constantinescu and R. Flume, *Phys. Lett.* **B257** (1991) 63.

[12] G. Moore and N. Seiberg, *Phys. Lett.* **122B** (1988) 451.

[13] K. H. Rehren and B. Schroer, *Nuc. Phys.* **B295** (1988) 229; *Commun. Math. Phys.* **116** (1988) 675.

[14] G. Felder, J. Fröhlich and G. Keller, *Commun. Math. Phys.* **124** (1989) 417; *Commun. Math. Phys.* **130** (1990) 1.

[15] E. Speer,Generalized Feynman Amplitudes, *Annals of Mathematical Studies* **Vol. 62** (1969).

[16] G. Felder, *Nucl. Phys.* **B317** (1989) 215.

[17] S. D. Mathur, *preprint HUTMP-90-B-299* (1990).

[18] F. Constantinescu and M. Lüdde, Braid Modules, *preprint Bonn HE-91-15* .

[19] J. Birman, "Braids, Links and mapping class groups", Princeton UP (1974).

[20] D. A. Kastor, E. J. Martinec and S. H. Shenker, *Nuclear Phys.* **316** (1989) 590.

[21] R. G. Pogossyan, Study of the vicinities of super conformal fixed points in two-dimensional field theory, *Yerevan preprint ePhN-1003* **53-87** (1987).

[22] A. Cappelli, J. I. Latorre *Nucl. Physics* **B340** (1990) 659.

[23] V. P. Yurov and Al. B. Zamolodchikov, *Int. J. Mod. Phys.* **A5** (1990) 3221.

[24] Al. B. Zamolodchikov, *Nucl. Phys.* **B342** (1990) 695.

[25] M. Henkel and J- Saleur, *J. Phys. A: Math. Gen.* **22** (1989) L513.

[26] G. Von Gehlen, *Nucl. Phys.* **B330** (1990) 741.

[27] D. Bernard, Quantum symmetries in 2-D massive field theories, *preprint Saclay SPh-T-91-124* (1991).

[28] E. Witten, *Commun. Math. Phys.* **92** 81994) 455.

[29] D. J. Gross and A. Neveu, *Phys. Rev.* **D10** (1974) 3235.

[30] S. Chauduri and J. A. Schwarz, *preprint CLNS 88/851* (1988).

[31] M. Lüscher and K. Pohlmeyer, *Nucl. Phys.* **B137** (1978) 46.

An alternative dynamical description of Quantum Systems

Benno Fuchssteiner
University of Paderborn, D 4790 Paderborn, Germany

For the analysis of finite-dimensional classical completely integrable Hamiltonian systems the representation by action-angle-variables is an essential tool. This representation can be carried over to the infinite-dimensional situation by use of mastersymmetries, thus leading to a suitable Viasoro algebra in the vector fields. For the quantum case, however, such a structure cannot exist in the corresponding operator algebra, due to a classical theorem of Kaplansky, although the concept of mastersymmetries can be used to give a formal description of the infinite-dimensional symmetry group of those "nonlinear" Quantum systems which are accessible by the Quantum Inverse Scattering Transform. In order to make that precise and to transfer classical notions and methods to the quantum case an alternative dynamical concept for quantum systems is proposed. We give two examples, in the discrete case we consider spin chains, like the Heisenberg anisotropic spin chain, and in the continuous case, where additional difficulties arise, we consider the quantization of the KdV.

1 Introduction

The Viasoro algebra of mastersymmetries ([11], [21]) is an essential tool for completely integrable systems. These time-dependent symmetries are closely connected to recursion operators and to the action-angle representation of the dynamics [15].

We give the necessary definition in the abstract case of an arbitrary Lie algebra $\mathcal{L}$. Let $\mathcal{L}_1$ be a sub-Lie algebra of $\mathcal{L}$. Recall that a map $d : \mathcal{L}_1 \to \mathcal{L}$ is said to be a *derivation* (on $\mathcal{L}_1$) if

$$d[A, B] = [d(A), B] + [A, d(B)] \text{ for all } A, B \in \mathcal{L}_1. \tag{1.1}$$

Special derivations are given by the adjoint $\hat{G}$ of elements $G \in \mathcal{L}$ (i.e. $\hat{G}A := [G, A]$ for all $A \in \mathcal{L}_1$). These derivations are called *inner*. A derivation on $\mathcal{L}_1$ is said to be an $\mathcal{L}_1$-*mastersymmetry* if it maps $\mathcal{L}_1$ into $\mathcal{L}_1$. The mastersymmetries are a sub-Lie algebra.

We briefly explain the use of mastersymmetries. For example, if we fix $K \in \mathcal{L}$ and define $\mathcal{L}_1 = K^{\perp} = \{G \in \mathcal{L} \mid [G, K] = 0\}$ to be the commutant of K, then a $K^{\perp}$-mastersymmetry d has the property that

165

R. Gielerak et al. (eds.), Groups and Related Topics, 165–178.

$d(K), d^2(K), \ldots, d^n(K)$ are elements of $K^\perp$. Hence, we are able to generate in a recursive way out of K infinitely many elements of $K^\perp$, maybe even all of $K^\perp$.

It looks as if for the construction of those mastersymmetries d which can be used to generate elements of $K^\perp$ we have to check how d acts on *all* of $K^\perp$. Fortunately this is not really necessary when $K^\perp$ is abelian:

Observation 1.1:   *Consider $\mathcal{L}_1 \subset \mathcal{B} \subset \mathcal{L}$, where $\mathcal{L}_1$ and $\mathcal{B}$ are sub-Lie algebras of $\mathcal{B}$ and $\mathcal{L}$, respectively. Fix $K \in \mathcal{L}_1$ such that $\mathcal{L}_1$ is equal to the commutant $K^\perp(\mathcal{B}) = \{A \in \mathcal{B} \mid [A, K] = 0\}$ of K in $\mathcal{B}$ (not in $\mathcal{L}$). Assume that $\mathcal{L}_1 = K^\perp(\mathcal{B})$ is abelian. Then an inner derivation $d : \mathcal{B} \to \mathcal{L}$ with $d(\mathcal{L}_1) \subset \mathcal{B}$ is a $K^\perp(\mathcal{B})$-mastersymmetry if and only if $d(K) \in K^\perp(\mathcal{B})$.*

Thus, in case of abelian structure we only have to try out how d acts on K, that means we only have to check if $d(K)$ commutes with K. The proof of this simple fact is mainly based on a successive application of the Jacobi identity, see [11]. Under very mild additional conditions, one can show that in this case the Lie algebra generated by K and d is a Viasoro algebra (see [15]).

We illustrate this crucial notion in different situations. First we concentrate on the classical situation. Let $\mathcal{M}$ be a C^∞-manifold, denote the variable on $\mathcal{M}$ by u and consider an evolution equation $u_t = K(u)$ where K is a C^∞-vector field on $\mathcal{M}$. Recall that the C^∞-vector fields are endowed with a Lie-algebra structure, namely the infinitesimal structure of the group of C^∞-diffeomorphisms on $\mathcal{M}$. Therefore the construction of $K^\perp$ via mastersymmetries amounts to the construction of the infinitesimal generators of the one-parameter symmetry groups. This way of construction works for all the popular completely integrable systems like KdV, mKdV, SG, BO, KP etc. (see [11] or [21]). Complete integrability in all these cases implies that $K^\perp$ is abelian. In addition to that, the $K^\perp$-mastersymmetries have a direct meaning in terms of time-dependent symmetry groups. To see this, consider $G(t) := \exp(t\hat{K})G_0$ for a mastersymmetry G_0. Then, due to the mastersymmetry property, the Taylor series reduces to a polynomial of first order in t. Thus $G(t)$ is a time-dependent symmetry generator.

In case our system is hamiltonian, i.e. the vector field $K = \Theta \nabla H$ is the image of a gradient field ∇H (H the Hamiltonian) under an invertible implectic (inverse symplectic operator or Poisson operator) map Θ [10]. Then in the space of zero-forms (scalar quantities on $\mathcal{M}$) one has a canonical Lie-algebra structure $\{,\}_\Theta$ induced by Θ (*Poisson brackets with respect to Θ*). A zero form is a conserved quantity with respect to the flow if and only if it commutes with H in the Lie algebra of Poisson brackets. Hence, we are able to construct out of H further conserved scalar fields via commutation with $H^\perp$-mastersymmetries. As above, if γ_0 is a mastersymmetry for H in the Lie algebra of Poisson brackets, then $\gamma(t) = \exp(t\hat{H})\gamma_0$ is a time-dependent scalar field, invariant under the flow, and a polynomial of first order in t; hence an angle variable.

For quantum systems the Lie-algebra under consideration are the operators on a suitable Hilbert space. The Lie-product is given by the usual commutation of operators (denoted in the following by $[\![\ ]\!]$). Given an operator H then, via $H^\perp$-mastersymmetries, we would be able to construct in a recursive way operators commuting with H. This would be most interesting, because if H is normal, then knowing a maximal abelian subalgebra of $H^\perp$ is the same as knowing the spectral resolution of H. Hence finding $H^\perp$-mastersymmetries is a big step forward towards the diagonalization of H. But at this point one gets disillusioned since there is a well known theorem of Kaplansky [17] (see [20] for the unbounded case), stating that for every continuous derivation d on an operator algebra, fullfilling $[H, d(H)] = 0$, the spectral radius of $d(H)$ must be equal to zero. So, mastersymmetries cannot exist in the proper sense. The way out of this is to consider mastersymmetries of outer type which can be represented by operators which are "very unbounded". This actually will be our concept for spin chains.

2   Canonical formulation of quantum systems

We embed the usual formulation of quantum mechanics into the frame of classical hamiltonian systems. The manifold $\mathcal{M}$ under consideration is the space of selfadjoint operators on some Hilbert space $\mathbf{H}$. Since $\mathcal{M}$ is a linear space we can identify $\mathcal{M}$ with the typical fiber of its tangent bundle. If some selfadjoint operator H is fixed, then a dynamic on $\mathcal{M}$ is described by the linear evolution equation

$$\dot{A}(t) = i[\![\ H, A(t)]\!]\ .\tag{2.1}$$

Symmetries of this system are given in the well-known way: Consider another flow of this type $\dot{A}(t) = i[\![\ H_1, A]\!]$, then it commutes with (2.1) if and only if $[\![\ H_1, H]\!]\ = 0$.

For the moment we restrict our attention to the special case of finite dimensional Hilbert space in order to avoid convergence difficulties. In that case we have a well defined *duality* on the tangent bundle given by $< A, B >:= \mathrm{trace}\{AB\}$. Via this duality we can identify tangent space $T\mathcal{M}$

and cotangent space $T^*\mathcal{M}$ and we can compute gradients. For example if $P(A) = (1/2) < A, A >$ then $\nabla P(A)$ is given by

$$< \nabla P(A), B >= \frac{1}{2}\frac{\partial}{\partial \epsilon}\Big|_{\epsilon=0} < A + \epsilon B, A + \epsilon B > = < A, B >.$$

Hence $\nabla P(A) = A$. It is easily seen that whenever a selfadjoint $H_1 \in E$ commutes with H, then $F(A) =< A, H_1 A >= trace\{AH_1 A\}$ is a conservation law for (2.1). So, from this aspect, for quantum mechanical systems there is no essential difference between symmetries and conserved quantities, or, in other words, all symmetries are hamiltonian.

We can use Noethers theorem (mapping conserved quantities to symmetry generators) to construct dynamical laws which are not of the type (2.1). We introduce a map $\Theta : T\mathcal{M}^* \to T\mathcal{M}$ by $\Theta(A) = i[\![H, A]\!]$. It is easily verified that Θ is skew-symmetric with respect to the duality introduced before. So Θ must be implectic because it is independent of the particular manifold point. The system (2.1) is hamiltonian in the classical sense because it can be rewritten as $\dot{A}(t) = \Theta \circ \nabla P(A)$. One should remark that this dynamical law is truly linear whatever the form of the hamiltonian is. This observation is puzzling insofar as, by the technique of the quantum inverse scattering transform, *nonlinear* methods really yield relevant results for certain quantum systems: Take some H_1 with $[\![H, H_1]\!] = 0$ then using the image (under Θ) of ∇F we find that the flow $\dot{A}(t) = i[\![H, AH_1 + H_1 A]\!]$ commutes with (2.1).

So, on the first view, there are additional symmetries for quantum systems. However, some of these additional symmetries may not really be relevant because either they destroy the commutation relations between canonical variables or they do not lead to new information. To some extent, this is due to the linear nature of the dynamics given by (2.1). Therefore, one of the aims of this paper is to see how some quantum systems can be considered as truly *nonlinear* systems. We illustrate this new viewpoint first in case of spin chains.

3   Hamiltonian Mastersymmetries for spin chains

We present a direct method for the computation of the commutants of the hamiltonians of the XYZ-model in ferromagnets. For details concerning the importance and the physical relevance of these spin chains we refer to the literature ([3]-[5], [1], [7], see also [2] where similar results can be found for the XYh-model).

At each point n of the lattice $\mathbb{Z}$ a three-component spin operator $\vec{S}_n = (S_n^X, S_n^Y, S_n^Z)$ is given. We assume spin-1/2 operators. i.e.

$$S_n^j S_n^k = \delta_{jk} + i \sum_l \varepsilon^{jkl} S_n^l \tag{3.1}$$

where ε^{jkl} is the cyclic totally antisymmetric tensor with $\varepsilon^{XYZ} = 1$. We either consider the unbounded case, where no periodicity of the lattice is assumed, i.e. where all spin operators at different places commute $[S_n^j, S_m^k] = 0$ for $n \neq m$, or we consider the periodic case where some N is given such that $S_n^k = \pm S_{n+N}^k$ and where the spin operators only commute for those $n \neq m$ which are different modulo N. The manifold of polynomials in the S_n^k fulfilling the constraints (3.1) we denote by $\mathcal{M}(1/2)$.

The Hamiltonian of the XYZ-model is

$$H_{XYZ} = \sum_{n,k} J_k S_n^k S_{n-1}^k \tag{3.2}$$

where the sum for k goes over $\{X, Y, Z\}$ and for n it either goes over all $n \in \mathbb{Z}$ (unbounded case) or from 1 to N (periodic case with periodicity N). The equation of motion is $\vec{S}_t = i[\![H, \vec{S}]\!]$ or explicitly:

$$\dot{S}_n^k = -2 \sum_{l,r} \varepsilon^{klr} J_r S_n^l (S_{n-1}^r + S_{n+1}^r) . \tag{3.3}$$

To describe the commutants of this hamiltonian we look for hamiltonian mastersymmetries. Here, hamiltonian mastersymmetries are operators (in some extended operator algebra) such that if they are commuted with the operator H, then we obtain again operators commuting with H.

For the two Systems (XYZ and XYh) such hamiltonian mastersymmetries can be found in the literature ([13], [2]). Indeed, by a computer algebra package [14], developed for this purpose, we were able to find such operators systematically.

For the XYZ-model we obtain that $M_0 = \sum_{n,k} n J_k S_n^k S_{n-1}^k$ is such a hamiltonian mastersymmetry. Indeed, commuting this operator formally with H_{XYZ} we find the well known operator $H_1 = \sum_{k,l,r} J_l J_r \varepsilon^{lkr} S_{n+1}^l S_n^k S_{n-1}^r$ (see [18], [13]), as first symmetry (or conservation law, if one likes). This process can be continued indefinitely and leads to an infinite sequence of hamiltonians which commute with H_{XYZ}. Of course, the quantity M_0 is not really an operator, it only defines an outer derivative.

One observes that M_0 does not give a translation invariant operator, so it is not compatible with the reductions leading to periodic lattices. Nevertheless one can also use this mastersymmetry in case of periodic lattices because a successive application of the commutator given by M_0 to the original H_{XYZ} always leads to translation invariant operators.

We did not find higher order mastersymmetries for the XYZ-model (contrary to the situation encountered for the XYh-model), although the one mastersymmetry already allows a simple construction of commuting hamiltonians. However, the XYZ-model is considered to be completely integrable, and in case of classical complete integrability we always can find a complete set of action-angle variables, and only the action variables are found, since these correspond to the commuting hamiltonians. The angle variables should correspond to mastersymmetries, and since these do not exist, some doubts seem to be cast upon the complete integrability of this system.

These doubts are not really justified, because the quantum mechanical formulation of the system only allows for hamiltonian vector fields, since all dynamical laws have to be given by operators, thus leading to hamiltonian structure. Even in the classical case of complete integrability, angle variables on infinite dimensional manifolds do not always exist, only nonhamiltonian mastersymmetries exist in these cases, which then yield the angle variables if finite dimensional reductions are taken. The point is, that the dynamical laws given by the quantum mechanical formulation are not a rich enough structure to allow for nonhamiltonian quantities. So, we have to look for an extension of the dynamics in such a way that for hamiltonian quantities the usual quantum mechanical dynamic prevails and nevertheless nonhamiltonian dynamical laws are possible.

4   An alternative description of the dynamics

Consider a vector operator $\mathbf{S}_j = (S_j^X, S_j^Y, S_j^Z)$ associated with every lattice point j. Let P(S) be the polynomials in S_j^n, where $n = X, Y, Z$ and $j \in \mathbb{Z}$. Define the space of *densities* (see [13] or [2]) to be the quotient of P(S) with respect to $Q(\mathbf{S}) :=$ linear span $\{AB - BA \mid A, B \in P(\mathbf{S})\}$. Equivalence classes will be denoted by [ ], and the equivalence by $\equiv$. The construction of density space is done in such a way that $Q(\mathbf{S})$ can be understood as the kernel of a tracelike operation; it is exactly that in case the operators are Hilbert-Schmidt.

Let $\mathbf{A}$ and $\mathbf{B}$ be three-component operator-valued vectors whose components are in P(S). Define for $\mathbf{A}, \mathbf{B} \in P(\mathbf{S})$ the inner product by

$$(\mathbf{A}, \mathbf{B}) :\equiv [\sum A_j^n B_j^n] \equiv \text{equivalence class of } \sum A_j^n B_j^n \ . \qquad (4.1)$$

Define the *directional derivative* of a density F in the direction of $\mathbf{B}$ by

$$F'[\mathbf{B}] = \frac{\partial}{\partial \epsilon}\bigg|_{\epsilon=0} [F(\mathbf{S} + \epsilon\mathbf{B})] \ . \tag{4.2}$$

Since cyclic permutations of factors are allowed in densities we obtain the result ([13] or [16]) that there is a unique operator ∇F in $P(\mathbf{S})$ such that $F'[\mathbf{B}] = (\nabla F, \mathbf{B})$. The operator ∇F is defined to be the *gradient* of F. For example, one obtains the gradient of $[H_{XYZ}]$, the equivalence class given by H_{XYZ}, to be $\nabla[H_{XYZ}]_j = (J_X(S^X_{j+1} + S^X_{j-1}), J_Y(S^Y_{j+1} + S^Y_{j-1}), J_Z(S^Z_{j+1} + S^Z_{j-1}))$ or for example $(\nabla S^l_m)^k_n = \delta_{lk}\delta_{mn}$. Next one introduces a vector product, whose k-th component on the n-th place is given as

$$(\mathbf{B} \times \mathbf{A})^k_n = \frac{1}{2}\sum_{rs} \epsilon^{rsk}(B^r_n A^s_n - A^r_n B^s_n) \ . \tag{4.3}$$

For example $\mathbf{S} \times \mathbf{S} = 0$. Now equation (3.3) can be written as

$$\dot{\mathbf{S}} = -2(\mathbf{S} \times \nabla[H]), \tag{4.4}$$

where $H = H_{XYZ}$. This suggests to consider this dynamical formulation as a flow on $\mathcal{M}(1/2)$. Observe that all flows of the form (4.4) leave this manifold $\mathcal{M}(1/2)$ invariant. We define a Lie algebra structure (Poisson brackets) by

$$\{[G], [H]\}_\Theta \equiv -2(\nabla[G], \mathbf{S} \times \nabla[H]) \tag{4.5}$$

which fulfills the Jacobi identity on this special manifold. A density G is *invariant* under the flow (4.4) if and only if $G_t + \{G, H\}_\Theta \equiv 0$.

Therefore, also in this formulation, H, or rather the density given by it, is said to be the *hamiltonian* of (4.4). Furthermore, the map $G \longrightarrow -2\mathbf{S} \times \nabla G$ is as usual a Lie algebra homomorphism from the Poisson brackets into the vector fields. Hence $\Theta = -2\mathbf{S}\times$ constitutes an implectic operator, i.e. an operator which can be used to define Poisson brackets not only for scalar fields but as well for covector fields. This yields in addition that for the system (4.4) Θ maps invariant covector fields onto invariant vector fields. These statements hold true for any flow of the type (4.4).

Observe, that we found a new hamiltonian structure for the given dynamical systems which drastically differs on the structural level from the one we had before. The main differences are:
•The manifold under consideration is not anymore the manifold of all selfadjoint operators but rather the manifold $\mathcal{M}(1/2)$ of suitable functions in the

spin variable S. Thus we have reduced the dynamics to a manifold which is considerably smaller.

• The dynamical system now truly is a nonlinear one, whereas in its canonical formulation if was linear.

This new approach, which completely fits into the classical formulation of hamiltonian systems, now allows us to look for flows on the new manifold which are not anymore of hamiltonian nature. It turns out that, at least for reductions to the periodic case, action angle variables can be found (a result which also follows from [7] or from Baxters work [3]-[5]).

5   Quantization of KdV

Recall that for the KdV $u_t = u_{xxx} + 6uu_x$ the Poisson bracket structure for scalar fields is defined by

$$\{F_1(u), F_2(u)\} := \int_{-\infty}^{+\infty} (\nabla F_1(u))(\nabla F_2(u))_x dx \qquad (5.1)$$

where ∇ denotes the usual gradient. We shall follow the rule that quantum brackets are operator generalizations of the classical Poisson brackets. First we rewrite (5.1) for the case $F_i(u) = \int_{-\infty}^{+\infty} \varphi_i(x)u(x)dx$ where $\varphi_i(x)$ are suitable test functions. For these special fields we find $\{F_1, F_2\} = \int_{-\infty}^{+\infty} \varphi_i(x)\varphi_2(x)_x dx$. Now, taking limits such that $\varphi_1(x) \to \delta(\hat{x})$ and $\varphi_2(x) \to \delta(x - \hat{x})$, then we obtain $\{u(\hat{x}), u(\tilde{x})\} = \delta_{\hat{x}}(\hat{x} - \tilde{x})$ and the Poisson bracket between field variables at different points is a derivative of the δ-distribution. So quantization of the KdV-field must lead to

$$[u(x), u(\tilde{x})] = i\delta_x(x - \tilde{x}) . \qquad (5.2)$$

Serious difficulties are to overcome in order to make this heuristic approach precise. To show that an algebra, fulfilling this relation, exists at all, we have to give an interpretation of terms like $(u(x)u(\tilde{x}) - u(\tilde{x})u(x))^2$ which would be equal to $\delta_x(x - \tilde{x})^2$, a quantity not yet defined. So, we first have to make some remarks about distribution multiplication. We follow closely the concept introduced in [12] and [8].

A distribution $\phi(x)$ is said to be *almost-bounded* if, for every $n \in \mathbb{N}$, its n-th derivative is of the form $\phi^{(n)}(x) = b(x) + \Delta(x)$ where b is a locally bounded function and where Δ is a distribution with discrete support without accumulation point.

A fundamental observation is that in the space of almost-bounded distributions, there is a canonical algebraic structure fulfilling associativity, product-rule of differentiaton, translation invariance with respect to x, and having the property that it extends the usual pointwise algebra of functions. The algebra is non-commutative, hence there must be two different algebras (interchange of order of factors). In that canonical algebra the product of two distributions with discrete support vanishes. The two product realizations are given by $\phi(x)\phi(\tilde{x}) = \lim_{\epsilon \downarrow 0} \phi(x + \epsilon)\tilde{\phi}(x)$ or $\phi(x)\phi(\tilde{x}) = \lim_{\epsilon \downarrow 0} \phi(x)\tilde{\phi}(x + \epsilon)$. To make the following considerations consistent we choose one of these realizations, say the last.

Now we denote by $u(x)$ a variable in the space of real almost-bounded distributions of degree 3 (third order derivatives of continuous functions). We define $\mathcal{F}(x)$ to be the algebra, generated in the space of almost-bounded distributions, by $u(x)$, all its translations $u(x+\tilde{x}), \tilde{x} \in \mathbb{R}$ and by the almost-bounded distributions. Observe that this is a non-commutative algebra. By $\otimes\mathcal{F}(x)$ we define the algebra of arbitrary tensor products. We will realize an algebra fulfilling (5.2) by taking suitable congruence classes in that algebra. Consider the ideal J generated in $\otimes\mathcal{F}(x)$ by the following relations $\simeq$:

$$\phi_1(x) \otimes \phi_2(\tilde{x}) \simeq \phi_1(x)\phi_2(\tilde{x}) \otimes 1 \simeq 1 \otimes \phi_1(x)\phi_2(\tilde{x})$$

$$u(x) \otimes \phi_1(\tilde{x}) \simeq u(x)\phi_1(\tilde{x}) \otimes 1 \simeq 1 \otimes u(x)\phi_1(\tilde{x})$$

$$\phi_1(\tilde{x}) \otimes u(x) \simeq \phi_1(\tilde{x})u(x) \otimes 1 \simeq 1 \otimes \phi_1(\tilde{x})u(x)$$

$$u(\hat{x}) \otimes u(\tilde{x}) - u(\tilde{x}) \otimes u(\hat{x}) \simeq i\delta_x(\hat{x} - \tilde{x}) \otimes 1$$

$$A \otimes 1 \simeq 1 \otimes A \simeq A \tag{5.3}$$

for ϕ_1, ϕ_2 arbitrary almost-bounded distributions, and $A \in \otimes\mathcal{F}(x)$.

Taking now the quotient $QF(x) = \otimes\mathcal{F}(x)/J$ of $QF(x)$ with respect to the ideal J we have found our quantum realization (named $QF(x)$, the *quantum fields* generated by $\mathcal{F}(x)$). In this new algebraic structure we then have $u(x) \cdot u(\tilde{x}) - u(\tilde{x}) \cdot u(x) = i\delta_x(x - \tilde{x})$. Since elements of $QF(x)$ may be considered as operators (by multiplication) on $QF(x)$ itself, we have found the required operator representation of the Poisson structure of the KdV. Now, we have the prerequisites to define the time evolution for quantum systems by taking suitable Hamiltonian operators. For example, taking

$$H = \int_{-\infty}^{+\infty} \left\{ \frac{1}{2}u_\xi(\xi)u_\xi(\xi) + u(\xi)u(\xi)u(\xi) \right\} d\xi \tag{5.4}$$

and defining the action of a commutator on an integral, as integral (in convolution sense) over the commutator with its integrand, then we find

$$u(x)_t = i[H, u(x)] = u_{xxx}(x) + 3u_x(x)u(x) + 3u(x)u_x(x) . \tag{5.5}$$

This is the quantum version of the KdV, it leaves the crucial relation (5.2) invariant. The main problem is to prove that this equation is completely integrable in the usual sense, i.e. that it has infinitely many commuting symmetry groups (or conserved quantities which are in involution).

6 Densities

In order to give a recursive description of the symmetries and the conserved quantities of the evolution (5.5), as before an alternative representation of its dynamics is introduced.

Define the space of *densities* to be the quotient of $QF(x)$, first with respect to $\mathcal{L}_1 = $ linear span $\{AB - BA \mid A, B \in QF(x)\}$ and then with respect to $\mathcal{L}_2 = $ linear span $\{DA - AD \mid A \in QF(x)\}$, where D denotes differentiation with respect to the variable x. The equivalence relation coming from the successive spaces $\mathcal{L}_1$ and $\mathcal{L}_2$ will be denoted by $\equiv$ and the class of A by $[A]$. The construction of density space is done in such a way that the factor spaces can be understood as the kernel of a trace operation such that formal integrals (from $-\infty$ to $+\infty$) over total derivatives vanish. Indeed, it were exactly that kernel, in case our operators were Hilbert-Schmidt operators vanishing rapidly at $x = \pm\infty$.

Let $\mathbf{A}$ and $\mathbf{B}$ elements in $QF(x)$. Define for $\mathbf{A}, \mathbf{B} \in QF(x)$ an *inner product* by

$$< \mathbf{AB} >:\equiv \text{equivalence class of } \int_{-\infty}^{+\infty} \mathbf{A}(x)\mathbf{B}(x)dx \ . \qquad (6.1)$$

Observe that, due to $\mathcal{L}_2$, the differential operator is antisymmetric with respect to that density-valued inner product. Let $F = F(u)$ be a density depending in some way on the field variable u, define the *directional derivative* of F in the direction of $\mathbf{B}$ by $F'[\mathbf{B}] := \partial/\partial\epsilon|_{\epsilon=0} F(u + \epsilon\mathbf{B})$. The equivalence relation yields the result that there is a unique operator ∇, mapping densities into density-valued linear functionals on $QF(x)$ such that $F'[\mathbf{B}] =< \nabla F, \mathbf{B} >$ for all $\mathbf{B} \in QF(x)$. The quantity ∇F is said to be the *gradient* of F. In case that the densities can be understood as kernels of traces, then the gradient defined this way is indeed the classical gradient of the corresponding scalar quantity given by the integral over the trace. For example, one obtains the gradient of $[H_1]$, where

$$H_1 = \int_{-\infty}^{+\infty} \left\{ \frac{1}{2}u_\xi(\xi)u_\xi(\xi) + u(\xi)u(\xi)u(\xi) \right\} d\xi \qquad (6.2)$$

as $u_{xxx}(x)+3u_x(x)u(x)+3u(x)u_x(x)$. The evolution equation (5.5) therefore can be rewritten as

$$u_t = D\nabla[H_1] \tag{6.3}$$

where H_1 is given above, and where D denotes the operator of taking the derivative with respect to x. This suggests, to consider this alternative dynamical formulation of the system (5.5). This is a flow not on all of $QF(x)$ but rather on the manifold given by those $u(x)$ which are realizations of (5.2). However, this now is a hamiltonian system in the classical sense since D is an implectic operator.

Given this implectic operator Θ, define a Lie algebra structure in the space of densities (*Poisson brackets*) by $\{G,H\}_\Theta :\equiv< \nabla G, \Theta \nabla H >$. These fulfill the Jacobi identity. A density $G(u,t)$ is then *invariant* under the flow $u_t = \Theta \nabla H$ if and only if $G_t + \{G,H\}_\Theta \equiv 0$. If G does not depend explicitly on t then G is invariant if and only if its Poisson bracket with H vanishes. Therefore, H is said to be the *classical hamiltonian* of $u_t = \Theta \nabla H$. The map $G \longrightarrow \Theta \nabla G$ is a Lie algebra homomorphism from the Poisson brackets into the vector fields.

Observe, that we found a new hamiltonian structure, for the given quantum system, which drastically differs from the one we had before in (5.5). The main differences are:
• The manifold under consideration is not anymore the manifold of all elements of $QF(x)$ but rather the manifold of all $u(x)$ fulfilling (5.2). Thus we have reduced the dynamics to a manifold which is considerably smaller.
• The dynamical system now is a truly nonlinear one, whereas in its canonical formulation $A_t = i[H, A]$, $A \in QF(x)$ it was a linear one.

This new approach, which completely fits into the classical formulation of hamiltonian systems, now allows to look for other implectic operators which generate the same dynamics. So, we may use the bi-hamiltonian formulation, given by that, to construct the recursion operator in the usual way.

7 Recursion operator and mastersymmetries

We are now able to obtain the second hamiltonian formulation of the quantum KdV. Denote by u the field variable and introduce $L(u)A := uA$ $R(u)A := Au$ where $a \in QF(x)$. These are the operators of multiplication with u from the left and from the right, respectively. Then set:

$$\Theta = D^3 + DL(u) + DR(u) + R(u)D + L(u)D + (L(u) - R(u))D^{-1}(L(u) - R(u))$$

which gives an operator being antisymmetric with respect to the inner product defined in the last section. We claim Θ satisfies, for arbitrary $A, B, C \in QF(x)$, the *implectic condition*: $< A, \Theta'[\Theta B]C > +$ cyclic permutations $= 0$. Hence we can conclude that Θ is an implectic operator and provides therefore the second hamiltonian formulation of (6.3). The actual verification of the implectic condition is tedious and elaborate. However, the operator Θ is formally, up to a change of sign, the same as the one considered in the general KP-theory presented in [19] or [6]. So, one may apply formally the structural arguments found in these papers, although the operator space considered there is quite different from the one considered here.

Since the operator Θ satisfies the implectic condition, we may now consider the conserved quantity: $H_0 := \frac{1}{2} \int_{-\infty}^{+\infty} u(\xi)u(\xi)d\xi$. Then from the rules laid down in the previous section we get $\nabla[H_0] = u$, whence, $u_t = \Theta\nabla[H_0]$ again results in the flow (5.5).

Thus we have two classical hamiltonian formulations for this quantum flow. This allows to apply the usual theory of hereditary operators ([9] or [10] in order to have a recursive generation of conserved densities and vector fields. We observe that replacing $u(x)$ by $u(x) + \alpha$ in the operator Θ preserves the implectic character of that operator (trivial Bäcklund transformation). But this transformation now yields the operator $\Theta + 4\alpha D$. Hence, Θ and D are compatible implectic operators, so $\Phi := \Theta D^{-1}$ is hereditary and generates out of the vector field, given by the right side of (5.5), a hierarchy of commuting flows. All these flows then constitute symmetry group generators for the quantum KdV, since that equation is among the members of the hierarchy. On the other hand, we may consider the density $[H_0]$ as a conserved quantity for (5.5), then recursive application of Φ^+ (transpose) yields other conserved densities. This is an immediate consequence of the fact that Φ^+ generates, because of its hereditaryness, a sequence of elements whose Poisson brackets commute.

The complete set of mastersymmetries is now easily found. Define, as in the commutative situation, a scaling symmetry by $\tau_0 := xu_x + 2u$. Then, if $K(u)$ is the right side of the quantum KdV, we find for the Lie derivatives $L_{\tau_0}K(u) = 3K(u)$ and $L_{\tau_0}\Phi = 2\Phi$. Using now the hereditary structure of Φ we find that the $\{\tau_n \mid \tau_n := \Phi^n\tau_0, \ n \in \mathbb{N}\}$ is the set of master-symmetries forming a Virasoro algebra together with the set of symmetries $\{K(u)_n \mid K(u)_n := \Phi^n K(u)_0, \ n \in \mathbb{N}\}$.

References

[1] E. Barouch: *On the Ising model in the presence of magnetic Field*, Physica, 1 D, p.333-337, 1980

[2] E. Barouch and B. Fuchssteiner: *Mastersymmetries and similarity Equations of the XYh-model*, Studies in Appl.Math., 73, p.221-237, 1985

[3] R.J. Baxter: *Eight-vertex model in lattice statistics and one-dimensional anisotropic Heisenberg chain, I*, Ann.Physics, 76, p.1-24, 1973

[4] R.J. Baxter: *Eight-vertex model in lattice statistics and one-dimensional anisotropic Heisenberg Chain, II*, Ann.Phys, 76, p.25-47, 1973

[5] R.J. Baxter: *Eight-vertex model in lattice statistics and one-dimensional anisotropic Heisenberg Chain, III*, Ann.Physics, 76, p.48-71, 1973

[6] I. Ya. Dorfman and A. S. Fokas: *Hamiltonian Theory over noncommutative Rings and Integrability in Multidimensions*, INS #181, Clarkson University, Potsdam New-York, p.1-20, 1991

[7] L. D. Faddeev and L. A. Takhtadzhan: *The Quantum Method of the Inverse Problem and the Heisenberg XYZ-model*, Russian Math. Surveys, 34, p.11-68, 1979

[8] B. Fuchssteiner: *Eine assoziative Algebra über einem Unterraum der Distributionen*, Math. Ann., 178, p.302 - 314, 1968

[9] B. Fuchssteiner: *Application of Hereditary Symmetries to Nonlinear Evolution equations*, Nonlinear Analysis TMA, 3, p.849-862, 1979

[10] B. Fuchssteiner and A. S. Fokas: *Symplectic Structures, Their, Bäcklund Transformations and Hereditary Symmetries*, Physica, 4 D, p.47-66, 1981

[11] B. Fuchssteiner: *Mastersymmetries, Higher-order Time-dependent symmetries and conserved Densities of Nonlinear Evolution Equations*, Progr. Theor.Phys., 70, p.1508-1522, 1983

[12] B. Fuchssteiner: *Algebraic Foundation of some distribution algebras*, Studia Mathematica, 76, p.439-453, 1984

[13] B. Fuchssteiner: *Mastersymmetries for completely integrable systems in Statistical Mechanics*, in: Springer Lecture Notes in Physics 216 (L. Garrido ed.) Berlin-Heidelberg-New York, p. 305-315 , 1985

[14] B. Fuchssteiner and U. Falck: *Computer algorithms for the detection of completely integrable quantum spin chains*, in: Symmetries and nonlinear phenomena (D. Levi and P. Winternitz ed.), Singapore, World Sc. Publishers, p.22-50, 1988

[15] B. Fuchssteiner and G. Oevel: *Geometry and action-angle variables of multisoliton systems*, Reviews in Math. Physics, 1, p.415-479, 1990

[16] B. Fuchssteiner: *Hamiltonian structure and Integrability*, in: Nonlinear Systems in the Applied Sciences, Math. in Sc. and Eng. Vol. 185 Academic Press, C. Rogers and W. F. Ames eds., p.211-256, 1991

[17] I. Kaplansky: *Functional Analysis* , in Surveys in Applied Mathematics, Wiley, New York, 4, p.1-34, 1958

[18] M. Lüscher: *Dynamical Charges in the quantized renormalized massive Thirring model*, Nuclear Physics, B117, p.475-492, 1976

[19] F. Magri and C. Morosi: *Old and New Results on Recursion Operators: An alebraic approach to KP equation*, in: Topics in Soliton Theory and Exactly solvable Nonlinear equations (eds: M. Ablowitz, B. Fuchssteiner, M. Kruskal) World Scientific Publ., Singapore, 1987 p.78-96,

[20] M Mathieu: *Is there an unbounded Kleinecke-Shirokov theorem*, Sem. Ber. Funkt. anal. Universität Tübingen, SS 1990, p.137-143, 1956

[21] W. Oevel: *A Geometrical approach to Integrable Systems admitting time dependent Invariants*, in: Topics in Soliton Theory and Exactly solvable Nonlinear equations (eds: M. Ablowitz, B. Fuchssteiner, M. Kruskal) World Scientific Publ., Singapore, p.108-124, 1987

On the solutions of the Yang-Baxter equations

Ladislav Hlavatý

Institute of Physics, Czechoslovak Academy of Sciences,
Na Slovance 2, 180, 40 Prague 8, Czechoslovakia

Abstract

Classification of triangular constant solutions of t he Yang-Baxter equation is done. The formulas for the baxterization of braid group representations are used for finding trigonometric solutions of the Yang-Baxter equations from the constant ones. Some of the obtained solutions correspond to knows solutions but several new noes are obtained as well. The calculations confirm that the baxterization formulas for more then two eigenvalues of the $\hat{R}$-matrix are not universal. The relationship between recently published representation of the coloured braid group and solutions found before is described.

1 Introduction

The Yang-Baxter equations (YBE) have appeared in many branches of theoretical physics and recently also in mathematics [1]. Therefore their solutions have rather wide field of applications.

Few years ago the author of the paper have published classification of 4×4 matrices of the eight - or - less vertex form.

$$R = \begin{bmatrix} q & 0 & 0 & d \\ 0 & r & c & 0 \\ 0 & b & s & 0 \\ a & 0 & 0 & t \end{bmatrix}, \qquad det\, R \neq 0 \tag{1.1}$$

that satisfy the Yang-Baxter equations (YBE)

$$R_{12}R_{13}R_{23} = R_{23}R_{13}R_{12} \tag{1.2}$$

and their "spectral-dependent" counterparts [2]. For reader's convenience and as they will be used in the following, the eight-or-less vertex constant solutions are listed in the Appendix.

In the next section we shall perform the classification of triangular constant solutions. The classification of general 4×4 solutions seems to be a rather complicated problem.

R. Gielerak et al. (eds.), Groups and Related Topics, 179–188.
© 1992 *Kluwer Academic Publishers.*

Recently, formulas for the trigonometric Yang-Baxterization of the constant R-matrices that satisfy the YBE (or equivalently braid groups relations) were found [3]. We are going to apply the formulas for baxterizing the constant solutions.

In [2], the spectral-dependent solutions $R(u, v)$ were obtained by the requirement that $R(u, u)$ be equal to a given constant solution R for some u. Most of the solutions were genuine functions of two variables (cf. [4]), not only of their difference or ratio or any other combination. Obviously the spectral-dependent solutions that we shall get by the baxterization formulas will differ from those obtained in [2] because the trigonometric solutions are of the form $R(u, v) = R(u/v)$ and the initial condition $R(0) = R$ used in [3] cannot be realized as $R(u, u) = R$.

An attempt to classify the eight-or-less vertex solutions dependent on the difference $u - v$ of spectral variables was made in [5]. Many of the solutions obtained by the baxterization formulas below belong to this list after the substitution $x = \exp(i\theta)$ but some new solutions will appear as well.

2   Constant triangular solutions

In this section we are going to classify the constant solutions of the equation (1.2) that are of the form

$$
R = \begin{bmatrix} q & 0 & 0 & 0 \\ x & r & 0 & 0 \\ g & y & s & 0 \\ p & h & z & t \end{bmatrix}, \qquad \det R = qrst \neq 0 \tag{2.1}
$$

Obviously there is a nontrivial intersection of the classes (1.1) and (2.1) so that many of the triangular solutions appeared already in the classification performed in [2]. For those reasons, here we shall be interested mainly in solutions where $x \neq 0$ or $g \neq 0$ or $h \neq 0$ or $z \neq 0$.

Besides that we shall exploit symmetries of the YBE

$$
R \mapsto k(A \otimes A)R(A \otimes A)^{-1}, \qquad A \in GL(2, \mathbb{C}), \ k \in \mathbb{C}\backslash\{0\}. \tag{2.2}
$$

Inserting (2.1) into (1.2) we get 18 independent equations for 10 entries of (2.1). They can be solved by brute force and we get a list of triangular solutions to the YBE. Most of them but not all can be transformed by (2.2) to the form (1.1) with $c = d = 0$.

The result is that up to the symmetries (2.2) there are just seven triangular solutions of the YBE. Five of them R_3, R_5, R_6, R_7, R_8 appeared in the classification of the eight-or-less vertex solutions. The remaining two are

$$
R_{10} = \begin{bmatrix} 1 & 0 & 0 & 0 \\ a & 1 & 0 & 0 \\ b & 0 & 1 & 0 \\ c & b & a & 1 \end{bmatrix}, \qquad R_{11} = \begin{bmatrix} 1 & 0 & 0 & 0 \\ -g & 1 & 0 & 0 \\ g & 0 & 1 & 0 \\ -gh & h & -h & 1 \end{bmatrix}, \tag{2.3}
$$

The solution R_{11} with $g = -h = 1$ appeared in [6] and in its general form was found by a different method in [7].

3   Baxterization of the constant solutions

Having a list of constant solutions we can look for their spectral dependent counterparts. A method for that provide the baxterization formulas.

First we shall baxterize the R-matrices that satisfy the (Hecke) condition

$$(PR - \lambda_1)(PR - \lambda_2) = 0 \qquad (3.1)$$

where P is the permutation matrix $P_{ij}^{kl} = \delta_j^k \delta_i^l$. In other words, the minimal polynomial of the matrix $\hat{R} := PR$, is of the second degree. They are the matrices $R_1 - R_6$ from the Appendix and R_{11}.

It was proved [3] that if R is a constant solution of the YBE then

$$R(x) = R + \lambda_1 \lambda_2 x P R^{-1} P \qquad (3.2)$$

is a solution of the "spectral-dependent" YBE

$$R_{12}(x) R_{13}(xy) R_{23}(y) = R_{23}(y) R_{13}(xy) R_{12}(x) \qquad (3.3)$$

It turns out that if x is replaced by $exp(i\theta)$ the solutions may be expressed in terms of the trigonometric functions.

Baxterization of the matrices R_1, R_2, R_5, R_6 yields solutions $8V(III)$, $8V(II)$, $6V(I)$, $6V(II)$ presented in [5]. (for R_2 see also [8]).

The minimal polynomial of $\hat{R}_3$ is of the second degree

$$\left(\hat{R}_3 - 1\right)\left(\hat{R}_3 + t\right) = 0 \qquad (3.4)$$

so that the eigenvalues are $\lambda_1 = 1$, $\lambda_2 = -t$ and the baxterization formula (3.2) gives the solution

$$R_3(x) = \begin{bmatrix} 1 - xt & 0 & 0 & 0 \\ 0 & s(1-x) & x(1-t) & 0 \\ 0 & 1-t & st(1-x) & 0 \\ (1-x) & 0 & 0 & x-t \end{bmatrix}. \qquad (3.5)$$

The matrix R_4 gives $PR_3(x)P$.

Due to the fact that $PR_{11}^{-1}P = R_{11}$, the baxterization formula (3.2) supplies only unessential factor $(1 - x)$ to the constant solution R_{11}.

There are unproved but usually working formulas for baxterization such R-matrices that $\hat{R}$ have the minimal polynomial of the degree three [3], [9].

$$R(x) = a(x-1)R + bxP + cx(x-1)PR^{-1}P \qquad (3.6)$$

where

$$a = -\lambda_3^{-1}, \qquad b = 1 + \lambda_1\lambda_2^{-1} + \lambda_1\lambda_3^{-1} + \lambda_2\lambda_3^{-1}, \qquad c = \lambda_1 \qquad (3.7)$$

or

$$a = -\lambda_1\lambda_2^{-1}\lambda_3^{-1}, \qquad b = 1 + \lambda_1\lambda_2^{-1} + \lambda_1\lambda_3^{-1} + \lambda_1^2\lambda_2^{-1}\lambda_3^{-1}, \qquad c = \lambda_1 \qquad (3.8)$$

where $\lambda_1, \lambda_2, \lambda_3$ are eigenvalues of $\hat{R}$. As the eigenvalues do not appear symmetrically in (3.7), (3.8) their permutations may yield different function.

Using these formulas we can baxterize constant solutions R_7, R_9 and special cases of R_8.

The roots of the minimal polynomial $\hat{R}_7$ are

$$\lambda_1 = -\lambda_2 = \lambda_3 = 1. \qquad (3.9)$$

All the solutions obtained form the formulas (3.6) - (3.8) are (up to a scalar factor) of the form

$$R_7(x) - \begin{bmatrix} 1 + \epsilon x & 0 & 0 & 0 \\ 0 & s(1 + \epsilon x) & 0 & 0 \\ 0 & 0 & s(1 + \epsilon x) & 0 \\ 1 - \epsilon x & 0 & 0 & 1 + \epsilon x \end{bmatrix}, \qquad \epsilon = \pm 1. \qquad (3.10)$$

The matrix $\hat{R}_8$ has four eigenvalues in general. However, if $t = q$ or $t^2 = rs$ or $q^2 = rs$ some of the eigenvalues coincide. For $t^2 = q^2 = rs$ we get the special cases of the matrices R_5 or R_6. For $q^2 \neq rs$ and $t = q$ or $t^2 = rs$, the minimal polynomial for R_8 is of the third degree and

$$\lambda_1 = -\lambda_2 = \sqrt{rs}, \qquad \lambda_3 = q. \qquad (3.11)$$

The baxterization formulas (3.6) - (3.8) then give three x-dependent R-matrices. Two of them are not very interesting as they are diagonal and any diagonal x-dependent matrix is a solution of the YBE. The third one given by

$$R(x) = p^{-1}(x-1)R_8 + x(pq^{-1} - qp^{-1}) + px(x-1)PR_8^{-1}P, \qquad p^2 = rs \quad (3.12)$$

is a solution of the YBE only if $t = q$. The trigonometric version of this solution is $6V(I)$ of [5]. If $t^2 = rs \neq q^2$, the formula (3.12) does not yield a solution of the YBE.

The minimal polynomial for $\hat{R}_9$ is

$$\left(\hat{R}_9 - I\right)\left(\hat{R}_9 + 1\right)\left(\hat{R}_9 - t\right) \tag{3.13}$$

and the baxterization formulas yield the solutions

$$R_{9a}(x) = \begin{bmatrix} 0 & 0 & 0 & 1 \pm xt \\ 0 & 0 & t \pm x & 0 \\ 0 & t \pm x & 0 & 0 \\ 1 \pm xt & 0 & 0 & 0 \end{bmatrix}, \tag{3.14}$$

and

$$R_{9b}(x) = \begin{bmatrix} x(t - t^{-1}) & 0 & 0 & 1 - x^2 \\ 0 & 0 & t - x^2 t^{-1} & 0 \\ 0 & t - x^2 t^{-1} & 0 & 0 \\ 1 - x^2 & 0 & 0 & x(t - t^{-1}) \end{bmatrix}. \tag{3.15}$$

The matrices $\hat{R}_8$ (in generic case) and $\hat{R}_{10}$ have minimal polynomial of the fourth degree. Then the baxterization formula is

$$R(x) = A(x)RPR + B(x)R + C(x)P + D(x)PR^{-1}P \tag{3.16}$$

where A, B, C, D are quadratic and cubic functions of x with coefficients dependent in a rather complicated way on the eigenvalues of $\hat{R}$ [3]. The coefficients are not invariant w.r.t. the permutation of the eigenvalues so that the formula (3.16) represents several prescriptions for the baxterization.

When applied to R_8, the formula (3.16) give either purely diagonal solutions of matrices that do not satisfy the YBE.

More interesting is the matrix R_{10}. The minimal polynomial for $\hat{R}_{10}$ is $(x - 1)^3(x+1)$ so that the eigenvalues are ± 1. The baxterization formula then give four different solutions of the YBE. Two of them are constant up to a scalar factor and the other two are of the form

$$R_{10}(x) = \begin{bmatrix} q(x) & 0 & 0 & 0 \\ a(x) & q(x) & 0 & 0 \\ b(x) & 0 & q(x) & 0 \\ c(x) & b(x) & a(x) & q(x) \end{bmatrix}, \tag{3.17}$$

where

$$\begin{aligned} a(x) &= a(2x + 1) - bx(x + 1), \qquad b(x) = b(2x + 1) - ax(x + 1) \\ c(x) &= c(1 + x - x^2) - 2(a^2 + b^2)x + 2abx^2, \qquad q(x) = x^2 + x + 1. \end{aligned} \tag{3.18}$$

or

$$\begin{aligned} a(x) &= a(2x + 1) - bx(x + 1), \qquad b(x) = b - ax(x + 1) \\ c(x) &= c(1 - x - x^2) - 2(a^2 + b^2)x + 2abx^2, \qquad q(x) = x^2 + x + 1 \end{aligned} \tag{3.19}$$

4   Coloured braid group representations

Recently two 4 × 4 solutions of the Yang-Baxter equation (YBE)

$$R_{12}(\lambda,\mu)R_{13}(\lambda,\nu)R_{23}(\mu,\nu) = R_{23}(\mu,\nu)R_{13}(\lambda,\nu)R_{12}(\lambda,\mu), \qquad (4.1)$$

that are equivalent to the braid group relations, were published [10]. The solutions are non-additive in the sense that $R(\lambda,\mu) \neq f(\lambda-\mu)$. The purpose of this comment is to generalize these solutions and describe their relationship to the solutions

$$R_V(u,v) = \begin{bmatrix} u/v & 0 & 0 & 0 \\ 0 & (u/v)^{-1} & 0 & 0 \\ 0 & 1-k & kuv & 0 \\ 0 & 0 & 0 & v/u \end{bmatrix}, \qquad k = const \qquad (4.2)$$

$$R_{VI}(u,v) = \begin{bmatrix} u_1/v_2 & 0 & 0 & 0 \\ 0 & (u_1v_2)^{-1} & 0 & 0 \\ 0 & W & -u_1v_2 & 0 \\ 0 & 0 & 0 & v_2/u_1 \end{bmatrix}, \qquad W = u_1/u_2 + u_2/u_1 \quad (4.3)$$

that are given in [2]. Table 1 together with other non-additive solutions to the YBE. Note that the variables u,v in R_{VI} are two - component quantities.

To solve the equation (4.1) the ansatz

$$R(\lambda,\mu) = \begin{bmatrix} u_+(\lambda,\mu) & 0 & 0 & 0 \\ 0 & p^{(+-)}(\lambda,\mu) & 0 & 0 \\ 0 & W(\lambda,\mu) & p^{(-+)}(\lambda,\mu) & 0 \\ 0 & 0 & 0 & u_-(\lambda,\mu) \end{bmatrix} \qquad (4.4)$$

was accepted (for easy orientation we use the notation of [10]). The ansatz can be justified either [10] by weight-conservation or [2] by the requirement that we look for solutions that for $\lambda = \mu$ are of the form

$$R = q \begin{bmatrix} 1 & 0 & 0 & 0 \\ 0 & r & 0 & 0 \\ 0 & 1-rt & t & 0 \\ 0 & 0 & 0 & s \end{bmatrix}, \qquad s = 1 \text{ or } s = -rt. \qquad (4.5)$$

Important fact exploited in the following is that the set of solutions of (4.1) is in general invariant under the transformations

$$R(\lambda,\mu) \mapsto \phi(\lambda,\mu)R(\lambda,\mu) \qquad (4.6)$$

$$R(\lambda,\mu) \mapsto [T(\lambda) \otimes T(\mu)] R(\lambda,\mu) [T(\lambda) \otimes T(\mu)]^{-1} \qquad (4.7)$$

$$R(\lambda,\mu) \mapsto R(f(\lambda), f(\mu)) \qquad (4.8)$$

where ϕ and f are scalar functions and T is a $GL(2)$-valued function.

We can exploit the symmetry (4.6) to set $\mu_+(\lambda, \mu) = 1$. Then we immediately get from (4.1) that $p^{(+-)}$, $p^{(-+)}$ are functions of one variable only

$$p^{(+-)}(\lambda, \nu) = p^{(+-)}(\lambda, \mu) = p^+(\lambda),$$ (4.9)

$$p^{(-+)}(\lambda, \nu) = p^{(-+)}(\mu, \nu) = p^-(\nu).$$ (4.10)

(cf. (11), (12) in [10]). For $W(\lambda, \mu)$ we get the equation

$$W(\lambda, \mu)W(\mu, \nu) = W(\lambda, \nu)\left[1 - p^+(\lambda)p^-(\lambda)\right]$$ (4.11)

general solution of which is

$$W(\lambda, \mu) = \left[1 - p^+(\lambda)p^-(\lambda)\right]\xi(\lambda)/\xi(\mu)$$ (4.12)

where ξ is an arbitrary function. The equations for $u_-(\lambda, \mu)$ then imply that

$$u_-(\lambda, \mu) = p^+(\lambda)q(\mu)$$ (4.13)

where

$$q(\mu) = 1/p^+(\mu), \qquad p^+(\mu)p^-(\mu) = k \in \mathbb{C} \backslash \{0\}$$ (4.14)

or

$$q(\mu) = -p^-(\mu)$$ (4.15)

The conclusion is that *there are just two solutions to the YBE (4.1) of the form (4.4)*. They are

$$R_1(\lambda, \mu) = \phi(\lambda, \mu) \begin{bmatrix} 1 & 0 & 0 & 0 \\ 0 & p^+(\lambda) & 0 & 0 \\ 0 & (1-k)\xi(\lambda)/\xi(\mu) & k/p^+(\mu) & 0 \\ 0 & 0 & 0 & p^+(\lambda)/p^+(\mu) \end{bmatrix}$$ (4.16)

$$R_2(\lambda, \mu) = \phi(\lambda, \mu) \begin{bmatrix} 1 & 0 & 0 & 0 \\ 0 & p^+(\lambda) & 0 & 0 \\ 0 & W(\lambda, \mu) & p^-(\mu) & 0 \\ 0 & 0 & 0 & -p^+(\lambda)p^-(\mu) \end{bmatrix}$$ (4.17)

where W is given by (4.12) and p^+, p^-, ϕ and ξ are arbitrary functions.

The solutions in [10] are particular cases of (4.16), (4.17) where

$$p^+(\lambda) = q^{-1}\eta Q^{\Sigma_{j=1}^m \tilde{a}_j \lambda^j}, \qquad p^-(\mu) = q^{-2}/p^+(\mu). \tag{4.18}$$

$$\phi(\lambda,\mu) = q\phi_+(\lambda,\mu), \qquad \xi(\lambda) = g(\lambda), \qquad k = q^{-2}, \tag{4.19}$$

The appearance of functions ϕ and ξ in (4.16), (4,17) is a consequence of the symmetries (4.6), (4.7). *The symmetry (4.8) enables to consider $p^\pm(\lambda)$ and $p^\pm(\mu)$ as independent variables of the solutions.* This attitude was accepted in [2]. Namely, denoting $p^+(\lambda) = u^{-2}$, $p^+(\mu) = v^{-2}$ and choosing $\phi(\lambda,\mu) = u/v$, $\xi(\lambda) = u$, $\xi(\mu) = v$ we get $R_1(\lambda,\mu) = R_v(u,v)$. Similarly, denoting $P^+(\lambda) = u_1^{-2}$, $p^-(\lambda) = u_2^{-2}$, $p^+(\mu) = v_1^{-2}$, $p^-(\mu) = -v_2^{-2}$ and choosing $\phi(\lambda,\mu) = u_1/v_2$, $\xi(\lambda) = u_2^{-1}$, $\xi(\mu) = v_2^{-1}$ we get $R_2(\lambda, mu) = R_{VI}(u,v)$.

Besides that, R_{VI} corresponds to $N = 2$ solution given in [4].

5   Conclusions

We have classified triangular 4×4 solutions of the YBE. In addition to R_3, R_5, R_6, R_7, R_8 (see Appendix) obtained before, we have found two new constant solutions R_{10}, R_{11}.

We have applied the baxterization formulas of [3] to the known constant solutions and obtained spectral-dependent solutions of the YBE (3.3) by this way. Several of them correspond to the trigonometric solutions obtained before in [5] but those displayed in this paper are now to the best knowledge of the author.

The calculations confirmed that the baxterization formula for three and four eigenvalues of $\hat{R}$ are not universal but nevertheless may provide nontrivial solutions.

It was shown that recently found solutions to the coloured braid group relations [10] have a generalization that correspond to solutions found in [2].

This work was supported in part by the grant CSAV 11 086.

References

[1] Braid Group, Knot Theory and Statistical Mechanics, editors C.N. Yang, M.L. Ge, World Scientific, 1989

[2] L. Hlavaty, J. Phys. **A20** (1987) 1661

[3] Ge M.L., Wu Y.S. and Xue Y., Preprint ITP - SB - 90 - 02 Stony Brook, 1990
    Y. Cheng, M.L. Ge, and Xue Y., Copmm. Math. Phys. **136** (1991) 195

[4] Y. Akutsu, T. Deguchi, Phys. Rev. Lett. **67** (1991) 777

[5] K. Sogo, M. Uchinami, Y. Akutsu, M. Wadati, Prog. Theor. Phys. **68** (1982) 508

[6] E.E. Demidov, Yu. I. Manin, E.E. Mukhin, D.V. Zhdanovich, Prog. Theor. Phys. Suppl. 102 (1990) 203

[7] H. Ewen, O. Ogievetski, J. Wess preprint MPI - PAE/PTh 18/91

[8] M.-L. Ge, L.-H. Gwa, H.-K. Zhao, J. Phys. **A23** (1990) L795

[9] M. Couture, Y. Cheng, M.L. Ge, K. Xue, Int. J. Mod. Phys. **A6** (1991) 559

[10] Ge. M. L., Xue K. 1991 J. Phys. **A24** (1991)L895

Appendix: The eight - or - less vertex solutions of the YBE:

$$
R_0 = \begin{bmatrix} 1 & 0 & 0 & 0 \\ 0 & 0 & 1 & 0 \\ 0 & 1 & 0 & 1 \\ 0 & 0 & 0 & 1 \end{bmatrix}, \quad
R_1 = \begin{bmatrix} 1 & 0 & 0 & i \\ 0 & 1 & 1 & 0 \\ 0 & 1 & -1 & 0 \\ i & 0 & 0 & 1 \end{bmatrix}
$$

$$
R_2 = \begin{bmatrix} 1+t & 0 & 0 & 1 \\ 0 & \sqrt{1+t^2} & 1 & 0 \\ 0 & 1 & \sqrt{1+t^2} & 0 \\ 1 & 0 & 0 & 1-t \end{bmatrix},
$$

$$
R_3 = \begin{bmatrix} 1 & 0 & 0 & 0 \\ 0 & \pm t & 0 & 0 \\ 0 & 1-t & \pm t & 0 \\ 1 & 0 & 0 & -t \end{bmatrix}, \quad R_4 = PR_3P
$$

$$
R_5 = \begin{bmatrix} q & 0 & 0 & 0 \\ 0 & 1 & 0 & 0 \\ 0 & q-t & qt & 00 & 0 & 0 & q \end{bmatrix}, \quad
R_6 = \begin{bmatrix} q & 0 & 0 & 0 \\ 0 & 1 & 0 & 0 \\ 0 & q-t & qt & 0 \\ 0 & 0 & 0 & -t \end{bmatrix},
$$

$$
R_7 = \begin{bmatrix} 1 & 0 & 0 & 0 \\ 0 & \pm 1 & 0 & 0 \\ 0 & 0 & \pm 1 & 0 \\ 1 & 0 & 0 & 1 \end{bmatrix}, \quad
R_8 = \begin{bmatrix} q & 0 & 0 & 0 \\ 0 & r & 0 & 0 \\ 0 & 0 & s & 0 \\ 0 & 0 & 0 & t \end{bmatrix},
$$

$$
R_9 = \begin{bmatrix} 0 & 0 & 0 & 1 \\ 0 & 0 & t & 0 \\ 0 & t & 0 & 0 \\ 1 & 0 & 0 & 0 \end{bmatrix}.
$$

These matrices represent all regular solutions of the form (1.1) to the YBE (1.2) up to the symmetry

$$\tilde{R} = k(T \otimes T)R(T \otimes T)^{-1}, \qquad k \in \mathbb{C}, \; T \in GL(2,\mathbb{C}).$$

State sum invariants of compact 3-manifolds with boundary and 6j-symbols *

M. Karowski1

W. Müller 2

R. Schrader[1]

---

* Supported in part by Deutsche Forschungsgemeinschaft and Akademie der Wissenschaften zu Berlin. Talk given by R. Schrader at the Proceedings of the First German-Polish Max Born Symposium (Wroclaw, Polen).

1 Institut für Theoretische Physik, Freie Universität Berlin, Germany

2 Max Planck Institut für Mathematik, Bonn, Germany

R. Gielerak et al. (eds.), Groups and Related Topics, 189–196.
© 1992 Kluwer Academic Publishers.

In a recent article Turaev and Viro [TV] have constructed nontrivial "quantum" invariants of closed compact 3-manifolds M^3 in the form of state sums (called partition functions in statistical physics and vacuum functionals in quantum field theory) associated with the quantized universal enveloping algebra $U_q(sl(2, \mathbf{C}))$ when q is a complex root of unity of a certain degree $2r > 4$. The state sum is first defined for a given triangulation X of M and then shown to be independent of the triangulation X of M thus giving rise to a well defined invariant of M, which thus depends on q. This result may be viewed as a rigorous mathematical construction of what is called a topological quantum field theory. In fact, in the language of physicists, a triangulation corresponds to the introduction of a high-energy cut-off. Now topological quantum field theories have trivial dynamics, in other words they only deal with the vacuum sector, are scale invariant and more generally independent of any metrics. Invariance under subdivision is just the statement that the associated renormalization group transformation is trivial. This result suggests that the familiar techniques from algebraic topology should become useful to construct and discuss other topological quantum field theories. Moreover, similar combinatorial techniques could become helpful in discussing particle structures of low dimensional quantum field theories, whose vacuum sector is given in terms of a topological quantum field theory.

Another interesting novel feature of the Turaev Viro approach is that the states in the state sum are labelled by elements of the dual of the (quantum) group, i.e. by irreducible representations. This contrast with the familar ansatz in lattice gauge theories, where the states are labelled by elements of the group.

In a recent article [KMS] the authors have extended the Turaev Viro construction to the case where M has nonempty boundary. This leads in particular to a simplified proof of the main result of Turaev and Viro, which is the invariance of the state sum under triangulations. Furthermore by this extension one has introduced observables into the Turaev-Viro approach in the form of certain closed (piecewise) smooth 2-submanifolds consist of several copies of 2-tori, these 2-tori may be viewed as blown up links, i.e. the boundaries of tubular neighborhoods of such links. Invariance is then just the statement that these 2-tori around the links may be chosen arbitrary "small".

To construct the state sum for manifolds with boundary, we first briefly recall the abstract set-up of Turaev and Viro. Let K be a commutative ring with unit. By K^* we denote the set of invertible elements in K. Let I be a finite set, $w \epsilon K^*$ a distinguished element and $i \mapsto \omega_i$ a map from I into K^*. We set

$$\tilde{\omega}^2 = \sum_{i \epsilon I} \omega_i^4. \tag{1}$$

We assume there is given a nonempty set of unordered triples $(i, j, k) \epsilon I$ called admissible. We set $\delta(i, j, k) = 1$ if (i, j, k) is admissible and zero otherwise. An ordered 6-tuple (i, j, k, ℓ, m, n) is called admissible if the 4 unordered 3-triples $(i, j, k), (k, \ell, m), (i, m, n)$ and (j, ℓ, n) are admissible. To each such admissible 6-tuple we assume there is associated an element of K, the abstract 6j-symbol, denoted by

$$\begin{vmatrix} i & j & k \\ \ell & m & n \end{vmatrix}$$

satisfying the following symmetry relations

$$\begin{vmatrix} i & j & k \\ \ell & m & n \end{vmatrix} = \begin{vmatrix} j & i & k \\ m & \ell & n \end{vmatrix} = \begin{vmatrix} i & k & j \\ \ell & n & m \end{vmatrix} = \begin{vmatrix} i & m & n \\ \ell & j & k \end{vmatrix}. \tag{2}$$

In addition we impose three conditions. First, the following "orthogonality" relations are supposed to hold

$$\sum_{k \epsilon I} \omega_k^2 \omega_n^2 \begin{vmatrix} i & j & k \\ \ell & m & n \end{vmatrix} \begin{vmatrix} i & j & k \\ \ell & m & n' \end{vmatrix} = \delta_{n,n'}. \tag{3}$$

The summation is such that all symbols are supposed to be defined, i.e. both 6-tuples (i, j, k, ℓ, m, n) and (i, j, k, ℓ, m, n') are admissible. Secondly, we assume that

$$\omega^2 = \omega_k^{-2} \sum_{ij} \omega_i^2 \omega_j^2 \delta(i, j, k) \tag{4}$$

holds for all $k \epsilon I$. Assume now I to be irreducible, i.e. to any $(i, j) \epsilon I$ there exists a sequence $(\ell_1, ... \ell_n) \epsilon I$ with $\ell_1 = i, \ell_n = j$ and $\delta(\ell_\nu, \ell_{\nu+1}, \ell_{\nu+2}) = 1$ for all $| \leq \nu \leq n - 2$. Then one may prove [KMS] that

$$\sum_i \omega_i^2 \delta(i, j, k) = \frac{\omega^2}{\tilde{\omega}^2} \omega_j^2 \omega_k^2. \tag{5}$$

For the special case of quantum 6j-symbols associated to $U_q(sl(2, C))$ with $q = \exp(i\pi s/r)$ (r and $s \epsilon Z$ relatively prime) one has

$$\omega_i = (\sqrt{-1})^{2i} \left(\frac{q^{2i+1} - q^{-2i-1}}{q - q^{-1}} \right)^{\frac{1}{2}} \qquad (i = 0, \frac{1}{2}, 1, ... \frac{r}{2} - 1) \tag{6}$$

and

$$\omega^2 = \tilde{\omega}^2 = \frac{-2r}{(q - q^{-1})^2}. \tag{7}$$

Finally we assume the following polynomial relation to hold in K (the abstract Bieden-harn-Elliot identity for 6j-symbols).

$$\sum_n \omega_n^2 \begin{vmatrix} i & j & k \\ \ell & m & n \end{vmatrix} \begin{vmatrix} i & n & m \\ D & A & C \end{vmatrix} \begin{vmatrix} j & \ell & n \\ D & C & B \end{vmatrix} = \begin{vmatrix} i & j & k \\ B & A & C \end{vmatrix} \begin{vmatrix} k & \ell & m \\ D & A & B \end{vmatrix} \tag{8_2}$$

again with restrictions similar to those in (3). This relation has a graphical representation in the form

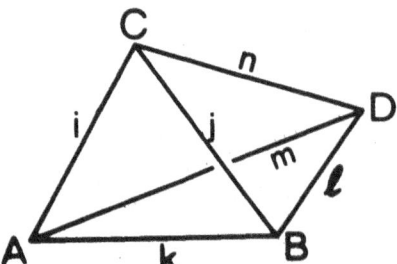

Using (2), (3) and (4), relation (8_2) gives four additional relations

$$\begin{vmatrix} i & j & k \\ \ell & m & n \end{vmatrix} = \omega^{-2} \sum_{A,B,C,D} \omega_A^2 \omega_B^2 \omega_C^2 \omega_D^2 \begin{vmatrix} i & j & k \\ B & A & C \end{vmatrix}$$
$$\cdot \begin{vmatrix} k & \ell & m \\ D & A & B \end{vmatrix} \begin{vmatrix} j & n & \ell \\ D & B & C \end{vmatrix} \begin{vmatrix} i & m & n \\ D & C & A \end{vmatrix}, \tag{8_0}$$

$$\begin{vmatrix} i & j & k \\ \ell & m & n \end{vmatrix} \begin{vmatrix} i & n & m \\ D & A & C \end{vmatrix} = \sum_B \omega_B^2 \begin{vmatrix} i & j & k \\ B & A & C \end{vmatrix} \begin{vmatrix} k & \ell & m \\ D & A & B \end{vmatrix} \begin{vmatrix} j & n & \ell \\ D & B & C \end{vmatrix}, \tag{8_1}$$

$$\sum_{\ell,m,n \atop D} \omega_\ell^2 \omega_m^2 \omega_n^2 \omega_D^2 \begin{vmatrix} i & j & k \\ \ell & m & n \end{vmatrix} \begin{vmatrix} i & n & m \\ D & A & C \end{vmatrix}$$
$$\cdot \begin{vmatrix} j & \ell & n \\ D & C & B \end{vmatrix} \begin{vmatrix} k & m & \ell \\ D & B & A \end{vmatrix} = \omega^2 \begin{vmatrix} i & j & k \\ B & A & C \end{vmatrix}, \tag{8_3}$$

$$\sum_{i,j,k,\ell,m,n \atop A,B,C,D} \omega_i^2 \omega_j^2 \omega_k^2 \omega_\ell^2 \omega_m^2 \omega_n^2 \omega_A^2 \omega_B^2 \omega_C^2 \omega_D^2 \begin{vmatrix} i & j & k \\ \ell & m & n \end{vmatrix} \begin{vmatrix} i & n & m \\ D & A & C \end{vmatrix}$$
$$\cdot \begin{vmatrix} j & \ell & n \\ D & C & B \end{vmatrix} \begin{vmatrix} k & m & \ell \\ D & B & A \end{vmatrix} \begin{vmatrix} i & k & j \\ B & C & A \end{vmatrix} = (\omega^2)^3 \tilde{\omega}^2. \tag{8_4}$$

To construct the state sum for a compact 3-manifold M with boundary we now proceed as follows. Let X be a triangulation of M which induces a triangulation ∂X of the boundary ∂M of M.

Definition: An edge colouring of X is a map $\underline{j} : \sigma^1 \to j(\sigma^1)$ from the set of nonoriented 1-simplexes σ^1 of X into I. A vertex colouring of ∂X is a map $\underline{J} : \sigma^0 \to J(\sigma^0)$ from the set of vertices of ∂X into I.

To a given edge colouring of X and to every nonoriented 3-simplex $\sigma^3 \epsilon X$ we associate the 6j-symbol

$$(6\underline{j})(\sigma^3) = \begin{vmatrix} j(\sigma_1^1) & j(\sigma_2^1) & j(\sigma_3^1) \\ j(\sigma_4^1) & j(\sigma_5^1) & j(\sigma_6^1) \end{vmatrix} \tag{9}$$

provided the 6 tuple $(j(\sigma_1^1), ... j(\sigma_6^1))$ is admissible. Here σ_i^1 and $\sigma_{i+2}^1 (i = 1, 2, 3)$ are opposite edges in $\partial \sigma^3$ and $[\sigma_1^1, \sigma_2^1, \sigma_3^1]$ is a two simplex on $\partial \sigma^3$.

To a given edge colouring $\underline{j}$ of X, vertex colouring $\underline{J}$ of ∂X and nonoriented 2-simplex $\sigma^2 \epsilon \partial X$ we associate the 6j-symbol

$$(6\underline{j}, \underline{J})(\sigma^2) = \begin{vmatrix} j(\sigma_1^1) & j(\sigma_2^1) & j(\sigma_3^1) \\ j(\sigma_1^0) & j(\sigma_2^0) & j(\sigma_3^0) \end{vmatrix} \tag{10}$$

provided the 6-tuple $(j(\sigma_1^1), j(\sigma_2^1), j(\sigma_3^1), j(\sigma_1^0), j(\sigma_2^0), j(\sigma_3^0))$ is admissible. Here σ_i^0 are the vertices opposite to the edges $\sigma_i^1 (i = 1, 2, 3)$. Given $\underline{j}$ and $\underline{J}$ we define the "Boltzmann-Gibbs" weight factor

$$W(X)(\underline{j}, \underline{J}) = \prod_{\sigma^0 \epsilon X} \omega^{-2} \prod_{\sigma^0 \epsilon \partial X} \omega^2_{J(\sigma^0)} \prod_{\sigma^1 \epsilon X} \omega^2_{j(\sigma^1)}$$
$$\prod_{\sigma^3 \epsilon X} (6\underline{J})(\sigma^3) \prod_{\sigma^2 \epsilon \partial X} (6\underline{j}, \underline{J})(\sigma^2). \tag{11}$$

For a given triangulation X of M we define the state sum to be given by

$$Z(X) = \sum_{\underline{j}, \underline{J}} W(X)(\underline{j}, \underline{J}) \tag{12}$$

with a summation over all $\underline{j}$ and $\underline{J}$ such that the weight factor (11) is defined. For $\partial M = \phi$ this state sum definition agrees with that of Turaev and Viro.

Theorem: $Z(X)$ is independent of the particular choice of the triangulation and hence defines an invariant $Z(M)$ of the manifold M.

The proof is based on the fact that the relations $(8_i)(0 \leq i \leq 4)$ have geometric interpretation in form of a local Stokes theorem.

We note that a local Stokes theorem is also responsible for the fact, that in the context of Regge calculus [R] the Lipschitz-Killing curvatures for piecewise linear manifolds are invariant under subdivisions [CMS]. In the present context, it is sufficient to consider invariance under local subdivisions like e.g. the Alexander moves. If this subdivision is localized in $Y \subset X$, then there are two cases

(i) If Y is near the boundary ∂X, we shift ∂X using invariance under isotopies of ∂X such that the state sum has no contribution from Y.

As an example consider the case where $\sigma^3 \epsilon X$ is such that $\partial \sigma^3$ has two 2-simplexes in common with ∂X. We now remove σ^3 from X. Then ∂X changes in the form that $\partial X \cap \partial \sigma^3$ is replaced by the two other 2-simplexes in $\partial \sigma^3$. Invariance of $Z(X)$ under this elementary isotopy then follows by using (8_2). If in general $\sigma^3 \epsilon X$ has $i(1 \leq i \leq 3)$ 2-simplexes in common with ∂X then one uses the relation (8_i).

(ii) If Y is not near the boundary ∂X, we first remove a $\sigma^3 \epsilon Y$ using (8_0), thus creating a hole in X with a new boundary component $\partial \sigma^3$. We then enlarge this new boundary using (i) to remove all of Y from X.

Invariance under a subdivision in Y is now obvious.

One may now use surgery techniques to show that e.g.

$$Z(D^3) = \frac{\tilde{\omega}^2}{\omega^2},$$
$$Z(S^3) = \frac{\tilde{\omega}^2}{\omega^4}, \tag{13}$$
$$Z(S^2 \times S^1) = 1.$$

Note that the formula for $Z(D^3)$ is just a consequence of (8_4). In the context of $U_q(sl(2,C)$ one has

$$\omega^{-2} = \tilde{\omega}^{-2} = \frac{2}{r} \sin^2 \frac{\pi s}{r} \tag{14}$$

This compares with the Chern-Simons topological quantum field theory at level k [Wi1]. Note that according to Reshetikhin and Turaev [RT] there is a rigorous definition $\tau_k(M)$ of the state sum for the Chern-Simons theory in case M is orientable such that

$$Z(M) = \tau_k(M)\tau_k(-M) = |\tau_k(M)|^2 \tag{15}$$

holds [T] with $k + 2 = r$.

Note added in proof:

Meanwhile the state sum has been calculated [KS] for the case $M = \Sigma_g \times S^1$ where Σ_g is a Riemann surface of genus g. One has

$$Z(\Sigma_g \times S^1) = \dim \mathcal{H}_{\Sigma_g}^{TV} \tag{16}$$

where $\mathcal{H}_{\Sigma_g}^{TV}$ is the Hilbert space for this topological quantum field theory. The calculation gives

$$\dim \mathcal{H}_{\Sigma_g}^{TV} = (\dim \mathcal{H}_{\Sigma_g}^{CS})^2 \tag{17}$$

where $\mathcal{H}_{\Sigma_g}^{CS}$ is the toplogical Hilbert space for the Chern Simons theory at level $k = r-2$. $\dim \mathcal{H}_{\Sigma_g}^{CS}$ is given by the Verlinde formula [V]

$$\dim \mathcal{H}_{\Sigma_g}^{CS} = \left(\frac{k+2}{2}\right)^{g-1} \sum_{1 \leq m \leq k+1} \sin^{2-2g}\left(\frac{m\pi}{k+2}\right) \tag{18}$$

(see e.g. [FG], [Wi2]). Note that relations (17) and (18) are compatible with (15).

References:

[CMS] Cheeger, J., Müller, W., Schrader, R., On the curvature of piecewise flat spaces, Commun. Math. Phys. **92**, 405-454 (1984).

[FG] Freed, D.S., Gompf, R.E., Computer calculation of Witten's 3-manifold Invariant, Commun. Phys. **141**, 79-117 (1991).

[KMS] Karowski, M., Müller, W., Schrader, R., State sum invariants of compact 3-manifolds with boundary and 6j-symbols, FU Berlin preprint 1991 (submitted to Journ. Phys. A).

[KS] Karowski, M., Schrader, S., in preparation.

[R] Regge T., General relativity without coordinates, Nuovo Cimento **19**, 558-571 (1961).

[RT] Reshetikhin, N. Yu., Turaev, V.G., Invariants of 3-manifolds via link polynomials and quantum groups, Inv. Math. **103**, 547-597 (1991).

[S] Segal, G., Two dimensional conformal field theories and modular functors, in: IXth International Conference on Math. Physics (Swansea, July 1988), Simon, B., Truman, A., Davies, I.M. (eds), Bristok, Adam Hilger (1989) 22, and preprint (to appear).

[T] Turaev, V.G., State sum models in low dimensional topology (preprint 1990).

[TV] Turaev, V.G., Viro, O.Y, State sum of 3-manifolds and quantum 6j-symbols (to appear in Topology).

[V] Verlinde, E., Fusion rules and modular transformations in 2d conformal field theory, Nucl. Phys. **B300**, 360-376 (1988).

[Wi1] Witten, E., Quantum field theory and the Jones Polynomial, Comm. Math. Phys., **121**, 351-399 (1989).

[Wi2] Witten, E., Quantum gauge theories in two dimensional dimensions, Commun. Math. Phys. **141**, 153-209 (1991).

SECTION IV
MISCELLANEOUS

Product of States

Klaus Fredenhagen

II. Institut für Theoretische Physik, Universität Hamburg

Abstract: On the set of asymptotically vacuumlike states there is a product which induces the composition of superselection sectors. Important concepts of the DHR theory of superselection sectors (statistical dimension, conjugation, fusion rules) are expressed directly in terms of states.

In quantum physics, observables can be multiplied. The physical meaning of this operation is by no means evident, but the fact that observables can be considered as selfadjoint elements of an operator algebra on Hilbert space is the constituting mathematical property of quantum theory.

States can be characterized as positive linear functionals on the algebra of observables. (The normalization condition which normally is imposed on states will be ignored within this note.) In the lecture of Professor Woronowicz we learned that an associative algebra with a coproduct is a natural generalization of the concept of a group. One may ask whether the algebra of observables in quantum theory also has a coproduct, so that observables may be interpreted as generalized symmetries.

Now a coproduct on an algebra immediately leads to a product on the space of linear functionals on the algebra, and it restricts to a product on the state space if the coproduct respects the *-structure. Actually, in quantum field theory there is a multiplicative structure on a certain subset of the state space. This product does not induce a coproduct on the whole algebra but it is indeed connected with the internal symmetries of the theory and their generalized group structure. It arises in the context of the theory of superselection sectors, and I will try to use it for a reformulation of this theory.

Let me first sketch the theory of superselection sectors following [1] (see [2] for the present stage of this field and [3] for a treatment in a textbook). Starting point is the description of a quantum field theory in terms of a Haag-Kastler net of local observable algebras

$$\mathcal{A} = (\mathcal{A}(\mathcal{O}))_{\mathcal{O} \in \mathcal{K}} \tag{1}$$

R. Gielerak et al. (eds.), Groups and Related Topics, 199–209.
© 1992 Kluwer Academic Publishers.

where $\mathcal{K}$ is the set of open double cones in Minkowski space. We consider the algebras $\mathcal{A}(\mathcal{O})$ as von Neumann algebras in a defining representation π_0 (the "vacuum representation") on a Hilbert space $\mathcal{H}_0$ which satisfies Haag duality

$$\pi_0(\mathcal{A}(\mathcal{O}'))' = \pi_0(\mathcal{A}(\mathcal{O})) \quad , \mathcal{O} \in \mathcal{K} \tag{2}$$

(here $\mathcal{O}'$ is the interior of the spacelike complement of $\mathcal{O}$, $\mathcal{A}(\mathcal{O}')$ is the C*-algebra generated by the algebras $\mathcal{A}(\mathcal{O}_1)$, $\mathcal{O}_1 \in \mathcal{K}$, $\mathcal{O}_1 \subset \mathcal{O}'$ and $\mathcal{M}'$ denotes for a subset $\mathcal{M} \subset B(\mathcal{H}_0)$ the commutant) and contains a vector Ω which is cyclic and separating for each algebra $\mathcal{A}(\mathcal{O})$, $\mathcal{O} \in \mathcal{K}$ (Reeh-Schlieder property), i.e. $\mathcal{A}(\mathcal{O})\Omega$ is dense in $\mathcal{H}_0$ and $A\Omega = 0$, $A \in \mathcal{A}(\mathcal{O})$ implies $A = 0$. The state ω_0 induced by this vector

$$\omega_0(A) = (\Omega, \pi_0(A)\Omega), \quad A \in \mathcal{A} \tag{3}$$

is called the vacuum. We will heavily rely on the well known fact that by the GNS construction every state ω gives rise to a representation π_ω with a cyclic vector Φ_ω inducing the state in the above sense.

The DHR theory of superselection sectors treats representations of the algebra of observables which satisfy the selection criterion

$$\pi|_{\mathcal{A}(\mathcal{O}')} \simeq \pi_0|_{\mathcal{A}(\mathcal{O}')}, \quad \mathcal{O} \in \mathcal{K}. \tag{4}$$

Due to Haag duality for each $\mathcal{O} \in \mathcal{K}$ there is an endomorphism ρ of $\mathcal{A}$ acting trivially on $\mathcal{A}(\mathcal{O}')$ such that

$$\pi \simeq \pi_0 \circ \rho. \tag{5}$$

Moreover, two such endomorphisms are related by an inner automorphism of $\mathcal{A}$. The encoding of representations in terms of endomorphisms induces a product of representations

$$\pi_0 \circ \rho_1 \times \pi_0 \circ \rho_2 = \pi_0 \circ \rho_1\rho_2 \tag{6}$$

which extends to a composition law for the equivalence classes. Due to locality of the underlying net the composition law is commutative on the level of classes. On the level of endomorphisms unitary intertwiners occur which induce a representation of the braid group (in $D = 2$ dimensions) or the symmetric group (in $D > 2$ dimensions) and which provide an intrinsic definition of particle statistics. The composition of irreducible representations may be reducible. This effect can be controlled in terms of a number $d(\pi) \in [1, \infty]$ (the "statistical dimension" of π). Products of representations with finite statistical dimension are completely reducible

$$\rho_i\rho_j \simeq \bigoplus N_{ij}^k \rho_k \tag{7}$$

with nonnegative integers N_{ij}^k, and the statistical dimensions satisfy

$$d(\rho_i)d(\rho_j) = \sum N_{ij}^k d(\rho_k), \tag{8}$$

hence they behave under product and decomposition like dimensions of group representations. Furthermore each irreducible representation with finite statistical dimension has a conjugate in the sense that the product contains a subrepresentation equivalent to the vacuum representation. The conjugate may be chosen to be irreducible, and it then is unique up to equivalence.

The sketched formalism is a beautiful description of the general structure of superselection sectors, and it is that structure which has been observed more recently in conformal field theory [4,5,6]. It has, however, a drawback, since it is often very difficult to apply the general formalism to specific examples. (See however [7,8, 9,10] for successful attempts.) It is the aim of the present note to start a reformulation of the theory in terms of states replacing the often nonaccessible endomorphisms. My hope is that this formulation will be useful for the analysis of models, and I will apply it to a simple model suggested by Rehren. Similar ideas have been presented by Buchholz [11] who aimed at a formulation in terms of Wightman fields.

We consider the set of positive linear functionals

$$\mathcal{S} = \{\omega \in \mathcal{A}_+^* \,|\, \pi_\omega \text{ satisfies the DHR criterion}\} \tag{9}$$

A norm dense subset $\mathcal{S}_0$ of $\mathcal{S}$ is formed by those functionals which are dominated at spacelike infinity by ω_0, $\mathcal{S}_0 = \bigcup_{\mathcal{O} \in \mathcal{K}} \mathcal{S}(\mathcal{O})$, where

$$\mathcal{S}(\mathcal{O}) = \{\omega \in \mathcal{S} \,|\, \exists \lambda > 0 \text{ such that } \omega(A') \leq \lambda \omega_0(A') \,\forall A' \in \mathcal{A}(\mathcal{O}'), A' \geq 0\} \tag{10}$$

As an example for a state in $\mathcal{S}(\mathcal{O})$ consider a partial intertwiner (a "vertex operator") for a DHR representation π which is localized in $\mathcal{O} \in \mathcal{K}$, i.e. an operator $S : \mathcal{H}_0 \to \mathcal{H}_\pi$ with

$$SA' = \pi(A')S \quad , A' \in \mathcal{A}(\mathcal{O}') \tag{11}$$

We have $S^*SA' = A'S^*S$, $A' \in \mathcal{A}(\mathcal{O}')$ and hence, by Haag duality, $S^*S \in \mathcal{A}(\mathcal{O})$, thus the state ω induced by $S\Omega$,

$$\omega(A) = (S\Omega, \pi(A)S\Omega) \quad , A \in \mathcal{A} \tag{12}$$

satisfies

$$\omega(A') = \omega_0(S^*SA') \leq ||S^*S||\omega_0(A') \tag{13}$$

for $A' \in \mathcal{A}(\mathcal{O}')$, $A' \geq 0$, i.e. $\omega \in \mathcal{S}(\mathcal{O})$. Actually, all states in $\mathcal{S}_0$ are obtained in this way. Namely, let $\omega \in \mathcal{S}_0$ and let $(\Phi_\omega, \mathcal{H}_\omega, \Phi_\omega)$ denote the GNS triple associated to ω. Let $\mathcal{O} \in \mathcal{K}$ such that $\omega \in \mathcal{S}(\mathcal{O})$. We then define a partial intertwiner S_ω localized in $\mathcal{O}$ by

$$A'\Omega \mapsto \pi_\omega(A')\Phi_\omega \quad , A' \in \mathcal{A}(\mathcal{O}') \tag{14}$$

Due to the Reeh-Schlieder property, S_ω is well defined on a dense domain. Moreover, we have

$$||S_\omega A'\Omega||^2 = ||\pi_\omega(A')\Phi_\omega||^2 = \omega(A'^*A') \leq \lambda\omega_0(A'^*A') = \lambda||A'\Omega||^2 \quad (15)$$

hence S_ω is bounded and may be extended to all of $\mathcal{H}_0$. In particular, S_ω is independent of the choice of $\mathcal{O} \in \mathcal{K}$ with $\omega \in S(\mathcal{O})$. Moreover, S_ω has the intertwining property (11). We now use this operator for the definition of a completely positive mapping on $\mathcal{A}$,

$$\chi_\omega(A) = S_\omega^* \pi_\omega(A)S_\omega \quad , A \in \mathcal{A} \quad . \quad (16)$$

χ_ω satisfies $\omega_0 \circ \chi_\omega = \omega$ and

$$\chi_\omega(A'BC') = A'\chi_\omega(B)C' \quad , A', C' \in \mathcal{A}(\mathcal{O}'), B \in \mathcal{A} \quad , \quad (17)$$

χ_ω is uniquely determined by these properties; namely, on the dense domain $\mathcal{A}(\mathcal{O}')\Omega$ the matrix elements of $\chi_\omega(A)$ are given by

$$(B'\Omega, \chi_\omega(A)C'\Omega) = (\Omega, \chi_\omega(B'^*AC')\Omega) = \omega(B'^*AC'). \quad (18)$$

As an easy further consequence we find that the association $\omega \mapsto \chi_\omega$ respects positive linear combinations. Actually, by Haag duality, χ_ω maps $\mathcal{A}(\mathcal{O}_1)$ into $\mathcal{A}(\mathcal{O}_1)$ for $\mathcal{O}_1 \supset \mathcal{O}$, hence $\mathcal{A}$ into $\mathcal{A}$. The argument is analogous to that which leads to endomorphisms in the DHR theory. Namely, let $A \in \mathcal{A}(\mathcal{O}_1)$, $\mathcal{O}_1 \in \mathcal{K}$ and $\mathcal{O}_1 \supset \mathcal{O}$ and let $B' \in \mathcal{A}(\mathcal{O}_1')$. Then by (16) and by locality

$$[\chi_\omega(A), B'] = \chi_\omega([A, B']) = 0, \quad (19)$$

hence by Haag duality $\chi_\omega(A) \in \mathcal{A}(\mathcal{O}_1)$. We conclude that the positive mappings associated to states in S_0 can be composed.

This fact will be used for a definition of a product of states. Let $\omega_1, \omega_2 \in S_0$. Then we set

$$\omega_1 \times \omega_2 := \omega_0 \circ \chi_{\omega_1}\chi_{\omega_2} \quad . \quad (20)$$

$\omega_1 \times \omega_2$ clearly is a positive linear functional. It is again an element of S_0. For, if $\omega_1, \omega_2 \in S(\mathcal{O})$ and $A' \in \mathcal{A}(\mathcal{O}')$ we have by (17)

$$\omega_1 \times \omega_2(A') = \omega_0(\chi_{\omega_1}\chi_{\omega_2}(1)A'). \quad (21)$$

But (17) also implies, in connection with Haag duality, that $\chi_{\omega_1}\chi_{\omega_2}(1) \in \mathcal{A}(\mathcal{O})$, hence for $A' \geq 0$

$$\omega_1 \times \omega_2(A') \leq ||\chi_{\omega_1}\chi_{\omega_2}(1)||\omega_0(A'), \quad (22)$$

thus $\omega_1 \times \omega_2 \in S(\mathcal{O})$. We also also observe the homomorphism property

$$\chi_{\omega_1 \times \omega_2} = \chi_{\omega_1}\chi_{\omega_2}. \quad (23)$$

We now want to compare this law of composition with the composition within the DHR theory. Let $\mathcal{O} \in \mathcal{K}$, $\omega \in \mathcal{S}(\mathcal{O})$ and $\rho \in \text{End}(\mathcal{A})$ such that ρ acts trivially on $\mathcal{A}(\mathcal{O}')$ and

$$\pi_\omega \simeq \pi_0 \circ \rho. \tag{24}$$

Let V be a unitary mapping from $\mathcal{H}_0$ to $\mathcal{H}_\omega$ which realizes this equivalence. Then

$$\Psi = V^* \Phi_\omega \tag{25}$$

is a vector in $\mathcal{H}_0$ which is cyclic for $\rho(\mathcal{A})$ and satisfies

$$(\Psi, \rho(A)\Psi) = \omega(A), \quad A \in \mathcal{A}. \tag{26}$$

Let $S = V^* S_\omega$. Then $S \in \mathcal{A}(\mathcal{O})$, $\chi_\omega(A) = S^* \rho(A) S$, $A \in \mathcal{A}$ and

$$\omega = \omega_{0,S} \circ \rho \tag{27}$$

where we use the notation

$$\phi_A(B) = \phi(A^* B A) \quad, \phi \in \mathcal{A}_+^*, A, B \in \mathcal{A}. \tag{28}$$

S and ρ are, in contrast to χ_ω, not uniquely determined by ω. If we require that $S\Omega$ is cyclic for $\rho(\mathcal{A})$ the freedom consists in replacing S by US and ρ by $\text{Ad}U \circ \rho$ for unitaries $U \in \mathcal{A}(\mathcal{O})$. ($\text{Ad}U$ means the adjoint action $\text{Ad}U(A) = U A U^{-1}$ on $\mathcal{A}$.)

Now let us compute the product of states $\omega_1, \omega_2 \in \mathcal{S}(\mathcal{O})$. We have

$$\omega_i = \omega_{0,S_i} \circ \rho_i \quad, i = 1, 2 \tag{29}$$

and thus

$$\omega_1 \times \omega_2(A) = \omega_{0,S_1} \circ \rho_1(S_2^* \rho_2(A) S_2) = \omega_{0,\rho_1(S_2)S_1} \circ \rho_1 \rho_2(A). \tag{30}$$

So $\omega_1 \times \omega_2$ is a state in the representation $\pi_0 \circ \rho_1 \rho_2$, thus the law of composition for states (21) leads to the DHR law of composition of sectors (6). It may happen, however, that the vector $\rho_1(S_2)S_1\Omega$ is not cyclic for the algebra $\rho_1 \rho_2(\mathcal{A})$, so $\pi_{\omega_1 \times \omega_2}$ is, in general, only equivalent to a subrepresentation of the DHR product $\pi_{\omega_1} \times \pi_{\omega_2}$.

In the next step we want to analyze the fusion rules (7) on the level of states. Let ρ_i be irreducible, localized in $\mathcal{O} \in \mathcal{K}$ and mutually inequivalent where i varies over the set of sectors. There are isometries $T_{ij}^{k,n} \in \mathcal{A}(\mathcal{O})$, $n = 1, \ldots, N_{ij}^k$ such that

$$\rho_i \rho_j(A) T_{ij}^{k,n} = T_{ij}^{k,n} \rho_k(A) \quad, A \in \mathcal{A} \tag{31}$$

satisfying the orthonormality and completeness relations [6]

$$(T_{ij}^{k,n})^* T_{ij}^{k',n'} = \delta_{kk'} \delta_{nn'} \tag{32}$$

and

$$\sum_{k,n} T_{ij}^{k,n}(T_{ij}^{k,n})^* = 1 \tag{33}$$

Hence we obtain the decomposition

$$\rho_i\rho_j(A) = \sum_{k,n} T_{ij}^{k,n}\rho_k(A)(T_{ij}^{k,n})^* \quad ,A \in \mathcal{A}. \tag{34}$$

Now let $\omega_i = \omega_{0,S_i} \circ \rho_i$. Then

$$\omega_i \times \omega_j = \omega_{0,\rho_i(S_j)S_i} \circ \rho_i\rho_j = \sum_{k,n}\omega_{0,(T_{ij}^{k,n})^*\rho_i(S_j)S_i} \circ \rho_k =: \sum_k \omega_k \tag{35}$$

where ω_k is a state in the representation ρ_k which is a sum of at most N_{ij}^k pure states. The decomposition of the product of states into the components ω_k is unique, hence the fusion rules may be obtained directly in terms of the states.

We now turn to the discussion of conjugates. Let $\rho, \bar{\rho}$ be irreducible, localized in $\mathcal{O}$ and conjugate to each other, i.e. there are isometries $R, \bar{R} \in \mathcal{A}(\mathcal{O})$ such that

$$\bar{\rho}\rho(A)R = RA, \quad \rho\bar{\rho}(A)\bar{R} = \bar{R}A. \tag{36}$$

According to the DHR theory, the phases of these isometries can be chosen such that

$$R^*\bar{\rho}(\bar{R})d(\rho) = 1 = \bar{R}^*\rho(R)d(\rho). \tag{37}$$

Now let $\omega = \omega_{0,S} \circ \rho$. We define a conjugate state $\bar{\omega}$ by

$$\bar{\omega} = d(\omega)\omega_{0,\rho(S^*)R} \circ \bar{\rho}. \tag{38}$$

The conjugate state is uniquely determined. This follows from the fact that the freedom in the choice of S, ρ, $\bar{\rho}$ and R consists in replacing S by US, ρ by $AdU \circ \rho$, $\bar{\rho}$ by $AdV \circ \bar{\rho}$ and R by $V\bar{\rho}(U)R$ for unitaries $U, V \in \mathcal{A}(\mathcal{O})$. Moreover, the double conjugate of ω coincides with ω. In fact, by (38)

$$\bar{\bar{\omega}} = d(\omega)^2\omega_{0,\rho(R^*\bar{\rho}(S))R} \circ \rho, \tag{39}$$

but by (36) and (37)

$$\rho(R^*\bar{\rho}(S))\bar{R} = \rho(R^*)\rho\bar{\rho}(S)\bar{R} = \rho(R^*)\bar{R}S = d(\omega)^{-1}S, \tag{40}$$

hence $\bar{\bar{\omega}} = \omega$.

We now look for a characterization of the conjugate which does not refer to the endomorphisms. Let us consider the products $\omega \times \bar{\omega}$ and $\bar{\omega} \times \omega$. For performing the straightforward calculations it is convenient to use the positive mappings ψ_A, $A \in \mathcal{A}$,

$$\psi_A(B) = ABA^* \quad ,B \in \mathcal{A}. \tag{41}$$

We note the rules

$$\psi_A \psi_B = \psi_{AB}, \ \psi_{\lambda A} = |\lambda|^2 \psi_A, \ \rho \psi_A = \psi_{\rho(A)\rho}, \ \psi_A = \mathrm{Ad}A \text{ for } A \text{ unitary.}$$
(42)

Now $\chi_\omega = \psi_{S \cdot \rho}$ and $\chi_{\bar\omega} = d(\omega) \psi_{R \cdot \rho(S)} \bar\rho$, hence

$$\chi_\omega \chi_{\bar\omega} = d(\omega) \psi_{S \cdot} \rho \psi_{R \cdot \rho(S)} \bar\rho = d(\omega) \psi_{S \cdot \rho(R \cdot) \rho \rho(S)} \rho \bar\rho.$$
(43)

By (36) $\rho \bar\rho \geq \psi_R$, hence

$$\chi_\omega \chi_{\bar\omega} \geq d(\omega) \psi_{S \cdot \rho(R \cdot) \rho \rho(S)} R$$
(44).

By the use of (40) and (42) we finally get

$$\omega \times \bar\omega \geq d(\omega)^{-1} \omega_{0,\chi_\omega}(1).$$
(45)

In a similar way

$$\bar\omega \times \omega \geq d(\omega)^{-1} \omega_{0,\chi_{\bar\omega}}(1).$$
(46)

The inequalities (45) and (46) are the desired relations which contain only the statistical dimension and the conjugate state. It is conceivable that they characterize the conjugate state as well as the statistical dimension completely in the following sense: let S_ω denote the set of all $\omega' \in S_0$ for which there exists some $\lambda > 0$ such that

$$\omega \times \omega' \geq \lambda \omega_{0,\chi_\omega}(1), \quad \omega' \times \omega \geq \lambda \omega_{0,\chi_{\omega'}}(1).$$
(47)

Let $\lambda(\omega')$ be the supremum of all numbers λ satisfying (47) for a given $\omega' \in S_\omega$. Then the conjecture is

(i) $d(\omega) = \inf_{\omega' \in S(\omega)} \lambda(\omega')^{-1}$.
(ii) $\bar\omega$ is the only state in S_ω with $\lambda(\bar\omega) = d(\omega)^{-1}$.

A proof of this conjecture is not available at the moment. I will instead illustrate the concepts discussed in this note on a simple model which was suggested by Rehren [12].

Let φ be the massless scalar free field in two dimensions, and let $j = \partial_0 \varphi - \partial_1 \varphi$. Then j depends only on the light cone variable $u = t - x$ and has the commutation relation

$$[j(u), j(u')] = 2i\delta'(u - u').$$
(48)

The 2-point function of j is

$$(\Omega, j(u)j(u')\Omega) = \frac{1}{\pi} \int_0^\infty dp\, p e^{-ip(u-u')} = \frac{-1}{\pi(u - u' - i\epsilon)^2}.$$
(49)

We introduce smeared fields

$$j(f) = \int du\, j(u) f(u)$$
(50)

with real valued test functions $f \in \mathcal{D}(\mathbb{R})$ and find

$$[j(f), j(g)] = -2i \int du \, f'(u) g(u) = 2i \int du \, f(u) g'(u) =: 2i\sigma(f, g) \qquad (51)$$

and

$$\|j(f)\Omega\|^2 = 2 \int_0^\infty dp \, p |\tilde{f}(p)|^2 =: 2\|f\|^2. \qquad (52)$$

For an analysis of this model in the framework of algebraic field theory (see e.g. [7]) it is convenient to introduce the Weyl operators

$$W(f) = e^{ij(f)} \qquad (52)$$

which satisfy the Weyl relations

$$W(f)W(g) = e^{-i\sigma(f,g)} W(f + g) \qquad (53)$$

as a mathematical precise version of the canonical commutation relations. Let $\mathcal{I}$ denote the set of bounded open intervalls on $\mathbb{R}$ and let $\mathcal{A}_0(I)$ denote the algebra generated by all Weyl operators $W(f)$ with $f \in \mathcal{D}(I)$. The vacuum is characterized by the expectation functional

$$(\Omega, W(f)\Omega) = \omega_0(W(f)) = e^{-\|f\|^2}, \qquad (54)$$

and the local algebras $\mathcal{A}(I)$, $I \in \mathcal{I}$ are defined as the weak closures of $\mathcal{A}_0(I)$ in the vacuum Hilbert space.

The Haag-Kastler net $(\mathcal{A}(I))_{I \in \mathcal{I}}$ has nontrivial outer automorphisms which act trivially on the complement I' of some $I \in \mathcal{I}$. Namely, let F be a smooth real function such that $F' \in \mathcal{D}(I)$. Then $\sigma(F, g)$ is well defined for all $g \in \mathcal{D}(I)$ and vanishes when $g \in \mathcal{D}(I')$. The automorphism is then defined on $\mathcal{A}_0$ by ·

$$\alpha_F(W(g)) = e^{2i\sigma(F,g)} W(g). \qquad (55)$$

For each intervall $J \in \mathcal{I}$ there is some test function $f_J \in \mathcal{D}$ such that $F' = f'_J$ on J, hence

$$\alpha_F|_{\mathcal{A}_0(J)} = \mathrm{Ad} W(f_J)|_{\mathcal{A}_0(J)}. \qquad (56)$$

This shows in particular that α_F can be uniquely extended to an automorphism of the von Neumann algebra $\mathcal{A}(J)$.

These automorphisms generate superselection sectors which are classified by the real number

$$q(F) = F(\infty) - F(-\infty), \qquad (57)$$

and because of the product rule $\alpha_F \alpha_G = \alpha_{F+G}$ they obey the fusion rules

$$q_1 \times q_2 = q_1 + q_2. \qquad (58)$$

As a model with nontrivial fusion rules Rehren proposed the restriction of this model to the even subalgebras $\mathcal{A}_e(I)$. These are the algebras which are invariant under the symmetry $\gamma : j(u) \mapsto -j(u)$; they can be generated by the hermitian operators

$$V(f) = \frac{1}{2}(W(f) + W(-f)) \tag{59}$$

which satisfy the relations

$$V(f)V(g) = \frac{1}{2}e^{-i\sigma(f,g)}V(f+g) + \frac{1}{2}e^{i\sigma(f,g)}V(f-g). \tag{60}$$

The sectors of $\mathcal{A}$ may split after restriction to $\mathcal{A}_e$. This is true for the vacuum, because there the automorphism γ is unitarily implemented by an operator Γ with $\Gamma^2 = 1$; the eigenspaces of Γ reduce the vacuum representation which splits into an even and an odd sector. It is not true for the charged sectors, since there the symmetry γ is spontaneously broken. Therefore there exists a sequence of odd elements of $\mathcal{A}$ which converges weakly to 1, thus each odd element can be obtained as a weak limit point of elements of $\mathcal{A}_e$, and $\mathcal{A}_e$ acts irreducibly on these sectors. As an example for such a sequence take

$$A_n = \frac{\lambda}{2i}(W(f_n) - W(-f_n)) \tag{61}$$

where $\lambda \in \mathbb{R}$ and $f_n(u) = f(\frac{u}{n})$. Then $[A_n, W(g)] \to 0$ for all $g \in \mathcal{D}$ and

$$\omega_0 \circ \alpha_F(A_n) = \lambda \sin 2\sigma(F, f_n)e^{-\|f_n\|^2} \to -\lambda \sin 2q(F)f(0)e^{-\|f\|^2}, \tag{62}$$

hence for a suitable choice of λ and f $A_n \to 1$ weakly.

We now want to study the fusion rules of this model. It seems to be difficult to exhibit the DHR endomorphisms, but it is very easy to give explicit formulae for the positive mappings. We find

$$\omega_F := \omega_0 \circ \alpha_F|_{\mathcal{A}_e} = \omega_0 \circ \chi_F \tag{63}$$

where

$$\chi_F(V(f)) = \cos \sigma(F, f)V(f) \quad , f \in \mathcal{D}. \tag{64}$$

We obtain the composition law

$$\chi_F\chi_G = \frac{1}{2}\chi_{F+G} + \frac{1}{2}\chi_{F-G}. \tag{65}$$

If $q(F+G)$ and $q(F-G)$ are nonzero, the corresponding decomposition of states,

$$\omega_F \times \omega_G = \frac{1}{2}\omega_{F+G} + \frac{1}{2}\omega_{F-G} \tag{66}$$

is a decomposition into pure states. If, however, $q(F-G) = 0$, say, there is some $f \in \mathcal{D}$ with $f = F-G$, so we have to investigate the state $\omega_f = \omega_0 \circ \chi_f$. Now χ_f has the decomposition

$$\chi_f = \chi_f^{(e)} + \chi_f^{(o)} \tag{67}$$

where $\chi_f^{(e)} = \psi_{V(f)}$ and

$$\chi_f^{(o)}(V(g)) = \frac{1}{2}\cos 2\sigma(f,g)V(g) - \frac{1}{4}V(g+2f) - \frac{1}{4}V(g-2f). \tag{68}$$

$\chi_f^{(o)}$ is not unit preserving, namely

$$\chi_f^{(o)}(1) = \frac{1}{2}(1 - V(2f)). \tag{69}$$

The statistical dimension may now be infered from the conjugation symmetry. We have

$$\omega_F \times \omega_F \geq \frac{1}{2}\omega_0 \tag{70}$$

which is consistent with $d(\omega_F) = 2$, and with $\omega_f^{(o)} = \omega_0 \circ \chi_f^{(o)}$

$$\omega_f^{(o)} \times \omega_f^{(o)} = \omega_{0,\frac{1}{2}(1-V(2f))}, \tag{71}$$

hence $d(\omega_f^{(o)}) = 1$. Actually, as Rehren showed, all statistical dimensions as well as the statistical phases of this model can be computed, up to one phase, once the fusion rules are known [12].

References

1.   S. Doplicher, R. Haag, J.E. Roberts: Local Observables and Particle Statistics. Commun.Math.Phys.**23**,199(1971) and **35**,49(1974)
2.   D. Kastler (ed.): The Algebraic Theory of Superselection Sectors. Introduction and Recent Results, World Scientific 1990
3.   R. Haag: Local Quantum Physics. Springer 1992
4.   J.Fröhlich: Statistics of Fields, the Yang-Baxter Equation, and the Theory of Knots and Links, in "Nonperturbative Quantum Field Theory", edited by G. 't Hooft, A. Jaffe, G. Mack, P. K. Mitter, and R. Stora, Plenum Publishing Corporation, 1988
5.   K.H. Rehren, B. Schroer: Einstein Causality and Artin Braids. Nucl. Phys. B **312**, 715 (1989)
6.   K. Fredenhagen, K.H. Rehren, B. Schroer: Superselection Sectors with Braid Group Statistics and Exchange Algebras. Commun.Math.Phys. **125**,201(1989)
7.   D. Buchholz, G. Mack, I. Todorov: The Current Algebra on the Circle as a Germ of Local Field Theories. Nucl. Phys. B (Proc. Suppl.) **5B**, 20 (1988)
8.   G. Mack, V. Schomerus: Conformal Field Algebras with Quantum Symmetry from the Theory of Superselection Sectors. Commun. Math. Phys. **134**, 139 (1990)

9.   J. Fuchs, A. Ganchev, P. Vecsernyes: Level 1 WZW Superselection Sectors. CERN-TH.6166/91 (preprint)

10.   A. Wassermann: Subfactors arising from positive energy representations of some infinite-dimensional groups (to appear)

11.   D. Buchholz: On Quantum Fields that Generate Local Algebras. J. Math. Phys. **31**, 1839 (1990)

12.   K. H. Rehren, private communication

Quantum Measurements and Information Theory.

K.-E. Hellwig.

Institut für Theoretische Physik,
Technische Universität Berlin.

January 31, 1991

1 Introduction.

Max Born [1] (1926) has proposed the statistical interpretation of quantum mechanics and there is no doubt that it well describes the statistics of pointer readings. Is this knowledge sufficient for to understand the physical reality of microparticles, i.e. can a physical theory be complete if nothing behind the statistics of pointer readings is assumed? Such questions arose shortly after Max Born's proposal in the Bohr-Einstein debate which culminated in the famous Einstein-Podolsky-Rosen paradox [2]. Later investigations have led to the Bell type inequalities [3]. The experimental tests by A. Aspect and his coworkers gave deep insight into the physical situation: If somebody believes in the physical reality of accidental properties of a single microparticle described by some kind of hidden variables, then he has to assume the existence of actions at a distance.

The occurrence of pointer positions can be described in terms of quantum mechanics and quantum statistics as a physical process. The description of this process can be understood as an analysis how we get knowledge about microsystems by their interaction with a measuring apparatus. The experiments to be analyzed have a very general scheme: There is one part of the experimental set-up that isolates a microparticle from its surroundings in a particular manner and another one that responds because of interaction with it by a certain pointer position. It is suggesting to assume the microparticle to carry some message from the first part of the set-up, the source of information, to the second, the receiver. This message, if it is any, can only be transmitted if it is coded into some property of the microparticle. If the language of the statistical theory of information can be used to describe this set-up one probably obtains new aspects about the nature of quantum systems.

In the following, I first give a short introduction to the theory of quantum measurements and then I describe the situation in terms of the statistical theory of

R. Gielerak et al. (eds.), Groups and Related Topics, 211–221.
© 1992 Kluwer Academic Publishers.

information.

2   The Theory of Quantum Measurements.

As already mentioned, quantum experimental set-ups usually are divided into two
parts. The first part is the means by which the quantum object is prepared under
control of relevant macroscopic parameters such that it belongs to a well defined
statistical ensemble described by a density operator W_O acting in the Hilbert space
$\mathcal{H}_O$. The second part is called the apparatus and consists of a macroscopic system in
a thermodynamically metastable state which can become instable because of interaction with the object. As a consequence of the interaction, one of several different
equilibrium states will arise and indicate the final result of a measurement [4], which
we call *"pointer position"*. As a many particle quantum system the metastable state
of the apparatus also will be described by a density operator, say W_A acting in a
Hilbert space $\mathcal{H}_A$. The preparations of the object and the apparatus are supposed to
be independent foeach other such that the initial state of the coupled system will be
the uncorrelated one, i.e. it is described by the density operator $(W_O \otimes W_A)$. This
is a highly simplified scheme and there are many features which I cannot describe
in full detail or evenmention here.

   As long as the irreversible dynamics towards the final equilibrium of the apparatus is not involved, the dynamics of the interaction process is described by a unitary
operator. Although it is very doubtful whether the final state is determined by the
reversible part of the process and no stochastics enters thereafter with respect to
the resulting final equilibrium we will confine ourselves to consider only the unitary
dynamics which may be given by the unitarian S acting in $\mathcal{H}_O \otimes \mathcal{H}_{A_1}$. It should be
mentioned that the influence of the irreversible part of the process on the final result
is not completely investigated up to now. We will consider here $S(W_O \otimes W_A)S^+$ as
the final state which determines the measurement result.

2.1   The Historical Scheme.

For the convenience of the reader to enter into this material, I will shortly remind
the historical scheme although it works only for such observables of the object that
have discrete spectra. Let the self-adjoint operator acting in $\mathcal{H}_O$ of the observable
to be measured be given by

$$B = \sum_{i=1}^{\infty} b_i |\psi_i >< \psi_i|.$$

The self-adjoint operator acting in $\mathcal{H}_A$ and corresponding to the macroscopic equilibrium quantity finally to be observed shall be given by

$$A = \sum_{i=1}^{\infty} a_j |\Psi_j >< \Psi_j|.$$

The eigenvalues of this operator can be identified with the final pointer positions. In a realistic situation as a macroscopic equilibrium the observable A should be highly degenerated. Here we will assume it in the contrary to be nondegenerated because this simplifies our reasoning without big loss of generality. By the same reason we assume W_A to represent a pure state, say $\Psi_1 \in \mathcal{H}_A$. Finally let also W_O represent a pure state, say $\phi \in \mathcal{H}_O$. Hence the final state of the coupled system will be $S(\phi \otimes \Psi_1)$. By a suitable choice of S and A, which formally is always possible, one gets a complete correlation of the pointer positions with the eigenvalues the eigenvalues of the operator A in the final state. This situation is given if for each $i \in \mathbb{N}$

$$S(\psi_i \otimes \Psi_1) = \psi_i \otimes \Psi_i$$

holds true. By the linearity of S we have immediately

$$S(\phi \otimes \Psi_1) = \sum_{i=1}^{\infty} < \psi_i|\phi > (\psi_i \otimes \Psi_i)$$

such that the probability to read the pointer position a_i from the apparatus is just given by $|< \psi_i|\phi >|^2$ as it should be.

There are some serious difficulties with this approach. The final state after the pointer position a_j has been fixed should be a mixture rather than a pure state. The simplest explanation is given by J. M. Jauch [5]. Since A is an observable of a many particle system that directly can be (classically) observed, it belongs to a commutative subset of all observables of this system. On the spectrum of any observable of this subset the pure state and the corresponding mixture give the same probability distributions. Hence the problem disappears. Problems arise, however, if the entropy of states is taken into consideration since the pure state has a lower entropy than the mixture. The first idea for to solve them is due to J. v. Neumann and states that this transition to a mixture is caused when the first person gets knowledge about the value by reading it from the apparatus. It has been assumed that a chain of succeeding processes of this kind happen until a human consciousness irreversibly decides about the final result [6]. Clearly, there arises a problem of objectivity since different persons watching the apparatus may decide differently. Although it seems much more natural that a final pointer position arises without interaction with a human brain the first idea is still alive. The Everett-Wheeler interpretation of many worlds [7] solves the objectivity problem at least formally. The realistic point of view is to assume that the final pointer position is the result of an irreversible process inside the apparatus as mentioned at the beginning.

Other difficulties are the following: This scheme only works for discrete spectra. Moreover, suitable choices for S and A do only exist if B commutes with each universally conserved quantity [8]. Since each component of angular momentum of a closed system is a universallly conserved quantity and spin-projections for different directions do not commute, the historical approach is too narrow for to explain measurements of one of them.

2.2   The Modern Approach.

A much more general and simple approach arose at the beginning of the sixties
[9]. In contrast to the historical approach where the operator B is assumed to
be given and the suitable choice of S and A has to be determined, the modern
approach assumes S and A to be given and it is asked what will be measured. This
formulation of the problem accounts for the obstructions mentioned at the end of
the latter subsection 2.1 and includes the problem of approximate measurements as
well. Let again A denote the self-adjoint operator that represents the observable
finally to be observed on the apparatus. The present approach does not require the
spectrum to be discrete. With the spectral resolution E in $\mathcal{H}_A$ the operator A can
be written in the form

$$A = \int adE(a).$$

Let W_A denote the density operator of the metastable equilibrium of the apparatus.
Let b be a Borel set on the real line such that

$$\mathrm{tr}(W_A E(b)) = 1.$$

b represents the set of pointer positions for which the experimenter would state
"zero". Let W_O denote the density operator of the ensemble of objects which are
prepared by the first part of the experimental set-up acting in $\mathcal{H}_O$. Then by the
rules of quantum mechanics the probability to find the pointer finally in the Borel
set a is given by

$$p(a, W_O) = \mathrm{tr}((W_O \otimes W_A)S^+(1 \otimes E(a)S).$$

Since this expression is trace norm continuous in W_O and bounded in between 0
and 1, there is exactly one self-adjoint operator $F(a)$ acting in $\mathcal{H}_O$ and fulfilling
$0 \leq F(a) \leq 1$ such that for each trace class operator W_O acting in $\mathcal{H}_O$ we have

$$p(a, W_O) = \mathrm{tr}(W_O F(a)).$$

Since $p(\cdot, W_O)$ is a probability measure, $F(\cdot)$ is a positive operator valued measure
and $F(\mathbb{R}) = \mathbb{1}$.

Given the self-adjoint operator A acting on $\mathcal{H}_A$ and the unitarian S acting in
$\mathcal{H}_O \otimes \mathcal{H}_A$, the observable measured on the object has values in the spectrum of A
and is represented by a positive operator valued measure F on the real line with
$F(\mathbb{R}) = \mathbb{1}$ which may be projection valued or not. If it is projection valued, then

$$B := \int aF(da)$$

is a self-adjoint operator acting in $\mathcal{H}_O$ with the usual interpretation in quantum
mechanics. If it is not, then one may think that it approximates an observable in
the usual sense in that it accounts for systematic errors. Since there is no primary

principle for to determine which self-adjoint operator in some ultraweak neighbour-hood of B is approximated, one usually generalizes the concept of observables in quantum mechanics to all positive operator valued measures on the Borel sets of the real line $\mathbb{R}$ with $F(\mathbb{R}) = \mathbb{1}$. Among them the projection valued measures are called decision observables by G. Ludwig [10].

From the probabilistic point of view and starting from what is given in nature, realizable metastable systems and their interactions with quantum objects, this scheme seems very natural. Moreover, no problems with continuous spectra arise.

2.3   Informational Completeness.

Let

$$F : B(\mathbb{R}) \longrightarrow [\mathbb{1},\mathbb{1}] \subset \mathcal{B}(\mathcal{H}_O),$$

where $B(\mathbb{R})$ denotes the algebra of Borel sets on the real line and $\mathcal{B}(\mathcal{H}_O)$ the set of bounded operators on $\mathcal{H}_O$, be a positive operator valued measure fulfilling $F(\mathbb{R}) = \mathbb{1}$. Then

$$W_O \longmapsto \mathrm{tr}(W_O, F(\,\cdot\,))$$

is an affine mapping from the density operators on $\mathcal{H}_O$ into the probability measures on $B(\mathbb{R})$. F is called "*informationally complete* " in case this map being injective. It is well known that there is no informational complete projection valued mea-sure. Moreover, the probability distributions defined by the spectral measures of the position and momentum operators together do not determine a density operator uniquely. But there are a lot of informationally complete positive operator valued measures.

Well known examples are given by the so called "*joint position-momentum ob-servables*" which can be generated in the following way: Let $\varphi \in L^2(\mathbb{R})$, $\|\varphi\| = 1$, $< \varphi | x \varphi >= 0$, and $< \varphi | \frac{1}{i} \frac{\partial}{\partial x} \varphi >= 0$. Consequently, for the Galilean shifts

$$\psi_{p,q}(x) := e^{-\frac{px}{2}} \varphi(x - q) \qquad (x, p, q \in \mathbb{R})$$

there hold $< \psi_{p,q} | x \psi_{p,q} >= q$, and $< \psi_{p,q} | \frac{1}{i} \frac{\partial}{\partial x} \psi_{p,q} >= p$. Now for any density operator W_O of the object the function $< \psi_{p,q} | W_O \psi_{p,q} >$ is integrable on the (p, q)-plane. Moreover, on the Borel sets of this plane a probability measure is given by

$$a \mapsto \frac{1}{2\pi} \int_a < \psi_{p,q} | W_O \psi_{p,q} > dp dq, \quad a \in B(\mathbb{R}).$$

This probability measure can also be written as

$$\mathrm{tr}(W_O\, F(a)), \quad F(a) := \frac{1}{2\pi} \int_a |\psi_{p,q} >< \psi_{p,q}| dp dq,$$

the latter defining a positive operator valued measure fulfilling $F(\mathbb{R}) = \mathbb{1}$. It has been shown by S. T. Ali and E. Prugovečki [11] that this measure is informationally complete whenever

$$< \varphi | \psi_{p,q} > \neq 0 \quad \text{a.e. on } \mathbb{R}^{\mathbb{1}}$$

holds true.

Finally, using informationally complete joint position-momentum observables and the related *"phase space representations"* W. Stulpe and M. Singer [12] have shown how far classical probability can approximate quantum probability. Let ρ_{W_O} be the probability density on the (p, q)-plane defined by

$$\int_a \rho_W \, dp dq = \text{tr}(W_O, F(a))$$

or. equivalently, by $\rho_{W_O} := frac12\pi < \psi_{p,q}|W_O\psi_{p,q}$. The mapping $W_O \mapsto \rho_{W_O} \in \mathbb{R}^2$ is affine and injective, the latter being a consequence of the informational completeness of F. Now consider a finite set of density operators $\{W_i\}_{i=1,2,3,\dots,n}$, $n \in \mathbb{N}$, and let $\epsilon > 0$. Then there is a mapping

$$B(\mathcal{H}_O) \ni A \longmapsto f \in L^\infty(\mathbb{R}^\varkappa)$$

such that for each $A \in B(\mathcal{H}_O)$

$$|\text{tr}(W_i A) - \int \rho_{W_i} f \, dp dq| < \epsilon.$$

This means: Given a finite set of density operators, the quantum mechanical expectation values can uniformly be approximated up to arbitrary accuracy by the corresponding classical expressions..

3   Quantum Measurements and Information.

As already mentioned in the introduction, the statistical theory of information is based on the probability distribution for the events to be observed, and one may ask whether it can help to get deeper insight into the nature of microsystems than one gets if only the statistical interpretation of quantum theory is taken into consideration. The hope is, that new knowledge may arise because new aspects of interpretation enter with the statistical theory of information.

3.1   Concepts of Classical Statistical Information Theory.

An informational set-up consists at least in three things: A *"source"* of information, a transmitting *"channel"* and a *"receiver"*. Let the source be able to send one of n *"letters"* D_k, $k = 1, 2, 3, \dots, n$, and let the receiver be equipped with m different lamps L_l, $l = 1, 2, 3, \dots, m$. Now let $p_l(D_k)$ be the probability that L_l will shine when D_k has been sent. The receiver will be called *"ideal with respect to $\{D_k\}_{k=1,2,3,\dots,n}$"* when $m \geq n$ and, say,

$$p_l(D_k) = \delta_{lk}, \quad l, k = 1, 2, 3, \dots, n.$$

For simplicity we have assumed here the channel to be ideal, too, i.e. the channel does not change the signals it is transmitting.

The informational value of a single message is the central point. It is defined by the value of the "*information function*" which depends on the probability by which the message is sent. This information function is assumed to be a mapping

$$I : [0, 1] \longmapsto [0, \infty]$$

and is determined by one further axiom which shall be described now: Enumerate the letters of the source by two numbers instead of one, i.e. write for them D_{ij} ($i = 1, 2, 3 \ldots, r$, $j = 1, 2, 3, \ldots, s_i$). Then assume a coarse receiver with r lamps L_l ($l = 1, 2, 3, \ldots, r$) such that

$$p_l(D_{i1} \vee D_{i2} \vee D_{i3} \vee \ldots \vee D_{is_i}) = \delta_{li}.$$

This receiver is ideal with respect to some decomposition into disjoint subsets of the set of letters, but it is not able to discern between individual elements of them. Let another receiver be equipped with $n = \sum_{i=1}^{r} s_i$ lamps L_{ij} and assume it to be ideal with respect to $\{D_{i1}, D_{i2}, D_{i3}, \ldots, D_{is_i}\}_{i=1,2,3,\ldots,r}$, i.e.

$$p_{ij}(D_{i'j'}) = \delta_{ii'} \, \delta_{jj'}.$$

The second receiver, obviously, gets more information than the first one. Let w_{ij} denote the probability by which the letter D_{ij} is sent. Hence $w_i = \sum_{j=1}^{s_i} w_{ij}$ is the probability that a letter from the i-th subset is sent. Now let $I(w_i)$ be the information obtained when some letter from the i-th subset has been received. For the surplus information which can only be obtained by the second receiver which detects, say that D_{ij} is sent, only the letters $\{D_{ij}\}_{j=1,2,3,\ldots,s_i}$ are in question. Therefore, it is natural to base it on the conditional probabilities

$$\mathbf{w}_{ij} := \frac{w_{ij}}{\mathbf{w}_i}.$$

This motivates the axiom

$$I(w_{ij}) = I(\mathbf{w}_i) + I(\mathbf{w}_{ij}),$$

which is equivalent to

$$I(\mathbf{w}_i \mathbf{w}_{ij}) = I(\mathbf{w}_i) + I(\mathbf{w}_{ij}).$$

It can be shown that I is determined up to a constant $c \in \mathbb{R}_+$ by

$$I(\lambda) = -c \log(\lambda).$$

it is called the information function.

3.2   Quantum Measurements.

We now will apply the concepts of statistical information theory to quantum measurements. The preparative part of an experiment shall be taken for the source and the registrative part for the receiver. Let the preparative part produce an ensemble described by the density operator W_O acting in $\mathcal{H}_O$. One is tempted to consider a decomposition into density operators W_i

$$W_O = \sum_i w_i W_i; \quad w_i \in (0,1), \quad \sum_i w_i = 1$$

and to call W_i a letter sent with the probability w_i. But such decomposition is in general not unique. The minimal requirement for to interpret the components as letters seems to be that there exists a receiver which can discern between them. This motivates the definition: A decomposition into density operators V_i

$$W_O = \sum_i v_i V_i; \quad v_i \in (0,1), \quad \sum_i v_i = 1$$

is called an "*admissible*" one iff there exists a positive operator valued measure F on the power set of $\{1, 2, 3, \ldots, n\}$, $F_i := F(\{i\})$, $\sum_i F_i = 1$, such that

$$\text{tr}(V_i F_k) = \delta_{ik}.$$

There may be many admissible decompositions of one and the same density operator.

Now the entropy of a density operator with respect to an admissible decomposition $(V_1, V_2, V_3, \ldots, V_n)$ is defined by the average information sent using V_i as letters:

$$H_W((V_1, V_2, V_3, \ldots, V_n)) := -c \sum_{i=1}^{n} v_i \log v_i, \quad v_i \neq 0.$$

For this expression it can be shown that

$$0 < H_W((V_1, V_2, V_3, \ldots, V_n)) \leq c \log n,$$

where the equality holds iff $v_i = \frac{1}{n}$. Moreover, consider (admissible) decompositions of the V_i, say

$$V_i = \sum_{j=1}^{m_i} u_{ji} U_{ji},$$

such that $(U_{11}, U_{12}, U_{13}, \ldots, U_{nm_n})$ is an admissible decomposition of W, then there holds

$$H_W((V_1, V_2, V_3, \ldots, V_n)) \leq H_W((U_{11}, U_{12}, U_{13}, \ldots, U_{nm_n})).$$

Finally, it can be shown that the usual entropy is just the supremum of the entropies with respect to the admissible decompositions

$$\mathbf{H}_W := \sup\{H_W(V_1, V_2, V_3, \ldots, V_n) \mid \text{admissible decompositions}\} = -c\,\text{tr}(W \log W).$$

3.3 Disturbed Transmission.

We now assume a nonideal channel which disturbs the signal. Such disturbance is most generally described by an affine map K operating on the density operators. If $W = \sum_{i=1}^{n} v_i V_i$ is an admissible decomposition of the density operator W of the source, the statistical ensemble arriving at the receiver will be described by $K(W) = \sum_{i=1}^{n} v_i K(V_i)$ when this decomposition of $K(W)$ is not admissible in general. Let G denote the positive operator valued measure on the power set of $\{1, 2, 3, \ldots, n\}$ by which the receiver analyses the signal and let $G_i := G(\{i\})$. Then this nonideality is expressed by

$$\mathrm{tr}(G_k K(V_l)) =: p_{kl} \neq \delta_{kl}.$$

Since p_{kl} is the probability that the lamp k will shine when the letter l is sent, we have $\sum_{k=1}^{n} p_{kl} = 1$. The conditional probability that V_l has been sent when the lamp k is shining is given by

$$q_{kl} := \frac{v_l \mathrm{tr}(G_k K(V_l))}{\mathrm{tr}(G_k K(W))}$$

where $\sum_{l=1}^{n} q_{kl} = 1$. Now we ask for the average loss of information when the non-ideal receiver is used instead of the ideal one. Under the hypothesis that the lamp k is shining the average information $-c \sum_{l=1}^{n} q_{kl} \log q_{kl}$ gets lost. If we average this loss over all lamps remembering that the k-th one is shining with the probability $\mathrm{tr}(G_k K(W))$, we get the "*equivocation with respect to $(V_1, V_2, V_3, \ldots, V_n)$*" as

$$\mathbf{E}_W(\{V_j\}) := -c \sum_{k=1}^{n} \mathrm{tr}(G_k K(W)) \sum_{l=1}^{n} q_{kl} \log q_{kl}.$$

The average information correctly transmitted is called the "*transinformation with respect to $(V_1, V_2, V_3, \ldots, V_n)$*" and is the average information produced by the source minus equivocation

$$\mathbf{T}_W(\{V_j\}) := H_W(\{V_j\}) - \mathbf{E}_W(\{V_j\}).$$

In the case that there exists a decomposition

$$K(W) = \sum_{k=1}^{n} u_k U_k$$

such that the operators G_k give rise to an ideal receiver with respect to $(U_1, U_2, U_3, \ldots, U_n)$, then one may also write

$$\mathbf{T}_W(\{V_j\}) = H_{K(W)}(\{U_j\}) - \mathbf{I}_{K(W)}(\{K(V_j)\}),$$

where

$$\mathbf{I}_{K(W)}(\{K(V_j)\}) := -c \sum_{l=1}^{n} v_l \sum_{k=1}^{n} p_{kl} \log p_{kl}$$

is called the noise or "*irrelevance*" produced by the nonideal channel.

4   Conclusion.

It has been demonstrated how concepts of the theory of statistical information can be fitted into the description of quantal measurements. For details and proofs as well as further literature I refer to the Dissertation of M. Singer (TU Berlin, 1989) [13]. Technical applications of such investigations are possible in quantum optical communication systems. I propose they may also give additional insight into the nature of microparticles, but results do not seem not to exist up to now.

Acknowledgement.

The author thanks to Werner Stulpe, Volker Perlick, Utz Grimmer, and Michael Keyl for critical readings of the manuscript.

References

[1] M. Born: *"Zur Quantenmechanik der Stoßvorgänge"*, Z. Physik **37**, 863 - 867 (1926).

[2] A. Einstein, B. Podolsky, N. Rosen: *"Can quantum mechanical description of physical reality considered complete"*, Phys. Rev. **47**, 777 - 780 (1935).

[3] J. S. Bell: *"On the Einstein-Podolsky-Rosen paradox"*, Physics **1**, 195 - 200 (1964).

[4] W. Weidlich: *"Problems of the quantum theory of measurement"*, Z. Physik **205**, 199 - 220 (1967).

[5] J. M. Jauch: *"The problem of measurement in quantum machanics"*, Helvetica Physica Acta **37**, 293 - 316 (1964).

[6] M. Jammer: *"The philosophy of quantum mechanics"*, New York (1974), see Ch. 11.2 for a review.

[7] B. DeWitt, N. Graham (eds.): *"The many worlds interpretation"*, Princeton (1973).

[8] E. P. Wigner: *"Die Messung quantenmechanischer Operatoren"* Z. Physik **133**, 101 - 108 (1952); E. P. Wigner, M. M. Yanase: *Information contents of distributions"*, Proc. N. A. S. **49**, 910 - 918 (1963).

[9] For a closed representation see K. Kraus: "States, effects, and operations", Lecture Notes in Physics **190**, Springer, Berlin (1983).

[10] G. Ludwig: Foundations of Quantum Mechanics I, Texts an Monographs in Physics, Springer, New York (1985).

[11] S. T. Ali, E. Prugovečki: *"Systems of imprimitivity and representations of quantum mechanics on fuzzy phase space"*, J. Math. Phys. **18**, 219 - 228 (1977); *"Classical and quantum statistical mechanics in the common Liouville space"*, Physica **89A**, 501 - 521 (1977).

[12] M. Singer, W. Stulpe: *"Phase space representations of general statistical physical theories"*, J. Math. 3Phys. **33**, 131 - 142 (1992).

[13] Matthias Singer: *"Zur Informationstheorie statistischer Experimente"*, Fachbereich Physik der Technischen Universität Berlin, D 83 (1989).

A COMMENT ON A 3-DIMENSIONAL EUCLIDEAN SUPERSYMMETRY

Jan Łopuszański*
Institute of Theoretical Physics, University of Wrocław,
Cybulskiego 36, 50-205 Wrocław, Poland

Abstract

We present a new nonrelativistic supersymmetric model, satisfying the correspondence principle with its relativistic counterpart.

1 Heuristic derivation of the super-Lie-algebra.

The standard relations between the momentum 4-vector and the energy and momentum 3-vector are

$$P_0 = \frac{E}{c}, \qquad P_j = P_j, \qquad j = 1,2,3,$$

where P_μ ($\mu = 0,1,2,3,$) denote the momentum 4-vector, E the energy of the system and c the light velocity. Thus the momentum 4-vector has the dimension

$$\left[g \, \frac{cm}{sec} \right].$$

All quantities are taken real, viz. $P_\mu = \bar{P}_\mu$. For a one particle system of rest mass m we have

$$P_0 = \frac{mc}{\left(1 - \dfrac{v^2}{c^2}\right)^{\frac{1}{2}}} = mc + \frac{m}{2} \frac{v^2}{c} + o(c^{-2}) \tag{1}$$

$$P_j = \frac{mv_j}{\left(1 - \dfrac{v^2}{c^2}\right)^{\frac{1}{2}}} = mv_j + o(c^{-1}) \qquad j = 1,2,3 \tag{2}$$

---

*This work was supported financially by the KBN 2 0095 91 01.

R. Gielerak et al. (eds.), Groups and Related Topics, 223–231.
© 1992 Kluwer Academic Publishers.

where v_j stands for the velocity 3-vector. For the dimensionless generators of the 3-dimensional rotations we have

$$S_{ij} = \frac{M_{ij}}{\hbar}$$

$$M_{ij} = x_i P_j - x_j P_i$$

or

$$L_k \equiv \sum_{ij=1}^{3} \frac{1}{2} \epsilon_{kij} S_{ij} = \bar{L}_k, \qquad i, j, k = 1, 2, 3.$$

The standard commutation relation for the generators of the group E_3 extended by P_o are

$$[L_i, V_j] = i\, \epsilon_{ijk} V_k$$

where V_l stands either for L_l or P_l

$$[L_i, P_0] = 0$$

$$[P_\mu, P_\nu] = 0.$$

In the sequel we shall consider these generators as abstract operators and shall use the notation X^+ for the hermitean conjugate of X, instead of $\bar{X}$ (complex conjugate quantity). So the standard commutation relation for the supersymmetric generators read

$$[L_i, Q_A] = -\frac{1}{2} \sum_{B=1}^{2} \left(\sigma^i\right)_{AB} Q_B \qquad A, B, \dot{A}, \dot{B} = 1, 2 \qquad (3)$$

$$\left[L_i, (Q^+)_{\dot{A}}\right] = \frac{1}{2} \sum_{\dot{B}=1}^{2} \left(Q^+\right)_{\dot{B}} \left(\sigma^i\right)_{\dot{B}\dot{A}}. \qquad (4)$$

Here $(\sigma^i)_{AB} = (\bar{\sigma}^i)_{BA}$ are the Pauli matrices. Formula (3) follows from the relativistic spinor relations

$$[S_{ij}, Q_A] = -\sum_{B=1}^{2} (\sigma_{ij})_A{}^B Q_B$$

where

$$(\sigma_{ij})_A{}^B \equiv \frac{1}{2}\, \epsilon_{ijk} \left(\sigma^k\right)_{AB}.$$

Further relations read

$$[P_\mu, Q_A] = 0$$

$$\{Q_A, Q_B\} = 0 \qquad (5)$$

$$\left\{Q_A, \left(Q^+\right)_{\dot{B}}\right\} = c(\sigma^\mu)_{A\dot{B}}\, P_\mu, \qquad \left(\sigma^0\right) \equiv \begin{pmatrix} 1 & 0 \\ 0 & 1 \end{pmatrix}.$$

In the last relation we inserted on the r.h.s. the light velocity c to make the l.h.s. have the dimension of energy. For a one particle system we have, taking into account (1) and (2),

$$\{Q_A, (Q^+)_{\dot{B}}\} = c^2 m (\sigma^0)_{A\dot{B}} + cm \sum_i (\sigma^i)_{A\dot{B}} v_i +$$
$$+ \frac{m}{2} (\sigma^0)_{A\dot{B}} v^2 + o(c^0). \tag{6}$$

Let us make the following ansatz, namely

$$Q_A = cG_A + R_A + o(c^{-1}). \tag{7}$$

Notice that we could as well assume

$$Q_A = cG_A + R_A + \frac{1}{c} S_A + o(c^{-2}). \tag{8}$$

The only motivation for (7) is that this ansatz leads to a simple closed super-Lie algebra, while the application of (8) results in a much more complicated scheme. If we insert (7) on the l.h.s. of (6), order the terms with respect to the powers of c and compare coefficients of the same power of c on both sides with each other we get

$$\left\{G_A, \left(G^+\right)_{\dot{B}}\right\} = m \left(\sigma^0\right)_{A\dot{B}} \tag{9'}$$

$$\left\{R_A, \left(G^+\right)_{\dot{B}}\right\} + \left\{G_A, \left(R^+\right)_{\dot{B}}\right\} = \sum_{i=1}^{3} \left(\sigma^i\right)_{A\dot{B}} m v_i \tag{10}$$

$$\left\{R_A, \left(R^+\right)_{\dot{B}}\right\} = \left(\sigma^0\right)_{A\dot{B}} \frac{1}{2} m v^2. \tag{11}$$

In a similar way we get from (5) taking into account (7)

$$\begin{aligned} \{G_A, G_B\} &= 0 \\ \{R_A, G_B\} + \{G_A, R_B\} &= 0 \\ \{R_A, R_B\} &= 0. \end{aligned} \tag{12}$$

Let us now forget about the derivation based upon the use of a one particle model of rest mass m and velocity v_i and redefine

$\frac{1}{2} m v^2$ to $E^{(0)} \equiv H =$ hamilton operator

$m v_i$ to $P_i^{(0)} = P_i =$ momentum 3-vector operator.

Then (10) and (11) read

$$\left\{R_A, \left(G^+\right)_{\dot{B}}\right\} + \left\{G_A, \left(R^+\right)_{\dot{B}}\right\} = \sum_i \left(\sigma^i\right)_{A\dot{B}} P_i \tag{13}$$

$$\left\{R_A, \left(R^+\right)_{\dot{B}}\right\} = \left(\sigma^0\right)_{A\dot{B}} H. \tag{14'}$$

Formulae (12) and (13) are not in a satisfactory form. The most general ansatz is in case of (13)

$$\{R_A, (G^+)_{\dot{B}}\} = \frac{1}{2} \sum_i (\sigma^i)_{A\dot{B}} P_i + i (\sigma^0)_{A\dot{B}} X + \frac{i}{2} \sum_i (\sigma^i)_{A\dot{B}} X_i \qquad (15')$$

where the operators X_μ are real. In a similar way we get in case of (12).

$$\{R_A, G_B\} = \epsilon_{AB} B. \qquad (16)$$

Here B does not need to be real.

In addition to the before mentioned relations we get from (3) and (4)

$$[L_i, S_A] = -\frac{1}{2} \sum_B (\sigma^i)_{AB} \, S_B \qquad S_A = G_A \text{ or } R_A$$

$$[L_i, (S^+)_{\dot{A}}] = \frac{1}{2} \sum (S^+)_{\dot{B}} (\sigma^i)_{\dot{B}\dot{A}} \qquad (17')$$

$$[P_i, S_A] = 0$$

$$[H, S_A] = 0$$

as well as

$$[L_i, H] = 0$$
$$[P_i, H] = 0.$$

The obvious consistency requirements following from the commutation relations are that:

$$[L_i, X_j] = i \sum_{ij} \epsilon_{ijk} X_k,$$

X_j being a vector, and

$$[P_i, X_j] = 0,$$
$$[L_i, X] = [P_i, X] = 0,$$
$$[L_i, B] = [P_i, B] = 0,$$

as R_A, G_B and P_j commute with P_j and X as well as B are scalars.

Before we enter into discussing the still missing commutator relations of our super Lie algebra let us make a side remark concerning the notion. We were using above the undotted as well as dotted spinor indices, reminiscent of the relativistic formulae of the $SL(2, C)$ group constituting the background of our approach. As we are at present concerned only with the 3–dimensional Euclidean group our tool will be exclusively the $SU(2)$ group. For this group, however, the undotted element

$$U = a_0 \sigma^0 + i \sum_{j=1}^{3} a_j \sigma^j$$

$$a_\mu = \bar{a}_\mu \qquad \sum_{\mu=0}^{3} a_\mu^2 = 1$$

is equivalent to its dotted counterpart, viz.

$$\sum_{\dot{B}\dot{C}} \left(\sigma^2\right)_{A\dot{B}} \bar{U}_{\dot{B}\dot{C}} \left(\sigma^2\right)^{-1}_{\dot{C}D} = U_{AD}$$

$$\bar{U} = a_0 \sigma^0 - i \sum_{j=1}^{3} a_j \bar{\sigma}^j.$$

This fact grants a good opportunity to use in the sequel undotted indices only. We define

$$\tilde{S}_A = \sum_{\dot{B}} \left(\sigma^2\right)_{A\dot{B}} \left(S^+\right)_{\dot{B}}$$

$$\left(S^+\right)_{\dot{A}} = \sum_{B} \left(\sigma^2\right)_{\dot{A}B} \tilde{S}_B.$$

In this notation formulae (9'), (14'), (15') and (17') read

$$\left\{G_A, \tilde{G}_B\right\} = i\,\epsilon_{AB}\,m \tag{9}$$

$$\left\{R_A, \tilde{R}_B\right\} = i\,\epsilon_{AB}\,H \tag{14}$$

$$\left\{R_A, \tilde{G}_B\right\} = -\frac{1}{2} \sum \left(\sigma^j \sigma^2\right)_{AB} (P_j + iX_j) - \epsilon_{AB} X \tag{15}$$

$$\left[L_i, \tilde{S}_A\right] = -\frac{1}{2} \sum_{i=1}^{3} \left(\sigma^i\right)_{AB} \tilde{S}_B \tag{17}$$

resp., where

$$\epsilon = i\sigma^2.$$

To close our super-Lie-algebra, presented above, we have to investigate the relations of the newly introduced operators X_i, X, B among themselves and the rest of (super)generators. We make ansatzes of the following form

$$[X_i, X_j] = \quad a_{ijk} P_k + b_{ijk} X_k + a_{ij} X + b_{ij} H + $$
$$+ c_{ij} B + d_{ij} B^+ + e_{ij}$$

$$[X_i, R_A] = \quad \sum_{B} \left(\alpha^i_{AB} R_B + \beta^i_{AB} G_B + \gamma^i_{AB} \tilde{R}_B + \delta^i_{AB} \tilde{G}_B\right)$$

$$[B, R_A] = \quad \sum_{B} \left(\alpha_{AB} R_B + \beta_{AB} G_B + \gamma_{AB} \tilde{R}_B + \delta_{AB} \tilde{G}_B\right)$$

$$\left[B, \tilde{R}_A\right] = \quad \sum_{B} \left(\tilde{\alpha}_{AB} R_B + \tilde{\beta}_{AB} G_B + \tilde{\gamma}_{AB} \tilde{R}_B + \tilde{\delta}_{AB} \tilde{G}_B\right).$$

Here the coefficients $a_{ikj}, a_{ij}, \alpha^i_{AB}, \alpha_{AB}$ are numbers satisfying obvious relations like $a_{ijk} = -\bar{a}_{ijk}$. To evaluate these coefficients we use the Jacobi and super - Jacobi identities, like

$$\{[B, F_1] F_2\} - \{[F_2, B] F_1\} + [\{F_1, F_2\}, B] = 0$$

$$[\{F_1, F_2\} F_3] + [\{F_2, F_3\} F_1] + [\{F_3, F_1\}, F_2] = 0.$$

2   Results and discussion

The result of our computations is as follows.

X_i commute with $X_k, P_k, R_A, G_B, H, B, B^+$. They do not commute with L_k. X commutes with all (super)generators. One may keep X_i and X but one can put them as well equal to zero. There is no reasonable motivation to introduce them into the algebra. As we want to keep the algebra as small as possible we assume

$$X_i = X = 0.$$

Other relations, not used before, are

$$[B, R_A] = \xi R_A$$

$$[B, \tilde{R}_A] = \xi \tilde{R}_A$$

$$[B, G_A] = -\xi G_A$$

$$[B, \tilde{G}_A] = -\xi \tilde{G}_A$$

$$[B, H] = 2\xi H$$

$$[B, B^+] = 0.$$

Here ξ is a number satisfying the constraint

$$\xi m = 0. \tag{18}$$

The choice $m = 0$ $\xi \neq 0$ (we do not discuss the trivial case $m = \xi = 0$) means

$$\{G_A, (G^+)_{\dot{B}}\} = 0. \tag{19}$$

This follows from (9). In quantum case, where G_A is regarded as an operator in a Hilbert space (19) would imply the vanishing of G_A (the Hilbert space has a positive definite metric). This, however, leads to a contradiction as relation (15) can no longer hold true. Even if we assume $X_j \neq 0$ this would not help us, as P_j and X_j are both real operators.

Thus the only choice, left to us, is, according to (18).

$$m \neq 0 \qquad \xi = 0.$$

In this case one super - subalgebra is spanned by $R_A, \tilde{R}_A$ and H with the algebra

$$\{R_A, R_B\} = \{\tilde{R}_A, \tilde{R}_B\} = 0 \tag{20}$$

$$\{R_A, \tilde{R}_B\} = i \, \epsilon_{AB} \, H \tag{21}$$

$$[R_A, H] = [\tilde{R}_A, H] = 0,$$

the other one by $G_A, \tilde{G}_B$ and $\mathbf{1}$ with the algebra

$$\{G_A, G_B\} = \{\tilde{G}_A, \tilde{G}_B\} = 0 \tag{22}$$

$$\{G_A, \tilde{G}_B\} = i \, \epsilon_{AB} \, m. \tag{23}$$

We may extend the supersymmetry to encompass both these subalgebras by adding the generators P_j, B and B^+ and the commutation relations

$$[P_i, P_j] = [P_i, H] = [P_i, B] = [B, B^+] = [H, B] = 0 \qquad \text{and} \quad \text{h.c}$$

$$\{R_A, \tilde{G}_B\} = -\frac{1}{2} \sum (\sigma^i \sigma^2)_{AB} P_j \qquad \text{and} \quad \text{h.c.} \tag{24}$$

$$\{R_A, G_B\} = \epsilon_{AB} \, B \qquad \text{and} \quad \text{h.c} \tag{25}$$

$$[P_i, R_A] = [P_i, G_A] = [H, G_A] = [B, G_A] = [B_i, R_A] = 0 \qquad \text{and} \quad \text{h.c.}$$

Notice that we do not need to this aim to use rotations induced by L_i. We may, of course, supplement the algebras presented above by adding the commutation relations

$$[L_i, V_j] = i \sum_{k=1}^{3} \epsilon_{ijk} V_k \qquad V_l = L_l \quad \text{or} \quad P_l$$

$$[L_i, S_A] = -\frac{1}{2} \sum_{B=1}^{2} (\sigma^i)_{AB} S_B \qquad S_C = G_C \quad \text{or} \quad R_C$$

$$[L_i, H] = [L_i, B] = 0$$

and h.c. involving L_i.

The central charges of the latter (largest) supersymmetry are H, B, B^+ and $\mathbf{1}$, viz.

$$H = -\frac{i}{2}\,\epsilon_{AB}\left\{R_A, \tilde{R}_B\right\}$$

$$\mathbf{1} = -\frac{i}{2m}\,\epsilon_{AB}\left\{G_A, \tilde{G}_B\right\}$$

$$B = \frac{1}{2}\,\epsilon_{AB}\left\{R_A, G_B\right\} \qquad \text{and h.c.}$$

They can be considered as Casimir operators in addition to

$$\sum_{i=1}^{3} P_i P_i. \tag{26}$$

Notice that

$$\sum_{i=1}^{3} L_i P_i$$

is no longer a Casimir operator, as it does not commute with R_A and G_A. The problem whether there exist other Casimir operators apart of (26) and the central charges remains open. Quantities, quadratic in the spinor generators and commuting with the generators of the E_3 group, are

$$\epsilon_{AB}\hat{R}_A\hat{R}_B, \qquad \epsilon_{AB}\hat{R}_A\hat{G}_B, \qquad \epsilon_{AB}\,\hat{G}_A\hat{R}_B, \qquad \text{and} \qquad \epsilon_{AB}\hat{G}_A\hat{G}_B$$

where $\hat{S}_A$ stands either for S_A or for $\tilde{S}_A$.
Unfortunately, there is no way to combine them in such a way as to make them commute with all $R_A, G_A, \tilde{R}_A$ and $\tilde{G}_A$.

By changing the notation to

$$G_A \equiv G_A^{(1)} \qquad\qquad R_A \equiv G_B^{(2)}$$

we may comprise formulae (23), (24), (21) and h.c. to them as well as (22) (25) and (20) resp. in the following way

$$\left\{G_A^{(L)}, \tilde{G}_B^{(M)}\right\} = \frac{1}{2}\left[\left(\sigma_{LM}^0\,i\,(m+H) + \sigma_{LM}^3\,i\,(m-H)\right)\epsilon_{AB} - \right.$$

$$\left. - i\left(\sigma^j\sigma^2\right)_{AB} P_j\sigma_{LM}^2\right] \tag{27}$$

$$\left\{G_A^{(L)}, G_B^{(M)}\right\} = -\epsilon_{AB}\,\epsilon_{LM}\,B. \tag{28}$$

Formula (28) confirms the suggestion that B can be viewed as a central charge. The relation (27) differs, however, somehow from the standard form used in supersymmetric theories.

3 Remark concerning a different approach to the problem under consideration.

As mentioned in section 1 the ansatz (8) for Q_A considered as a power series in the light velocity c seems to be more natural than that used by us and given by formula (7). The formulae one gets after ordering the powers of c are

$$\{G_A, (G^+)_{\dot{B}}\} = m \, (\sigma^0)_{A\dot{B}}$$

$$\{R_A, (G^+)_{\dot{B}}\} + \{G_A, (R^+)_{\dot{B}}\} = (\sigma^i)_{A\dot{B}} \, P_i \qquad (29)$$

$$\{S_A, (G^+)_{\dot{B}}\} + \{R_A, (R^+)_{\dot{B}}\} + \{G_A, (S^+)_{\dot{B}}\} = (\sigma^0)_{A\dot{B}} \, H$$

as well as corresponding relations for $\{G_A, G_B\}$ etc. While the first two relations above do not differ from those, used by us, the formula (29) entails

$$\left\{R_A, \left(R^+\right)_{\dot{B}}\right\} = \left(\sigma^0\right)_{A\dot{B}} (H + Y) + \sum_{i=1}^{3} \left(\sigma^i\right)_{A\dot{B}} Y_i,$$

with $Y = Y^+$, $Y_i = Y_i^+$, otherwise arbitrary, as well as

$$\left\{S_A, \left(G^+\right)_{\dot{B}}\right\} = \left(\sigma^0\right)_{A\dot{B}} Z + \sum_{i=1}^{3} \left(\sigma^i\right)_{A\dot{B}} Z_i,$$

where Z and Z_i are such that

$$Z = -\tfrac{1}{2}Y + \tfrac{i}{2}U$$

$$Z_i = -\tfrac{1}{2}Y_i + \tfrac{i}{2}U_i \quad U = U^+, \quad U_i = U_i^+ \quad \text{otherwise arbitrary.}$$

From (29) does not also follow any restriction upon the expression $\{S_A, (S^+)_{\dot{B}}\}$ which has to be reasonably chosen by introducing a new bosonic operator.

These few remarks show that the scheme based upon the ansatz (8) contains many new parameters, absent in the model based upon the ansatz (7). This makes the task to construct the super - Lie - algebra for (8) more complicated and tedious.

Acknowledgment:

The author is grateful to Dr. Marek Mozrzymas and Dr. Jan Sobczyk for checking the calculations as well as for critical remarks. He is also thankful to Professor Helmut Reeh for extending to him a warm hospitality at the Institute of Theoretical Physics of the University of Göttingen where a part of this work was accomplished.

Chiral Nets And Modular Methods*

B. Schroer

Freie Universität Berlin, Institut für Theoretische Physik

Arnimallee 14, D-1000 Berlin 33

Abstract

We derive space-time covariances from modular properties of von Neumann algebras for chiral nets. The ensuing euclidean theory is shown to be non-commutative.

1. Motivation

For a majority of physicist quantum field theory has become synonymous with euclidean action and functional integrals. Yet if one looks at the necessary prerequisites for correlation functions to admit a euclidean functional integral representation, one finds that they are very restrictive and probably exclude many physically vital cases. The relevant theorem if the following.

Theorem: (Klein, Landau):[1]

Stochastically positive quantum field theories (A, C, α_t, ω) are in one-to-one correspondence with symmetric faithfull stochastic processes $\{X_t\}_{t \in R}$ which are O.S. positive and have their spectrum in C.

The intuitive content of this mathematical theorem is the following. Stochastically positive means that the noncommutative C^* algebra A contains a commutative subalgebra C which is big enough so that its time translates $\cup_t \alpha_t(C)$ are dense in the total algebra $\bar{A}$. Clearly this property forces theories which permit a stochastic Feynman-Kac representation to be very close to canonical theories. Indeed among the bosonic theories the ϕ^4 selfinteraction model (say in $d \leq 3$) is the prototype model for the working of this theorem. Theories with more geometrical

---

* talk contributed to the German-Polish Max-Born Symposium in theoretical Physics, Wroclaw, Poland, Sept. 27-29, 1991

R. Gielerak et al. (eds.), Groups and Related Topics, 233–245.
© 1992 Kluwer Academic Publishers.

structure have formal commutation relations (or Poisson brackets) which significantly deviate from
the Heisenberg-Weyl commutation relations. An example is a WZNW-model or a Chern-Simons
model. In addition to the difficulty in exhibiting a maximal abelian set, there is no nontrivial
translation α_t, a feature which is shared by all toplogical field theories. But even for much simpler
conventional models as the zero mass current model which has a Schwinger term, the prerequists
are violated. A zero mass potential for the current either does not exist (infrared-divergencies),
or if one enlarges the Hilbertspace by quantum mechanical degrees of freedom (zero modes) the
subalgebra generated by the field at one time fail to have the prerequisites of the theorem. To
be sure, even in those cases in which one can write down functional integrals, they may do not
represent a quantum theoretical object. For ordinary gauge theories where the algebra $\mathcal{A}$ is of
course generated by gauge invariant fields it is questionable whether a (gauge invariant) $\mathcal{C}$ with
the required property can be found. In this case physicist play the "trick" of enlarging the system
by gauge dependent indefinite metric degrees of freedom which formally enforces the prerequisites
at the expense of loosing the good mathematical controll through von Neumann algebras (in
perturbation theory one may not need this controll nor the theorem). Then with the formalism
of BRST (and a prayer), one hopes to descend to a reasonable physical quantum theory. Clearly
such tricks are guided by geometric principles, but the (classical) geometry of fibre bundles has
not led to a substantial quantum theoretical non-perturbative insight.

It is noteworthy that non of the interesting soluable two-dimensional QFT have been discovered
in this way.

The situations not covered by the theorem are those for which the euclidean theories exist, but
cannot be described in terms of a commutative von Neumann algebra. If they are to be described
in terms of stochastics, it should be "non-commutative stochastics". A class of euclidean theories
which are accessible to analysis are the chiral conformal field theories whose most characteristic
feature is braid group statistics. We will construct the "euclideanisation" of their observable
algebras.

The change of paradigma which makes this analysis possible is the replacement of (Lagrangian) quantization by the classification of causal nets as expressed in the following dictum:

Dictum: All local quantum physics is contained in a net of von Neumann algebras indexed by space-time regions. There is no physical information in an individual algebra. Net theory is not (unlike Quantum Mechanics) about material content but rather about relations and inclusions.

When the pioneers of algebraic QFT formulated this principle they tested it in the derivation of scattering theory (insensitivity of the S-matrix against changes in the field representatives of a local first class) and in the generalization of the Wigner-Wick-Wightman superselection rules.[2] As a result of the creation of the Tomita-Takasaki modular theory and its extension by Connes and Araki,[3] as well as the recent Jones inclusion theory,[4] the above dictum also got its mathematical underpinnings.

In these notes we will consider one-dimensional nets and show how they lead to conformal field theories. The symmetry transformations beyond the translation come from theTomita-Takesaki modular theory. Conformal field theories have interesting associated noncommutative euclidean theories.

2. The Modular Method for Chiral Nets

We first recall a recent result of Borchers.[5]

Theorem (Borchers) Let M be a von Neumann algebra in a Hilbertspace $\mathcal{H}$ with a translation invariant vacuum vector Ω. The translation operator $U(a)$ for $\alpha > 0$ is assumed to act on M as a one-sided compression: $U(a)MU^+(a) \subset M$. Then the modular objects (Δ^{it}, J) of (M, Ω) fulfill the geometric relations:

$$\Delta^{it} U(a) \Delta^{-it} = U(e^{2\pi t} a)$$

$$JU(a)J = U(-a)$$

(1)

For the proof we refer to the orignal paper. Note that $\Delta^{i \frac{t}{2\pi}} = D(t)$ and $U(a)$ form a geometric pair (dilation, translation) whereas the J behaves as a one-dimensional version of the TP (or TCP) operator.

The Tomita-Takesaki theory[3] will not be explained in these notes. For a physicist it is important to note that characteristic properties of algebras and states (including geometric aspects) become encoded into domains of unbounded "master" operators and the art of modular theory is to decode this encoding.

From M one can now construct a net indexed by intervalls (a, b):

$$M_{ab} = U(a)MU^+(a) \cap U^+(b)JMJU(b)$$

It was shown by Borchers[5] that this net is covariant under the restricted Möbiusgroup (leaving ∞ invariant) and that, assuming the standard cyclicity property of algebraic QFT (the Reeh-Schlieder property) $\overline{M_{ab}\Omega} = \mathcal{H}$, one also obtains (essential) Haag duality:

$$M'_{ab} = M_{-\infty,a} \vee M_{b,\infty}$$

(3)

In algebraic QFT one usually starts from a net generated by von Neumann algebras say $A(a, b)$

and defines a von Neumann algebra M belonging to the semiinfinite interval $(0, \infty) = I_\infty$

$$M = \bigvee_{(a,b) \in I_\infty} A(a, b) \tag{4}$$

In that case the $\{M_{a,b}\}$ net constructed above from M is contained in the originally given A-net.[5] They are equal only if the original net was a dual net.

In order to obtain the missing part of the Möbius-covariance we assume a slightly strengthened form of (essential) duality for $M_{-1,1}$. Let $(\tilde{\Delta}^{it}, \tilde{J})$ be the modular objects for $(M_{-1,1}, \Omega)$. Then in addition to

$$\tilde{J} M_{-1,1} \tilde{J} = M_{-\infty,-1} \vee M_{1,\infty} \tag{5}$$

we assume something about the localization of "quadrant algebras":

$$\tilde{J} M_{0,1} \tilde{J} = M_{1,\infty}, \quad \tilde{J} M_{-1,0} \tilde{J} = M_{-\infty,-1} \tag{6}$$

Note that this strengthened form does not rule out the possibility that

$$M_{-1,0} \vee M_{0,1} \subset M_{-1,1}, \quad \text{properly contained.} \tag{7}$$

i.e. that there may be nonlocal "defect" operators in $M_{-1,1}$ which cannot be found in the left hand algebra.[6]

In order to study properties of the product $R = J\tilde{J}$ of modular conjugations, we first note that

$$J\tilde{\Delta}^{it} J = \tilde{\Delta}^{-it}$$

$$[J, \tilde{J}] = 0 \tag{8}$$

as a result of the commutation relation

$$\tilde{S} J A \Omega = (JAJ)^* \Omega = JA^* \Omega = J\tilde{S} A \Omega \tag{9}$$

Here we use the standard notation $\tilde{S}$ for the unbounded involutive operator:

$$\tilde{S} A \Omega = A^* \Omega, \quad A \epsilon M_{-1,1} \tag{10}$$

whose polar decomposition gives the modular objects $(\tilde{\Delta}^{it}, \tilde{J})$. This yields $R^2 = 1$ as well as the relation

$$R \tilde{\Delta}^{it} R^+ = \tilde{\Delta}^{-it} \tag{11}$$

In order to derive the same commutation relation between R and Δ^{it} one uses the strengthened duality assumption on $\tilde{J}$.

$$R \Delta^{it} R = \Delta^{-it} \tag{12}$$

The strengthened assumption yields:

$$\tilde{J} S A \Omega = \tilde{J} A^* \Omega = S \tilde{J} A \Omega, \quad A \epsilon M_{0,1} \tag{13}$$

Since the $M_{0,1}$ algebra creates a dense set of states one obtains the commutation relations of $\tilde{J}$ with Δ^{it} from the polar decomposition as before.

With the help of the dilation Δ^{it} we may now generate a net from the four quadrant algebras $M_{\infty,1}, M_{1,0}, M_{0,-1}, M_{-1,-\infty}$. We obtain a subnet of the original translation net. The hermitean operator R acts on this net as

$$R M_{ab} R = M_{-1/a,-1/b} \tag{14}$$

If $1 < a < b$ the $M_{a,b}$ dilation net inside the first quadrant $(a = e^{2\pi t_a}, b = e^{2\pi t_b})$ is defined by:

$$M_{ab} = \Delta^{it_a} M_{\infty,1} \Delta^{-it_a} \cap (\Delta^{it_b} M_{\infty,1} \Delta^{-it_b})' \tag{15}$$

the other cases are analogous. The formula for the net transformation follows directly from the commutation relation of R with Δ^{it} and the crosswise action of R on the quadrant algebras.

The final conclusion of this section is that chiral nets, i.e. nets indexed by intervalls with reasonable quantum physical properties (duality properties) are automatically conformal. To be more precise, one can, under those stated assumption, always construct an associated conformally covariant net.

Note that there exist nets, which only fulfill the Borchers property but lack the rigid rotation invariance.[6] They have been obtained from a full conformal net by the removal of the point infinity. Our conclusions suggest that this may be the only mechanism by which one obtains non-fully conformal covariant one-dimensional nets.

Instead of only showing that $U(a)$, Δ^{it} and R (corresponding to $x \rightarrow -\frac{1}{x}$) generate the full representation theory of the Möbiusgroup, one can also directly construct from the modular operators Δ and $\tilde{\Delta}$ the rotation generator. This will be a side result of the next section.

Note that the Tomita Takesaki modular theory achieves the reduction of geometric invariances to intrinsic algebraic properties. Even the translations from which we started can be interpreted as the product of two algebraic reflections the so called canonical endomorphism related to a pair of von Neumann algebras where one is contained inside the other.[7]

3. Euclideanization and Modular Theory

It is well-known that the analytic continuation of real-time correlation functions defines under favourite conditions a commutative euclidean theory with a Feynman-Kac representation. It is therefore interesting to notice that chiral conformal nets have two non-commutative euclideanization, the "cartesian" and the "radial euclideanzization". In the following I will only scetch some results which have been obtained in collaboration with H.W. Wiesbrock[8]. For more details I refer to a forthcoming joint publication.

Having a von Neumann algebra A and a cyclic seperating Ω vector in X, one can always use the modular objects (Δ^{it}, J) in order to define a different inner product and a different *- algebra

$\hat{A}$ on a Hilbertspace. This new inner product <> is defined as

$$A, B\epsilon\mathcal{A} \rightarrow < A, B >= (A\Omega, \Delta^{1/2}B\Omega) \tag{16}$$

Modular theory yields the strict positive-definiteness of this sesquilinear form on $\mathcal{A}$. $\mathcal{A}$ operates on the ensuing Hilbertspace completion $\hat{\mathcal{H}}$ as:

$$\pi(A)B = AB \quad B, AB\epsilon\hat{\mathcal{H}} \tag{17}$$

Since

$$\|B\|_{\hat{\mathcal{A}}}^2 = \|\Delta^{1/4}B\Omega\|_{\mathcal{H}}^2 \tag{18}$$

and

$$\|\pi(A)B\|_{\hat{\mathcal{A}}}^2 \leq \|\Delta^{1/4}A\Delta^{-1/4}\|_{\mathcal{H}}\|B\|_{\hat{\mathcal{A}}}^2 \tag{19}$$

we obtain bounded operators $\pi(A)$ by restricting $\mathcal{A}$ to the domain of $\sigma_{i/4}$

$$D(\sigma_{i/4}) := \{A\epsilon\mathcal{A}|\Delta^{1/4}A\Delta^{-1/4}\epsilon\mathcal{L}(\mathcal{H})\} \tag{20}$$

This dense subalgebra of $\mathcal{A}$ is also a subalgebra of $B(\hat{\mathcal{H}})$. We define

$$\hat{\mathcal{A}} = \text{von Neumann algebra genrated by } \pi(D(\sigma_{i/4}) \tag{20}$$

where the underlying Hilbertspace $\hat{\mathcal{H}}$ has a distinguished vector $\hat{\Omega} = Id$. In this way we obtain again a modular situation since $\hat{\Omega}$ turns out to be cyclic and seperating. In fact we have:

$$\hat{\Delta} = D_1 \quad, \mathcal{J} = C \tag{21}$$

$$\text{with} :D_\lambda : A\epsilon\hat{\mathcal{H}} \rightarrow \sigma_\lambda(A)\epsilon\hat{\mathcal{H}} \tag{22a}$$

$$\text{and } C : A\epsilon\hat{\mathcal{H}} \rightarrow A^*\epsilon\hat{\mathcal{H}} \tag{22b}$$

i.e. $\hat{\Delta}$ originates from the old modular automorphism which becomes a unitary operator on $\hat{\mathcal{H}}$

and the $\hat{J}$ from the old *-involution which now becomes a conjugation on $\hat{\mathcal{H}}$. The interchange

of the *-operation with the modular conjugation is a general feature of modular euclideanization.

However, nothing is known about the precise nature the local euclidean algebras (factors of type

III_1 or nontrivial center?).

The first situation to which we apply this formalism is that of Borchers which we explain in

the previous section. Since

$$\Delta^{1/4}U(a)\Delta^{1/4} = U(ia) \tag{23}$$

we expect on the basis of the formal calculation with point-like fields:

$$\phi(ix) = \Delta^{1/4}U(x)\Delta^{-1/4}\phi(0)\Delta^{1/4}U(x)\Delta^{-1/4}$$
$$= \Delta^{1/4}\phi(x)\Delta^{-1/4} \tag{24}$$

that the euclidean theory is related to the euclidean analytic continuation. But it is somewhat

surprising that the euclidean translation

$$\tau^a : A\epsilon\mathcal{A}(0,\infty) \rightarrow U(a)AU^+(a)\epsilon\mathcal{A}(0,\infty) \quad a > 0 \tag{25}$$

which defines a contractive semigroup in $\hat{\mathcal{H}}$, has a generator $\hat{H}$ which again fulfills the spectrum

condition:

$$\tau^a = e^{-a\hat{H}}, \quad \|\tau^a A\|_{\hat{\mathcal{H}}} \leq \|A\|_{\hat{\mathcal{H}}} \tag{26}$$

The verification is indentical to the calculation of Barata and Fredenhagen.[9] The euclidean trans-

lation is defined by τ^{ia} and fulfills the spectrum condition. With

$$\mathcal{A}_E(0,\infty) := \hat{\mathcal{A}}(0,\infty) \tag{27}$$

we obtain

$$r^{ia} \mathcal{A}_E(0,\infty) r^{-ia} \subset \mathcal{A}_E(0,\infty) \quad a \geq 0. \tag{28}$$

In complete analogy to Borchers definition of a real time translation net, we now define the euclidean translation net

$$\mathcal{A}_E(a,b) = r^{ia} \mathcal{A}_E(0,\infty) r^{-ia} \cap (r^{ib} \mathcal{A}_E(0,\infty) r^{-ib})' \tag{29}$$

So we obtain two nets which although having different inner products (and stars) both admit a notion of localization. The common algebra $D(\sigma_{i/4})$ and its product structure is shared but note that $D(\sigma_{i/4})$ is not a star subalgebra of the real time algebra. It is now easy to see that the analytically continued real time correlation functions with operators from $D(\sigma_{i/4})$ are representable in the euclidean theory by using the $<>$ inner product and the operator $\sigma_{i/4}(A)$ with $A \epsilon \mathcal{A}$. Since the latter are, as a result of the modular invariance of $\hat{\Omega}$, equal to the untransformed operator A in the real time inner product, we obtrain equality between the corresponding correlation function of the two theories i.e. the real time correlation functions of point-like covariant fields are identical to their analytic continuation to the euclidean points. This suggests that the two nets are isomorphic, a feature which was already foreshadowed when we realized that both H and $\hat{H}$ fulfill the spectral condition.

Formally we have

$$\hat{H} = -iH \tag{30}$$

where $\hat{H}$ is positive in the euclidean inner product.

A much more interesting pair of theories is obtained from $\mathcal{A}(-1,1)$ and the modular objects $(\tilde{\Delta}, \tilde{J})$. An operator which plays a similar role to the translation in this new situation is the old dilation Δ^{it} which acts as a two sided compression on $\mathcal{A}(-1,1)$. Instead of the Borchers formula one now obtains

$$\tilde{\Delta}^{1/2}\Delta^{it}\tilde{\Delta}^{-1/2} = \Delta^{it} \quad \text{for } t \geq 0 \tag{31}$$

and $\tilde{\Delta}^{1/2}\Delta^{1/2} = \tilde{\Delta}^{1/2}\Delta^{-1/2}$, on a subspace of $\mathcal{H}_E$ generated from the algebra $\mathcal{A}(0,1)$. The last property is the periodicity property of the euclidean contraction $D_\lambda A = \Delta^{i\lambda} A \Delta^{-i\lambda}$

$$D^{i/2} = D^{-i/2} \tag{32}$$

The contraction semigroup therefore leads upon analytic continuation to a periodic translation with spectral positive property and defines a euclidean rigid rotation:

$$D^{it} =: e^{2\pi it L_\circ} \tag{33}$$

Rewriting the euclidean inner product of the $\mathcal{A}(-1,1)$ theories as

$$< A, B >= (A\Omega, \tilde{\Delta}^{1/2} B\Omega) = (\tilde{\Delta}^{1/4} A\tilde{\Delta}^{-1/4}\Omega, \tilde{\Delta}^{1/4} B\tilde{\Delta}^{-1/4}\Omega) \tag{34}$$

the modular expression for the rigid rotation at imaginary angles is

$$e^{-2\pi t L_\circ} = \tilde{\Delta}^{1/4}\Delta^{it}\tilde{\Delta}^{-1/4} \quad t \geq 0 \tag{35}$$

If we would have taken instead the algebra $\mathcal{A}(0,\infty)$, we would have obtained the formula:

$$e^{-2\pi t L_\circ} = \Delta^{1/4}\tilde{\Delta}^{it}\Delta^{-1/4} \quad t \geq 0 \tag{36}$$

One can now show the following two facts

1) The nuclearity condition[10] of algebraic QFT is related to the trace-class property of $e^{-2\pi L_\circ}$.

2) The original real time net generated by rigid rotations and the euclidean net generated by the euclidean rigid rotations are related in a very non-trivial manner, if one includes besides the vacuum state the lowest energy states of the real time rigid rotation generator L_0.

This last remark is related to the existence of hitherto very elusive modular object, the so called "chiral corner transfer operator", an object which is in turn at the root of Verlinde's observation[10] and which in the case of temperatur $T = 2\pi$ (symmetric torus) is a genuine symmetry involving the ground states in all sectors. This corner transfer operator and its properties are expected to replace the notion of stochastic positivity of the commutative euclidean theory. In contradistinction to the old $\Delta^{1/4}$ operator, the new chiral corner transfer operator cuts across all sectors and links "charge measures" to "charge transporters". Presently we are studying how the global information about charge sectors on the real time side manifests itself in properties of the local noncommutative euclidean algebras. The reader is referred to a forthcoming preprint with H.W. Wiesbrock.

Conclusions

As long as Lagrangian QFT based on quantization promised to give rich physical insight, only a few physicist were interested in the classification scheme of "local nets". This changed somewhat with the realization that the classification of conformal QFT_2 (and perhaps also the overcoming of stagnation in the development of gauge theories) requires different novel methods.

Wheras in Lagrangian field theory geometrical concepts can be directly employed, geometric aspects of the approach buildt on local nets have to be derived from the Tomita-Takesaki modular theory applied to subalgebras in the net. Internal symmetries are related to inclusions with conditional expectations (i.e. Jones inclusions) and space-time symmetries to modular groups. The space-time modular structures are also intimatly related to Haag duality, a sharpening of Einstein causality.

In this article we scetched some theorems of Tomita-Takesaki modular theory for chiral nets. We also constructed non-commutative euclidean theories. The aim is to eventually obtain a complete classification of chiral nets and conformal QFT from first principles.

Since all admissable charge composition structures (superselection rules) have precisely one realization in chiral conformal field theory (like topological theories, chiral conformal theories are essentially kinematical), the use of such a classification would by far transcend its immediate physicsl application for critical phenomena. It is designed to shed light on a new symmetry concept which is dual to braid group statistics.

I acknowledge numerous discussions with H.W. Wiesbrock with whom I have been collaborating chiral nets and I am indepted to H.J. Borchers for some remarks which led to improvements in the second section.

References

1) A. Klein and L.J. Landau, Journal of Functional Analysis **42** (1981) 368 and references therein.

2) H.J. Borchers, Commun. Math. Phys. **1** (1963) 37,

   R. Haag and D. Kastler, J. Math. Phys. **5** (1964) 848,

   S. Doplicher, R. Haag and J.E. Roberts, Commun. Math. Phys. **23** (1971) 199 and **33** (1974) 49.

3) The reader finds all the relevant theorem in "Le algebra C^* e le loro applicazioni alla meccanica statistica ed alla teoria quantistical dei campi", Proceedings of the 1976 Enrico Fermi School in Varenna, North Holland Publ. .

4) V. Jones, Inv. Math. **72** (1983) 1

5) H.J. Borchers "The CPT Theorem in Two-Dimensional Theories of Local Observables" University of Göttingen.

6) D. Buchholz and H. Schulz-Mirbach, Reviews in Mathematical Physics **2** (1990) 105.

7) R. Longo, Commun. Math. Phys. **126** (1989) 217 and **130** (1990) 285.

8) Joint work with H.W. Wiesbrock, to be issued as a FU preprint.

10) D. Buchholz and F. Wichmann, CMP 106 (1986).

Chiral Symmetry Breaking – Rigorous Results*)

M. SALMHOFER

Mathematics Department
University of British Columbia, Vancouver, BC, V6T 1Z1 (Canada)

and

E. SEILER

Max-Planck-Institut für Physik
– Werner-Heisenberg-Institut –
P.O.Box 40 12 12, Munich (Fed. Rep. Germany)

ABSTRACT

We prove the occurrence of chiral symmetry breaking in a class of lattice gauge models with infinite strength of the gauge coupling.

---

*)Talk given by E. Seiler at the first Max-Born-Symposium held at Wojnowice castle (Poland), September 1991

R. Gielerak et al. (eds.), Groups and Related Topics, 247–257.
© 1992 *Kluwer Academic Publishers.*

1.Introduction

I have to apologize for lecturing about this subject at a workshop on quantum groups. But at least the system studied contains a quantum (Planck's constant, being set =1) and also some groups.

Chiral symmetry breaking (χSB) has played a central role in the physics of strong interactions ever since Y.Nambu in 1960 [1] first introduced the idea that the pion can be understood as an approximate Goldstone boson (the name did not exist at the time) for the spontaneous breaking of chiral symmetry. This idea was an essential ingredient in the development of current algebra ($PCAC$) and has survived the tremendous changes in the description of strong interactions that have occurred since 1960: Nowadays it is believed to be a property of QCD.

Chiral symmetry is a continuous symmetry group, and spontaneous breaking of such symmetries has first been shown to occur in Statistical Mechanics by the method of the so-called Infrared (IR) bounds [2]. Now it is well known that bosonic quantum field theory (at least if it is CP invariant) can be fitted into the framework of Statistical Mechanics via the functional integral representation. But the problem with chiral symmetry is that it involves relativistic fermions, which are represented by so-called Grassmann numbers in the functional integral and therefore falls outside the framework of Statistical Mechanics, specifically there is no probabilistic description of the system.

Nevertheless it is possible to extend the method of IR bounds to infinitely strongly coupled lattice gauge theories with fermions and thereby establish the occurrence of χSB. The basic reason why this is possible (albeit with some technical effort) is the reflection positivity (RP) enjoyed by these models, reflecting the existence of an underlying quantum theory with a bona fide Hilbert space.

The work reported appeared in [3]. Whoever is interested in all the details should consult [4].

2.Models and Actions

Formally in the continuum the models we are studying are described by actions of the type

$$S = \frac{1}{g_{YM}^2} S_{YM} + \int \bar{\psi} \not{D}_A \psi + m \int \bar{\psi}\psi \tag{1}$$

where $\bar{\psi}\psi$ etc. is a shorthand notation implying also summation over color and flavor indices and S_{YM} is the Yang-Mills action. For $m = 0$ S is invariant under the chiral transformations

$$\psi \to e^{i\alpha\gamma_5}\psi, \quad \bar{\psi} \to \bar{\psi}e^{i\alpha\gamma_5} \tag{2}$$

or their flavor changing generalizations. The following two related notions are important:

—Chiral symmetry breaking (χSB): $\lim_{m\to 0}\langle\bar{\psi}\psi\rangle \neq 0$.

—Long range order (l.r.o.): For $m = 0$ $\lim_{|x|\to\infty}\langle\bar{\psi}\psi(0)\bar{\psi}\psi(x)\rangle \neq 0$.

It is generally believed, but not really proven in our context, that the two notions are equivalent; certainly the first implies the second.

Since we want to prove something rigorously, we have to adopt a scheme in which everything is well-defined, and for this we will choose the lattice with so-called staggered fermions. Furthermore we will limit the discussion to infinitely strong self-coupling of the gauge field which amounts to dropping the term $\frac{1}{g_{YM}^2} S_{YM}$. The lattice action is then given by

$$S_F = \frac{1}{2}\sum_{x,\mu}\left[\bar{\psi}(x)\Gamma_\mu(x)U_\mu(x)\psi(x + e_\mu) - \bar{\psi}(x + e_\mu)\Gamma_\mu(x)U_\mu(x)^*\right]$$
$$- m\sum_x \bar{\psi}(x)\psi(x) + \lambda\sum_{x,\mu}\bar{\psi}\psi(x)\bar{\psi}\psi(x + e_\mu) \tag{3}$$

$U_\mu(x)$ is a $U(N)$ matrix acting on the color degrees of freedom, expectation values are obtained as usual by integrating with the 'weight' $\exp(-S)$ and dividing by the partition function

$$Z = \int \prod_x (d\bar{\psi}(x)\psi(x)) \prod_{x,\mu} dU_\mu(x) e^{-S} \tag{4}$$

the 'integration' over the fermionic variables means the Berezin prescription and $dU_\mu(x)$ stands for the Haar measure on $U(N)$. As usual one first works in a finite volume Λ and then takes the thermodynamic limit.

The staggered fermions are characterized by the fact that at each lattice point there is only one spin component of the fermions $\bar{\psi}$ and ψ, whereas the Dirac matrices are replaced by

$$\Gamma_\mu(x) = (-1)^{\sum_{\rho=1}^{\mu-1} x_\rho}. \tag{5}$$

Symbols like $\bar{\psi}\psi$ are shorthand for $\sum_a \bar{\psi}_a\psi_a$ (a subsumes the color and possibly a flavor index). Chiral $U(1)$ rotations are defined by

$$\psi(x) \rightarrow e^{i\alpha\epsilon(x)}\psi(x)$$
$$\bar{\psi}(x) \rightarrow e^{i\alpha\epsilon(x)}\bar{\psi}(x) \tag{6}$$

$$\epsilon(x) = (-1)^{\sum_{\mu=1}^{D} x_\mu} \tag{7}$$

where D is the dimension of the lattice and as in the formal continuum version these chiral rotations leave the action invariant for $m = 0$.

Carrying out the integration over the gauge fields produces an effective fermionic theory with only nearest neighbor coupling, as has been noticed by various authors (see for instance [5]). With the gauge group $U(N)$ which we are using the effective action of the resulting fermionic theory is of the form

$$S_{eff} = \sum_{x,\mu} \tilde{W}\left(\frac{1}{4}\bar{\psi}\psi(x)\bar{\psi}\psi(x+e_\mu)\right) \tag{8}$$

Note that S_{eff} depends only on the invariant bilinears $\bar{\psi}\psi(x)$. For this reason one can use the following bosonization trick (see [5]): Let f be a polynomial, $\bar{\psi}\psi \equiv \sum_{a=1}^{N} \bar{\psi}_a\psi_a$. Then

$$\int d\bar{\psi}d\psi f(\bar{\psi}\psi) = N!\oint \frac{dz}{2\pi i z} z^{-N} f(z) \tag{9}$$

where the line integral can be taken for instance over the unit circle. This formula is now used for every lattice point. It is convenient to rescale the complex variables z, introducing instead $\sigma_x = \frac{1}{2N}z_x$ and integrate again over the unit circle. We obtain thereby a complex spin model with partition function

$$Z = \prod_x \left(\oint \frac{d\sigma_x}{2\pi i \sigma_x} \sigma_x^{-N} e^{\frac{2Nm}{\sigma_x}} \right) \exp\left(N \sum_{x,\mu} W(\sigma_x \sigma_{x+e_\mu}) \right) \qquad (10)$$

where $W(t) = \tilde{W}(N^2)/N$. We note that in the special case $N = 1$ (QED) the function W becomes linear: $W(t) = (1 + gN)t$.

Rossi and Wolff [5] also noticed that the complex spin system described by (10) can be mapped into a generalized monomer-dimer (M-D) system, for $N = 1$ into the standard M-D system. So in spite of what I said before, there is a statistical mechanics system behind all this. It should be noted, however, that in this monomer-dimer representation one cannot see the chiral symmetry, much less its possible spontaneous breakdown. For $N = 1$ the transformation to the monomer-dimer system allows to make use of a number of properties that have been proven for this system, such as the absence of any phase transition for $\mathcal{R}e\, m \neq 0$. The case $N = 1$ can be thought of as (infinitely coupled) lattice QED but for $\lambda \neq 0$ it also represents a gauged version of the Nambu–Jona-Lasinio (NJL) model [6].

3.General Results

a) Schwinger-Dyson equations

As usual, the Schwinger-Dyson equations (SDE) can be obtained through integration by parts; it is easiest to do this in the complex spin version of the models. Let us consider the typical observable $\sigma^L \equiv \prod_x \sigma_x^{L_x}$. One obtains the SDE

$$\frac{N - L_x}{N}\langle \sigma^L \rangle = 2m\langle \sigma_x \sigma^L \rangle + \sum_{y:|y-x|=1} \langle \sigma_x \sigma_y W'(\sigma_x \sigma_y) \sigma^L \rangle \qquad (11)$$

In particular one obtains for the special case $N = 1$ and $L = 0$ (after a trivial rescaling if $\lambda \neq 0$ — NJL models):

$$1 - 2m\langle \sigma_x \rangle = \sum_{y:|y-x|=1} \langle \sigma_x \sigma_y \rangle. \qquad (12)$$

This implies a lower bound on the nearest neighbor 2-point function (assuming cubic symmetry):

$$\langle \sigma_x \sigma_{x+e_\mu} \rangle \geq \frac{1}{2D}. \tag{13}$$

b) *Reflection Positivity*

Because the model is a reformulation of a lattice gauge theory, it inherits the reflection positivity (RP) from that model. It has the following form: Consider a finite lattice Λ symmetric under 'time reflections' r in a plane lying halfway between lattice planes. Let Θ be the antihomomorphism of the algebra of polynomials in the variables σ_x that maps σ_x into σ_{rx}. Then for any polynomial A depending only on variables at 'positive time'

$$\langle A\Theta A \rangle \geq 0 \tag{14}$$

c) *Analyticity*

For the special case $N = 1$ one can use the information that was obtained many years ago for the equivalent standard M-D models to obtain:

Theorem 1: For $N = 1$ all correlation functions are analytic functions of the mass m in the whole m-plane except for the interval $\{\mathcal{R}e\ m = 0, |\mathcal{I}m\ m| \leq \sqrt{8D}\}$, and they cluster exponentially there.

Analyticity follows from the results of [7,8], exponential clustering is then obtained by the method of Penrose and Lebowitz [9].

d) *'Mermin-Wagner theorems'*

In Quantum Field Theory it is impossible to have spontaneous breaking of continuous symmetries in dimension $D = 2$ ('Coleman's theorem' [10]); in Statistical Mechanics the analogous fact, known as the Mermin-Wagner theorem, also holds under quite general conditions (see for instance [11, 12, 13, 14, 15]) in dimension $D \leq 2$. So one might expect that a similar result is true also for lattice gauge models studied here. There is however a surprise:

In $D = 1$ the models defined by (3) for $N = 1$ show spontaneous breaking of chiral symmetry and long range order. This can be seen by explicit calculation and only expresses the trivial fact that a fermionic oscillator of frequency 0 has a degenerate ground state.

This result should serve as a warning against blind extrapolation of what is known about spontaneous symmetry breaking in Statistical Mechanics to the case at hand. The fact that there is a representation as the Statistical Mechanics of monomers and dimers is of no help: In this representation there is no chiral symmetry left. Nevertheless for $D = 2$ we can use the M-D representation to obtain a Mermin-Wagner like result:

In $D = 2$ there is no l.r.o. in the model defined by (3) for $N = 1$. This follows from an old result by Fisher and Stephenson [16] for the M-D system. Translated to our system, they show that the 2-point function of $\bar{\psi}\psi$ decays like $|x|^{-1/2}$ for large distances.

e) Universal Bounds

We note the following inequalities that hold for all the systems considered:

$$0 \le \langle \sigma^L \rangle \le 1 \tag{15}$$

The first inequality is a consequence of the equivalence to a generalized M-D sytem, the second one relies for its proof on the SD equations, RP in the form of chessboard bounds and special positivity properties of the interaction function W, see [3, 4].

4.Mean Field Theory

Beginning with the pioneering paper of Nambu and Jona-Lasinio [6] χSB was always only studied in the mean field approximation. We can show that mean field theory is not just an (uncontrolled) approximation, but provides actually an upper bound on the 'condensate' $\langle \bar{\psi}\psi \rangle$, at least for $N = 1$ or more generally the NJL class of models.

Mean field theory is as usual obtained by neglecting the spatial fluctuations in the order parameter σ; replacing in the effective action $\sum_{y:|y-x|=1} \sigma_x \sigma_y$ by $2D\sigma_x \langle \sigma_x \rangle_{MF}$ one obtains the consistency condition

$$2D\langle \sigma_x \rangle^2_{MF} + 2m\langle \sigma_x \rangle_{MF} - 1 = 0 \tag{16}$$

which has the positive solution

$$\langle \sigma_x \rangle_{MF} = \frac{m}{2D}\left(\sqrt{1 + \frac{2D}{m^2}} - 1\right) \tag{17}$$

(17) has the nonvanishing limit $1/\sqrt{2D}$ for $m \to 0+$. So mean field theory predicts χSB in any dimension, and fails drastically in $2D$. But it provides an upper bound:

$$\langle \sigma_x \rangle \le \langle \sigma_x \rangle_{MF} \tag{18}$$

(18) is an easy consequence of RP, which implies $|\langle \sigma_x \rangle|^2 \le \langle \sigma_x \sigma_{x+e_\mu} \rangle$, and the SDE (12).

5.Control of Fluctuations

The method of IR bounds pioneered by Fröhlich, Simon and Spencer [2] can be adapted to the case at hand, with some technical difficulties and some changes. Again RP plays the crucial role. We refer to [3, 4] for a proof and just quote the main result:

Theorem 2: Let $\hat{G}(k) \equiv \sum_x e^{ikx} \langle \sigma_0 \sigma_x \rangle$. Then for any $m \in \mathbb{R}$ $\hat{G}$ is a signed measure; for $k \ne 0, \hat{\pi}$ ($\hat{\pi} = (\pi, ...\pi)$) it is absolutely continuous with respect to Lebesgue measure and satisfies

$$-\frac{1}{2ND(k + \hat{\pi})} \le (2\pi)^D \hat{G}(k) \le \frac{1}{2ND(k)} \tag{19}$$

with $D(k) \equiv \sum_\mu (1 - \cos k_\mu)$.

6.Chiral Symmetry Breaking
and Long Range Order

The idea that goes back to [2] is to exploit the potential conflict between the IR-bound (19) and the lower bound (13) that can only be resolved by the presence of δ-function contributions at $k = 0$ and/or $k = \hat{\pi}$. More explicitly for $N = 1$ we have, using (13)

$$1 \leq \sum_{y:|y-x|=1} \langle \sigma_x \sigma_y \rangle = \frac{2}{(2\pi)^D} \int \hat{G}(k) \sum_{\mu=1}^{D} \cos k_\mu dk \qquad (20)$$

Now by (19) we can write $\hat{G}(k) = \hat{g}(k) + c_o \delta(k) + c_{\hat{\pi}} \delta(k - \hat{\pi})$ where $\hat{g}$ is absolutely continuous with respect to Lebesgue measure and bounded by a function that is integrable in dimension $D \geq 3$. If the integral is sufficiently small, (20) then requires $c_o > 0$, expressing either l.r.o. or a nonvanishing $\langle \sigma_x \rangle$.

For $N = 1$ it follows from exponential clustering (Theorem 1) that for $m > 0$ $c_{\hat{\pi}} = 0$. On the other hand for $m = 0$ we have $c_o = c_{\hat{\pi}}$, provided we have constructed a chiral invariant state (for instance by using periodic b.c.). So what remains to do is to examine quantitatively the bound obtained from (19). Unlike the case of ferrmognets, where one has a small parameter (temperature) to force the presence of l.r.o., the only parameter to help us here is $1/D$.

In detail we obtain the following results:

a) $N = 1$, $m \rightarrow 0$:

$$r \langle \bar{\psi}\psi \rangle_{MF} \leq \langle \bar{\psi}\psi \rangle \leq \langle \bar{\psi}\psi \rangle_{MF} \qquad (21)$$

with $r = \sqrt{max\{0, 1 - 2S(D)\}}$,

$$S(D) = \frac{1}{(2\pi)^D} \int \frac{d^D k}{D(k)} \sum_{\mu=1}^{D} \cos k_\mu \qquad (22)$$

It is not hard to see that $S(D) \rightarrow 0$ as $D \rightarrow \infty$, so in that limit mean field theory becomes exact, as expected.

More importantly in $D \geq 4$ we have $2S(D) < 1$, hence $r > 0$ and therefore (21) implies χSB and l.r.o..

b) $U(N), N > 1$:

In this case the lower bound (12) has to be replaced by something much more complicated because of the more complicated $SDEs$. The IR bound (19) remains unchanged, however. Another drawback of this case is that for the corresponding generalized M-D models much less is known, in particular there is no proof of analyticity and clustering for $m > 0$. But a detailed analysis of the $SDEs$ yields at least a proof of l.r.o. for $m = 0$, provided $N \leq 4$ and $D \geq 4$. It may be gratifying that the 'physical' case $D = 4$ and $N = 3$ is covered by this result.

But we should not hide that there are a lot of open questions. To name but a few:

—*What happens in $D = 3$?*

—*What happens if the gauge coupling is not infinite?*

—*Can one prove absence of χSB for weak gauge coupling?*

—*What happens for the gauge groups $SU(N)$?*

—*Is there a regime in which the symmetry is effectively enhanced from*

$U(1)_{ax} \times SU(N)_{flavor}$ *to, say,* $SU(N)_{left} \times SU(N)_{right}$ *as required by phenomenology?*

Let me close by saying that at least it has been possible to control the fluctuations in a simple case, thereby offering some hope that the chiral symmetry breaking that so far has always just been postulated, may in fact be a property of models of strong interactions such as QCD.

REFERENCES

[1] Y.Nambu, *Phys. Rev. Lett.* **4** (1960) 380.

[2] J.Fröhlich, B.Simon and T.Spencer, *Commun. Math. Phys.* **50** (1976) 79.

[3] M.Salmhofer and E.Seiler, *Commun. Math. Phys.* **139** (1991) 395.

[4] M.Salmhofer, Ph.D. thesis, MPI-PAE-PTh 80/90

[5] P.Rossi and U.Wolff, *Nucl. Phys.* **B 248** (1984) 105; U.Wolff, *Nucl. Phys.* **B 280** (1987) 680.

[6] Y.Nambu and G. Jona-Lasinio, *Phys.Rev.* **122** (1961) 345.

[7] C.Gruber and H.Kunz, *Commun. Math. Phys.* **22** (1971) 133.

[8] O.Heilmann and E.Lieb, *Commun. Math. Phys.* **25** (1972) 190.

[9] O.Penrose and J.Lebowitz, *Commun. Math. Phys.* **39** (1974) 165.

[10] S.Coleman, *Commun. Math. Phys.* **31** (1973) 259.

[11] N.D.Mermin and H.Wagner, *Phys. Rev. Lett.* **17** (1966) 1133.

[12] N.D.Mermin, *J. Math. Phys.* **8** (1967)1061.

[13] R.L.Dobrushin and S.B.Shlosman, *Commun. Math. Phys.* **42** (1975) 31.

[14] C.Pfister, *Commun. Math. Phys.* **79** (1981) 181.

[15] J.Fröhlich and C.Pfister, *Commun. Math. Phys.* **81** (1981) 277.

[16] M.E.Fisher and J.Stephenson, *Phys.Rev.* **132** (1963) 1411.

On a twistor shift in particle and string dynamics

V.A. Soroka, D.P. Sorokin, V.I. Tkach, D.V. Volkov

Kharkov Institute of Physics and Technology,
310108, Kharkov, USSR

Abstract

A generalization of relativistic particle and string dynamics based on a notion of twistor shift and containing a fundamental length constant is considered, which results in a modification of particle (or string) interactions with background fields.

1. Starting with the works by Penrose et al. (see, for example, [1] and references there in) twistors are widely used to describe massless relativistic (super)particles, strings and superstrings [2-9]. Being an alternative to the conventional coordinate approach twistor one gives, in several cases, more economical description of constraints and more transparent representation of symmetry properties. It clarifies, for example, the physical and geometrical meaning of field equations [10] as well as an origin of a local fermionic symmetry in superparticle and superstring theories [7,9,11].

Though a twistor program has been developed for a dozens of years, as we know, there has not been proposed any essential modifications of modern theoretical ideas which would be based on twistors.

In this note we would like to point out that a generalization of twistor formulation of relativistic particle and string dynamics is possible which leads to appearance of a fundamental length in a generalized theory and results in a modification of particle (or string) interactions with background fields.

As a simple example, we shall demonstrate the idea of the proposed generalization in the case of relativistic massless particle theory in $D = 3$ space-time dimensions, though, the results obtained can be generalized to higher dimensional particle mechanics in $D = 4, 6$ and 10 as well as to the case of string theory which is the most interesting from the physical point of view.

R. Gielerak et al. (eds.), Groups and Related Topics, 259–266.
© 1992 *Kluwer Academic Publishers.*

2. Massless relativistic particle dynamics in $D = 3$ space-time dimension can be described by the Lagrangian of the following form (3.7).

$$L_0 = \lambda_\alpha \lambda_\beta \dot{X}^{\alpha\beta}, \tag{1}$$

playing the role of a bridge between space-time and twistor formulations [7]. $\lambda_\alpha(\alpha, \beta, \ldots = 1, 2)$, is a commuting Majorana spinor, $X^{\alpha\beta}(\tau) = \frac{1}{\sqrt{2}} \gamma_m^{\alpha\beta} X^m(\tau)$ ($m = 0, 1, 2$) is a coordinate of particle trajectory parametrized by τ and $\dot{X}^{\alpha\beta}(\tau) \equiv \frac{dX^{\alpha\beta}(\tau)}{d\tau}$ being a particle velocity ($\gamma_m^{\alpha\beta}$ are $D = 3$ Dirac matrices). Metric signature is chosen to be $(-, +, +)$. The mass-shell condition $P_m P^m = 0$ (where P_m being a momentum) for a particle with $m = 0$ arises from (1) as a consequence of constraint

$$P_{\alpha\beta} = \lambda_\alpha \lambda_\beta \tag{2}$$

and the identity $\lambda_\alpha \lambda_\beta \epsilon^{\alpha\beta} \equiv \lambda_\alpha \lambda^\alpha = 0 \quad (\epsilon^{12} = -\epsilon^{21} = 1)$.

The absence of $\dot{\lambda}_\alpha$ - derivative in (1) makes λ_α auxiliary variable of the twistor type, which can be eliminated by the solution of corresponding constraints and transition to the conventional dynamics description in terms of X^m and P_m (3.7). Nevertheless apriori one can not neglect such a term in the Lagrangian. It could be added in classical action by hand, or arise, for example, as a result of particle interaction with some kind of quantum fields. Here we would like to investigate the minimal generalization of ex. (1) of the following form

$$L = L_0 + L_1 = L_0 + l\lambda_\alpha \dot{\lambda}_\beta \epsilon^{\alpha\beta}, \tag{3}$$

where l is an arbitrary parameter of length dimension.

At the first sight the presence of L_1 in the Lagrangian (3) "revives" the spinor variables[1] and enreaches the dynamics by additional bosonic degrees of freedom but as we shall see below the shift in twistor space and corresponding redefinition of X^m (the formulae (10) and (11)) eliminate L_1 term in (3) and lead to the equivalence of the free particle theories (1) and (3). The difference arises when minimal interaction of the particles with external (gauge or gravitational) fields is switched on and manifests itself in appearance of non minimal interaction terms in Lagrangian (3) after the twistor shift of dynamical variables has been performed.

Free particle dynamics described by ex. (3) is subjected to the constraints (2) and

$$\phi_\alpha \equiv \pi_\alpha + l\lambda_\alpha = 0 \tag{4}$$

(π_α is a momentum canonically conjugate to λ^α). One of the constraints (2). $P_{\alpha\beta} \lambda^\alpha \lambda^\beta = 0$, is of the first class and corresponds to the local reparametrization

---

[1]Kinetic terms of this kind were used previously to construct spinning particle models based on commuting spinors (see, for example, [12] and refs. therein).

$(\tau \rightarrow \tau' = \tau - a(\tau))$ invariance of the theory under the following (infinitesimal) transformations of $\lambda_\alpha, X^{\alpha\beta}$:

$$\delta \lambda_\alpha = a(\tau)\dot{\lambda}_\alpha, \quad \delta X^{\alpha\beta} = a(\tau)\dot{X}^{\alpha\beta}.$$

Two constraints remained in (2) and two constraints (4) are of the second class. Thus, one can check that Lagrangians (1) and (3) possess the equal number of independent canonical variables with λ_α being auxiliary ones in both cases. To extract two second class constraints from eq. (2) let us introduce a spinor $\mu^\alpha = x^{\alpha\beta}\lambda_\beta$ (which together with λ_α form a twistor) [1,3,7] and make a projection of eqs (2) and (4) on $\lambda_\alpha, \mu^\beta$ directions:

$$\Phi^1 = P_{\alpha\beta}\mu^\alpha\mu^\beta - (\lambda\mu)^2 - (\mu_\alpha\phi^\alpha)(\lambda\mu) = 0,$$

$$\Phi^2 = P_{\alpha\beta}\lambda^\alpha\mu^\beta - (\lambda_\alpha\phi^\alpha)(\lambda\mu) = 0, \tag{5}$$

$$(\lambda\mu \equiv \lambda_\alpha\mu^\alpha).$$

The Poisson brackets of the constraints (4) and (5) have the following form:

$$\left[\phi^\alpha, \phi^\beta\right] = 2l\epsilon^{\alpha\beta}, \tag{6a}$$

$$\left[\Phi^i, \Phi^k\right] = \frac{1}{l}(\lambda\mu)^3\epsilon^{ik}, \quad (i, k = 1, 2) \tag{6b}$$

$$\left[\phi^\alpha, \Phi^i\right] = 0. \tag{6c}$$

The transition to the Dirac brackets

$$[f, g]^* = [f, g] - \frac{1}{2l}[f, \phi^\alpha]\epsilon_{\alpha\beta}[\phi^\beta, g] -$$

$$-\frac{l}{(\lambda\mu)^3}[f, \Phi^i]\epsilon_{ik}[\Phi^k, g] \tag{7}$$

(where f, g are arbitrary functions of dynamical variables) does not alter canonical relations between P and X

$$[P_{\alpha\beta}, P_{\gamma\delta}]^* = 0. \tag{8a}$$

$$\left[X^{\alpha,\beta}, P_{\gamma\delta}\right]^* = \frac{1}{2}\left(\delta^\alpha_\gamma\delta^\beta_\delta + \delta^\beta_\gamma \delta^\alpha_\delta\right) \tag{8b}$$

but breaks the commutativity of $X^{\alpha\beta}$

$$\left[X^{\alpha\beta}, X^{\gamma\delta}\right]^* = \frac{l}{2(\lambda\mu)^2}\left(\epsilon^{\alpha\delta}\mu^\gamma\mu^\beta + \epsilon^{\beta\gamma}\mu^\alpha\mu^\delta\right) \tag{8c}$$

When the gauge $X^0 = \tau$ is chosen, the commutation relation (8c) takes the simple form

$$[X_1, X_2]^* = -\frac{l}{\sqrt{2}\,E}$$

with E being a particle energy.

The Dirac brackets of λ^α, μ^β are as follows

$$\left[\lambda^\alpha, \lambda^\beta\right]^* = 0, \quad \left[\mu^\alpha, \lambda^\beta\right]^* = -\frac{1}{2}\,\epsilon^{\alpha\beta}, \tag{9a}$$

$$\left[\mu^\alpha, \mu^\beta\right]^* = \frac{1}{2}\,\epsilon^{\alpha\beta}. \tag{9b}$$

For eqs. (9) to be a conventional twistor commutation relations μ^α should commute with each other [1]. To restore the canonical character of Dirac brackets for twistor variables λ, μ let us make a twistor shift of μ^α in λ^α direction:

$$\hat{\mu}^\alpha = \mu^\alpha + \frac{l}{2}\,\lambda^\alpha. \tag{10}$$

Note that $\hat{\mu}^\alpha$ can be obtained heuristically by means of the following reasoning course: up to a full derivative term, $L_0 + L_1$ (3) can be represented in the twistor form [3]

$$L_0 + L_1 = -2\dot{\lambda}_\alpha\lambda_\beta X^{\alpha\beta} + l\lambda_\alpha\dot{\lambda}^\alpha = 2\dot{\lambda}^\alpha \left(X^\beta_\alpha\,\lambda_\beta + \frac{l}{2}\,\lambda_\alpha \right) \equiv 2\hat{\mu}_\alpha\dot{\lambda}^\alpha$$

where $\hat{\mu}_\alpha$ thus determined is the momentum conjugate to λ^α.

As a result the r.h.s. of (9b) vanishes for $\hat{\mu}^\alpha$. Considering $\hat{\mu}^\alpha$ as a function of λ_β and a new variables $\hat{X}^{\alpha\beta}$ ($\hat{\mu}^\alpha = \hat{X}^{\alpha\beta}\,\lambda_\beta$) one can find a relation between $\hat{X}^{\alpha\beta}$ and $X^{\alpha\beta}$ coordinates:

$$\hat{X}^{\alpha\beta} = X^{\alpha\beta} + \frac{l}{2\lambda\mu}\left(\lambda^\alpha\mu^\beta + \lambda^\beta\mu^\alpha\right) \tag{11}$$

Note that as a consequence of constraints (2) and the definition of μ^α the twistor shift of $X^{\alpha\beta}$ (11) is proportional to a particle orbital momentum. One can check that $\hat{X}^{\alpha\beta}$ commute with respect to brackets (7), but relations (8b) are changed. To restore them $P_{\alpha\beta}$ has to be modified by adding a term proportional to the first-class constraint $P_{\alpha\beta}\lambda^\alpha\lambda^\beta$ as follows

$$\hat{P}_{\alpha\beta} = P_{\alpha\beta} + \frac{l}{2\lambda\mu}\left(\lambda_\alpha\mu_\beta + \lambda_\beta\mu_\alpha\right)P_{\gamma\delta}\lambda^\gamma\lambda^\delta \tag{12}$$

After $X^{\alpha\beta}$ redefinition (11) the Lagrangian (3) turns into the Lagrangian (1) which signifies the equivalence of the corresponding free particle theories. Thus, when interaction is absent one cannot determine which of the variables, $X^{\alpha\beta}$ or $\hat{X}^{\alpha\beta}$ are the physical coordinates of a particle. To make a choice particle interactions with external fields ought to be taken into account, and depending on either X or $\hat{X}$ is the argument of external physical fields one has two different physical theories.

3. Let us consider, for example, the dynamics of a particle (3) minimally interacting with electromagnetic field $A_{\alpha\beta}(x)$

$$L_{int} = -\dot{X}^{\alpha\beta} A_{\alpha\beta}(x) \tag{13}$$

The structure of constraints (2), (4) and (5) remains as in the case of the free particle with only difference that $P_{\alpha\beta}$ is replaced by covariant momentum $D_{\alpha\beta} = P_{\alpha\beta} + A_{\alpha\beta}$ and the electromagnetic stress tensor $F_{mn} = \partial_{[n} A_{n]}$ appears in r.h.s. of (6c):

$$\left[\Phi^i, \Phi^k\right]^* = \frac{(\lambda\mu)^3}{l}\, \epsilon^{ik}\left(1 - \frac{l}{(\lambda\mu)^3}\,\mu^\alpha\mu^\beta F_{\alpha\beta,\gamma\delta}\lambda^\gamma\mu^\delta\right) \equiv \frac{(\lambda\mu)^3}{l}\,(1 - lF) \tag{14}$$

This leads to the modification of the Dirac brackets (7) of dynamical variables:

$$\left[X^{\alpha\beta}, X^{\gamma\delta}\right]^* = \frac{l}{2(\lambda\mu)^2(1 - lF)}\left(\epsilon^{\alpha\delta}\mu^\gamma\mu^\beta + \epsilon^{\beta\gamma}\mu^\alpha\mu^\delta\right), \tag{15a}$$

$$[D_{\alpha\beta}, D_{\gamma\delta}]^* = \left(\frac{1}{2}\,F_{mn} - \frac{l}{4(\lambda\mu)^2(1 - lF)}\,F_{mm'}F_{nn'}\epsilon^{m'n'l}\mu\gamma_l\mu\right)\gamma^m_{\alpha\beta}\gamma^n_{\gamma\delta}, \tag{15b}$$

$$\left[X^{\alpha\beta}, D_{\gamma\delta}\right]^* = \frac{1}{2}\left(\delta^\alpha_\gamma\delta^\beta_\delta + \delta^\beta_\gamma\delta^\alpha_\delta\right) +$$

$$+ \frac{l}{2(\lambda\mu)^2(1 - lF)}\left(\epsilon^{\alpha\rho}\mu^\beta\mu^\lambda + \epsilon^{\beta\lambda}\mu^\alpha\mu^\rho\right)F_{\lambda\rho,\gamma\delta}. \tag{15c}$$

Note that nonconventional terms in (15) caused by electromagnetic interaction are accompanied by factor l. The commutation relations (9) for twistor variables appear to be more complicated as well but we will not reproduce them explicitly concentrating our study on space-time description of the dynamics.

A transition to the representation where $\hat{X}^{\alpha\beta}$ commute can be fulfilled by the same twistor shift (11) as in the free particle case. As a result the Lagrangian (3)+(13) of the theory, with variables subjected to the twistor shift, acquires an infinite l^n series of nonminimal interaction terms of the definite structure containing $F_{mn}(\hat{x})$ and its higher derivatives. In the first order of l's power such a Lagrangian can be represented in the following form:

$$L = \hat{D}_{\alpha\beta}\dot{\hat{X}}^{\alpha\beta}\left(A_{\alpha\beta}(\hat{x}) + \frac{l}{2}\,F_{\alpha\beta}(\hat{x})\right) - \frac{e(\tau)}{2}\,\hat{D}_{\alpha\beta}\hat{D}^{\alpha\beta}, \tag{17}$$

where $F_{\alpha\beta} = \frac{1}{\sqrt{2}}\,\gamma^l_{\alpha\beta}\epsilon_{lmn}F^{mn}$, $\hat{D}_{\alpha\beta}$ is a redefined covariant momentum proportional to $\lambda_\alpha\lambda_\beta$ and, hence, satisfying the mass-shell condition $\hat{D}_{\alpha\beta}\hat{D}^{\alpha\beta} = 0$. This condition is taken into account in the last term of eq. (17) with $e(\tau)$ being a Lagrangian multiplier.

Thus, in the first order of $l's$ power the twistor shift leads to the theory of a particle "minimally" interacting with electromagnetic field described by an effective potential $\hat{A}_{\alpha\beta} = A_{\alpha\beta} + \frac{1}{2} F_{\alpha\beta}$.

Though the classical theory has been considered here the use of the Hamilton analysis and Dirac procedure provides a direct transition to the quantum case.

4. The consideration fulfilled admits a generalizations to $D = 4, 6$ and 10 space-time dimensions.

Let us make some comments on peculiarities of the generalized particle mechanics in $D = 4$.

The Lagrangian

$$L_0 + L_1 = \lambda_A \bar{\lambda}_{\dot{A}} \dot{X}^{A\dot{A}} + l\left(\lambda_A \dot{\lambda}^A + \bar{\lambda}_{\dot{A}} \dot{\bar{\lambda}}^{\dot{A}}\right),$$

where λ_A is a commuting Weyl spinor, describes particle dynamics restricted by 6 constraints. Among them two ones belong to the I class and generate reparametrization and local $U(1)$ symmetries of the Lagrangian, while remaining four constraints belong to the II class. The local $U(1)$ symmetry is not manifest in the case considered and its generator acts nontrivially on $X_{A\dot{A}}$ variables:

$$\delta\lambda_A = i\phi(\tau)\lambda_A,$$

$$\delta X_{A\dot{A}} = -\frac{il\phi(\tau)}{\lambda\mu}\left(\lambda_A \bar{\mu}_{\dot{A}} - \bar{\lambda}_{\dot{A}}\mu_A\right),$$

where $\mu_A = iX_A^{\dot{A}}\bar{\lambda}_{\dot{A}}$.

In the case of particle interaction with external electromagnetic field the following nonminimal term (being the first order in $l's$ power) arises in the Lagrangian reformulated in terms of $\hat{X}^m$:

$$L_{\text{int.}}^{\text{nonmin.}} \sim l\sqrt{F_{mn}(\hat{X})F_l^n(\hat{X})\dot{\hat{X}}^m \dot{\hat{X}}^l},$$

which transforms into $lF_{mn}\epsilon^{mnl}\dot{\hat{X}}_l$ after dimensional reduction from $D = 4$ to $D = 3$.

Note that the presence of additional term L_1 has to influence the structure of external field equations of motion. This could manifest itself, for example, in integrability conditions for superparticle interaction with background superfields, as well as in string and superstring theories. In the latter case interaction is intrinsically connected with topological configurations of strings and the vertex function responsible for interaction has much in common with the external field vertexes which have been considered above.

5. In conclusion, let us cite some general remarks on possible modification of classical string action in $D = 4$:

$$S = \int d\tau \, d\sigma (det\ e_\nu^a) \left\{ \bar\lambda^{\dot A} e_a^\mu \rho^a\ \lambda^A \left(\partial_\mu X_{A\dot A} - \frac{L^2}{2}\ \bar\lambda_{\dot A}\ e_\mu^b \rho_b \lambda_A \right) \right.$$

$$\left. + \frac{l}{2} \left(\lambda^A \rho^a e_a^\mu \partial_\mu \lambda^B \epsilon_{AB} + c.c. \right) \right\} \tag{18}$$

where ρ^a ($a = 0, 1$) are world-sheet Dirac matrices: $e_\nu^a(\tau, \sigma)$ is a zweinbein; λ_A, $\bar\lambda_{\dot A}$ ($A, \dot A$ are $D = 4$ Weyl indices) being spinors both in $D = 4$ and $d = 2$ world-sheet, parametrized by (τ, σ); L is a string length parameter. The first term in (18) is equivalent to classical bosonic string action[2] while the second one is the generalization of L_1 in (3). Both terms possess $2d$ local Lorentz, reparametrization and Weyl invariances with Weyl transformation of the $2d$ fields e_μ^a, $X_{A\dot A}$, λ_A looking as follows:

$$e_\mu^{'a} = \Omega(\tau, \sigma) e_\mu^a, \quad X'_{A\dot A} = X_{A\dot A}, \quad \lambda'_A = \Omega^{-\frac{1}{2}}(\tau, \sigma) \lambda_A.$$

In the case of free string it is possible to solve equations of motion so that $X = X(\sigma_+) + X(\sigma_-)$ and $\lambda_A^1 = \lambda_A^1(\sigma_+)$, $\lambda_A^2 = \lambda_A^2(\sigma_-)$. Then the Virasoro constraints $(\partial_\pm X)^2 = 0$ arise as a consequence of

$$\partial_+ X_{A\dot A} = \lambda_A^1 \bar\lambda_{\dot A}^1, \quad \partial_- X_{A\dot A} = \lambda_A^2 \bar\lambda_{\dot A}^2$$

For such a case transformation analogous to (10, 11) exists which transforms (18) into standard action. However, there is no such transformation for interacting strings and the modified action may result in new physical features of string theory.

The study of the model based on action (18) is in progress.

Acknowledgments

The authors are thankful to V.D. Gershun, A.I. Pashnev and A.A. Zheltukhin for useful discussion. One of the authors, D.P.S. is grateful to organizers of the Leipzig-Wrocław Seminar, and in particular prof. J. Lukierski and Dr A. Frydryszak, for their warm hospitality during his stay in Wrocław.

References

[1] Penrose R., M.A.H. MacCallum, Phys.Repts. **6** (1972)241.

[2] A. Ferber, Nucl.Phys. **B132** (1978)55.

---

[2]This type of string action has been considered independently by A.I. Pashnev and V.A. Chekalov and by A.A. Zheltukhin.

[3] T. Shirafuji, Progr.Theor.Phys. **70** (1983) 18.

[4] P. Budinich, Comm.Math.Phys. **107** (1986) 455.

[5] A.K.H. Bengtsson et al. Phys.Rev. **36D** (1987) 1766; I. Bengtsson, M. Ceder-
wall, Nucl.Phys. **B302** (1988) 81.

[6] Y. Eisenberg, S. Solomon. Nucl.Phys. **B309** (1988) 709: Y. Eisenberg, Mod.
Phys.Lett. **A4** (1989) 195; M. Plyushchay in Proc. XII Workshop on High
Energy Physics and Field Theory. Moscow: Nauka 1990.
J. Lukierski in "Selected Topics in QFT and Mathematical Physics" (Liblice
1989), World Sc. (Singapore).

[7] D.P. Sorokin, V.I. Thach, D.V. Volkov. Mod.Phys.Lett. **A4** (1989) 901;
V.I. Gumenchuk, D.P. Sorokin, Yad.Fiz. **51** (1990) 549; D.P. Sorokin, Fort-
schr.Phys. **38** (1990) 33.

[8] D.V. Volkov, A.A. Zheltukhin, Lett.Math.Phys. **17** (1989) 141, Nucl.Phys.
B335 (1990) 723.

[9] N. Berkovits, Rutgers University Preprints RU-89-30, 48 (1989), Phys.Lett.
241B (1990) 497; ibitd **247B** (1990), 45.

[10] E. Witten, Nucl.Phys. **B266** (19886) 245; L.-L. Chau, M.-L. Ge, C.-S. Lim.
Phys.Rev **33** (1986) 1056.

[11] D.P. Sorokin et al. Phys.Lett. **216B** (1989) 302.

[12] A.O. Barut, M. Pavšič, Phys.Lett. **216B** (1989) 297
R. Marnelius, V. Martensson, Göteborg Preprint 89-55, 1989. International
Journal of Modern Physics A in press.

The Metric of Bures and the Geometric Phase.

Armin Uhlmann
University of Leipzig,
Dept. of Physics

After the appearance of the papers of Berry [1], Simon [2], and of Wilczek and Zee [3], I tried to understand [4], whether there is a reasonable extension of the *geometric phase* - or, more accurately, of the accompanying phase factor - for general (mixed) states. A known recipe for such exercises is to use *purifications*: One looks for larger, possibly fictitious, quantum systems from which the original mixed states are seen as reductions of pure states. For density operators there is a standard way to do so by the use of Hilbert Schmidt operators (or by Hilbert Schmidt maps from an auxiliary Hilbert space into the original one).

Thus let

$$\mathbf{c} : \quad t \mapsto \varrho_t, \qquad 0 \leq t \leq 1 \tag{1}$$

be a path of density operators. A *standard purification* of (1) is a path

$$t \mapsto W_t, \qquad \varrho_t = W_t W_t^* \tag{2}$$

sitting in the Hilbert space of Hilbert Schmidt operators with scalar product

$$< W_1, W_2 >:= \operatorname{tr} W_1^* W_2 \tag{3}$$

The construction of standard purifications is by no means unique. Indeed, not only (2) but every *gauged* path

$$W_t \to W_t U_t, \qquad U_t \quad \text{unitary,} \tag{4}$$

is a purification of the same path of density operators.

The problem is, therefore, to distinguish within all purifications of the curve (1) of mixed states exceptional ones. In [4] this has been achieved as following. Let $W_1, ..., W_m$ be a *subdivision* of (2), i.e. a time-ordered subset of operators (2).

267

R. Gielerak et al. (eds.), Groups and Related Topics, 267–274.

These operators are of norm one since the density operators have trace one. Now the expression

$$\xi = <W_m, W_{m-1}> \ldots <W_3, W_2><W_2, W_1> \tag{5}$$

will be considered according to (4) for all gauges

$$\xi \mapsto \tilde{\xi} \quad \text{by} \quad W_j \mapsto W_j U_j, \qquad U_j \quad \text{unitary}, \tag{6}$$

and it will be looked within the set of gauged $\tilde{\xi}$ for choices with

$$|\tilde{\xi}| = \text{maximum!} \tag{7}$$

The necessary and sufficient condition for (7) reads [5], [6]:

$$|<\tilde{W}_{j+1}, \tilde{W}_j>| = \text{tr}\,(\varrho_j^{1/2}\varrho_{j+1}\varrho_j^{1/2})^{1/2} \quad \text{for} \quad j = 1, \ldots, m-1 \tag{8}$$

If (8) and hence (7) is fulfilled, the remaining arbitrariness is in a regauging $\tilde{W}_j \to \epsilon_j U \tilde{W}_j$ of the subdivision by numbers of modulus one and by an independent of j unitary U - provided the rank of the density operators (1) remains constant.

This, however, means the gauge invariance of the quantity

$$X \mapsto \nu_c^{subdivision}(X) = \xi\ <\tilde{W}_1, X\tilde{W}_m> \tag{9}$$

and it depends therefore only on the ordered set of the density operators $\varrho_k = W_k W_k^*$. In the limit of finer and finer subdivisions,

$$X \mapsto \nu_c(X) := \lim \nu_c^{subdivision}(X), \tag{10}$$

one obtains a gauge invariant linear form depending only on the original path (1). For closed loops of pure states the number $\nu_c(1)$ is exactly Berry's phase factor.

(10) defines a certain noncommutative product integral. For curves of faithful density operators it can be conveniently expressed by the help of the geometric mean of two positive operators

$$a\#b := a^{\frac{1}{2}}\,(a^{-\frac{1}{2}}ba^{-\frac{1}{2}})^{\frac{1}{2}}\,a^{\frac{1}{2}} \tag{11}$$

To this end one introduces the holonomy $V(c)$ of c by

$$\nu_c(X) = \text{tr}V(c)\varrho_0 X \tag{12}$$

to find [8] (- in [8] the exponents are not correctly assigned -)

$$V(c) = \lim_{subdivisions} (\varrho_m\#\varrho_{m-1}^{-1})(\varrho_{m-1}\#\varrho_{m-2}^{-1})\cdots(\varrho_2\#\varrho_1^{-1}) \tag{13}$$

My next aim is to obtain expressions of the above procedure which are more manageable. One idea is to use an infinitesimal variant of (8). Indeed one may sharpen (8) by adding the requirement

$$\tilde{W}_{j+1}^* \tilde{W}_j \geq 0 \qquad (14)$$

which in turn implies (8) for faithful density operators. Going to finer and finer subdivisions - and removing the tilde - (14) results in ($\dot{W}$ denotes the t-derivation of W)

$$W^* \dot{W} = \dot{W}^* W, \qquad (15)$$

the so-called *parallelity condition* [4] : A lift (2) of (1) fulfilling (15) is called a (standard) *parallel purification* or a *parallel lift*. Thus choosing a parallel purification of (1), it is

$$\nu_c(X) = < W_0, X W_1 > \qquad (16)$$

where W_0 and W_1 are the starting and the end point of a parallel lift.

Though the word *parallel* points to a parallel transport governed by a connection form (described later on), a more elementary explanation is possible. The scaler products of the subdivision attain their maximal possible value if (8) is true. The vectors W_j have norm one and hence the scalar product is the cosine between neighbouring vectors. Therefore (8) indicates the the angles between neighbouring vectors is as small as possible. Hence for infinitesimal neighbouring they are parallelly directed.

Note that from (15) it follows for parallel lifts

$$W^* \ddot{W} = \ddot{W}^* W, \qquad (15a)$$

Another idea is already indicated in a paper of Fock [7], who tried to minimize the arbitrariness in the transport of phases of degenerate eigenstates of Hamiltonians.

The observation [8] is as following: After choosing appropriate phases in (5) the scalar products $< W_{j+1}, W_j >$ can be made real and positive. But then ξ in (7) attains its maximum if and only if

$$\| W_m - W_{m-1} \| + ... + \| W_3 - W_2 \| + \| W_2 - W_1 \| \qquad (17)$$

attains its minimum. On the other hand, in going to finer and finer subdivisions, (17) tends to the length of the curve (2) in the metric given by (3). Therefore a purification (2) is a parallel one iff it solves the variational problem

$$\int \sqrt{< \dot{W}, \dot{W} >} dt = \text{Min !} \qquad (18)$$

However, the Euler equations of this variational problem are nothing else than the parallelity condition (15) !

One can calculate the minimal length (18), which, indeed, is the *Bures length* [9] of the path (1) of density operators. To do so one has to solve the parallelity condition. According to Dabrowski and Jadczyk [10], and to [11], this is done by an ansatz

$$\dot{W} = GW, \qquad G^* = G \qquad (19)$$

which gives easily the equation

$$\dot{\varrho} = G\varrho + \varrho G \qquad (20)$$

for the unknown G. G is gauge invariant, and depends only on the pair $\{\varrho, \dot{\varrho}\}$. This reflects the fact that the Bures length of the path (1) can be expressed without using lifts (2) : Inserting (19) into (18) one gets

$$L^{\text{Bures}}(c) = \int \sqrt{< GW, GW >}\, dt \qquad (21)$$

and a straightforward calculation shows

$$dt^2_{\text{Bures}} = < GW, GW > = \operatorname{tr} \varrho G^2 = \frac{1}{2} \operatorname{tr} G\dot{\varrho} \qquad (22)$$

There is a formal solution of (20) which reads for faithful density operators

$$G = \int_0^\infty (\exp -s\varrho)\dot{\varrho}(\exp -s\varrho)\, ds \qquad (23)$$

and which implies for the metric form (22) the expression

$$\frac{1}{2}\operatorname{tr} \int_0^\infty (\exp -s\varrho)\dot{\varrho}(\exp -s\varrho)\, \dot{\varrho}\, ds \qquad (24)$$

Now, switching to density operators of finite dimension n, one may choose a base E_k , where $k = 1, \ldots, n^2 - 1$, of traceless hermitian matrices, and write

$$\varrho = \frac{1}{n}\mathbf{1} + \sum x^k E_k \qquad (25)$$

to get from (24)

$$\frac{1}{2}\operatorname{tr} \dot{\varrho} G = \sum g_{jk}\dot{x}^j \dot{x}^k \quad \text{with} \quad g_{jk} = \frac{1}{2}\operatorname{tr} \int_0^\infty (\exp -s\varrho)E_j(\exp -s\varrho)E_k\, ds \qquad (26)$$

Therefore one has for the "moments conjugate to the coordinates", x_k,

$$p_k = 2\sum g_{kj}\dot{x}^j = \operatorname{tr} GE_k \qquad (27)$$

Example 1.

Here I show the simplest possible case, the Bures metric for $n = 2$. That this case can be solved is due to the following: Let δ be a derivation, $X > 0, Y$, 2-by-2 matrices, then

$$\delta X = YX + XY \tag{28}$$

is solved by

$$Y \operatorname{tr} X = \delta X + \frac{1}{2} X^{-1} \delta \det X - 1 \frac{1}{2} \operatorname{tr} X \tag{29}$$

which is easily derived by δ-differentiating the characteristic equation of X. Describing now the density operators by

$$\varrho = \frac{1}{2}(1 + x_1 \sigma_1 + x_2 \sigma_2 + x_3 \sigma_3) \tag{30}$$

which is a variant of (25), the metric space

$$\{\varrho > 0, \quad \operatorname{tr}\varrho = 1, \quad dt^2_{\text{Bures}}\} \tag{31}$$

can be isometrically imbedded into a sphere S^3 given by

$$1 = x_1^2 + x_2^2 + x_3^2 + x_4^2 \tag{32}$$

where x_4 is defined by

$$x_4 \geq 0, \qquad x_4^2 = 4 \det \varrho \tag{33}$$

and which is equipped with the metric

$$\frac{1}{4}(dx_1^2 + dx_2^2 + dx_3^2 + dx_4^2) \tag{34}$$

This example shows that the Bures metric turns the set of all 2-by-2 density matrices into a piece of a symmetric space, i.e. into half of a 3-sphere. see also [12]

Example 2.

Here the restriction of the Bures metric to maximal commutative submanifolds will be described. In a suitable base such a submanifold can be given by diagonal density matrices.

$$\varrho = (\lambda_j \delta_{jk}), \qquad G = (g_j \delta_{jk}) \tag{35}$$

Now (20) yields

$$g_j = \frac{\dot{\lambda_j}}{2\lambda_j} \tag{36}$$

Introducing the new variables

$$\lambda_j = y_j^2 \tag{37}$$

the metric of Bures reads

$$dt^2_{\text{Bures}} = \sum \dot{y}_j^2 \tag{38}$$

Hence the restriction on a maximal commutative subset of the Bures metric is isometrically isomorph to a piece of a sphere, i.e. of a symmetric space.

The set of density operators, equipped with the Bures metric, is metrically incomplete. One may ask whether there is a completion in which all geodesics close for dim > 2. To support this question let us consider

Example 3.

The geodesic connecting two faithful density operators, ϱ_j, $j = 1, 2$, within the space of density operators can be described as follows. Let $\varrho_j = W_j W_j^*$. Then the geodesic in the W-space connecting W_1 with W_2 is part of a large circle of the unit sphere. Its equation is

$$W = \lambda_1 W_1 + \lambda_2 W_2, \qquad < W, W >= 1 \tag{39}$$

where

$$a := \text{Real} \ < W_1, W_2 > \tag{40}$$

$$\lambda_1 = \cos \vartheta - \frac{a}{\sqrt{1 - a^2}} \sin \vartheta \tag{41}$$

$$\lambda_2 = \frac{\sin \vartheta}{\sqrt{1 - a^2}} \tag{42}$$

The (oriented) length is hence the arc ϑ_0 given by

$$\cos \vartheta_0 = \text{Real} < W_1, W_2 > \quad \text{with} \quad -\frac{\pi}{2} < \vartheta_0 < \frac{\pi}{2} \tag{43}$$

Clearly, the length attains its minimum if we choose lifts such that a is of maximal value. This can be achieved if (14), and hence (8), is valid for $j = 1$.

Thus the Bures length ϑ_0 of the geodesic joining the two density operators is given by

$$\cos \vartheta_0 = \text{tr} \, (\varrho_1^{1/2} \varrho_2 \varrho_1^{1/2})^{1/2}, \quad \text{with} \quad 0 < \vartheta_0 < \frac{\pi}{2} \tag{44}$$

The metric on the unit sphere of the Hilbert Schmidt W-space can be decomposed into a horizontal and a vertical part by an ansatz

$$< \dot{W}, \dot{W} >= < GW, GW > + < WA, WA >, \qquad A^* = -A \tag{45}$$

Then A can be defined equally well by [13]

$$W^* \dot{W} - \dot{W}^* W = A W^* W + W^* W A \tag{46}$$

This can be seen as follows. Going into (20) with an ansatz [13]

$$\dot{W} - WA = GW \tag{47}$$

and with $\varrho = WW^*$, it follows that A is antihermitian. Knowing this and the hermiticity of G one easily recovers (45). On the other hand, substituting (47) into the left side of (46), one arrives at the right side of this equation.

A is the restriction on the given lift of a connection 1-form, **A**, for the gauge transformations (4), and one has

$$W^*dW - dW^* W = \mathbf{A} W^*W + W^*W \mathbf{A} \tag{48}$$

Introducing the **A**-covariant derivation of an expression X transforming as W by

$$DX = dX - X\mathbf{A} \tag{49}$$

another form of (47) is

$$DW = GW \tag{50}$$

I now rewrite (48) in a form similar to (28). As a complex linear space defines a complex analytic structure, the total differential is decomposed naturally into $d = \partial + \bar{\partial}$. Using this one may rewrite (48) as

$$(\partial - \bar{\partial})(W^*W) = \mathbf{A} W^*W + W^*W \mathbf{A} \tag{51}$$

This may be contrasted to

$$d(WW^*) = (\partial + \bar{\partial})(WW^*) = WW^*G + GWW^* \tag{52}$$

Thus we have

$$W^*dW = \mathbf{A}^{1,0} W^*W + W^*W \mathbf{A}^{1,0} \tag{53}$$

$$dW W^* = WW^*\mathbf{G}^{1,0} + \mathbf{G}^{1,0}WW^* \tag{54}$$

Remark: In the case of 2-by-2 density operators (45) can be solved effectively by (29) using $\delta = \partial - \bar{\partial}$, $X = W^*W$, and $Y = \mathbf{A}$. The first explicit expression for **A** was obtained in [14], see also [15].

For $\mathrm{rank}(\varrho) = 1$ one falls back to the Berry case, and **A** describes the monopole structure. For $\mathrm{rank}(\varrho) = 2$ one gets instanton structures [16]. It is unknown what is with $\mathrm{rank}(\varrho) > 2$.

* * *

Note added in proof: In a recent preprint [17] some of the constructions are generalized and examined for C*-algebras. It is further indicated how possibly to proceed if the states (or density operators) have mutually inequivalent supports.

References

1) M. V. Berry, Proc. Royal. Soc. Lond. A 392 (1984) 45

2) B. Simon, Phys. Rev. Lett. 51 (1983) 2167

3) F. Wilczek, A. Zee, Phys. Rev. Lett. 52 (1984) 2111

4) A. Uhlmann, Rep. Math. Phys. **24**, 229, 1986

5) H. Araki, RIMS-151, Kyoto 1973

6) A. Uhlmann, Rep. Math. Phys. **9**, 273, 1976

7) V. Fock, Z. Phys. 49 (1928) 323

8) A. Uhlmann, Parallel Transport and Holonomy along Density Operators. In: "Differential Geometric Methods in Theoretical Physics", (Proc. of the XV DGM conference), H. D. Doebner and J. D. Hennig (ed.), World Sci. Publ., Singapore 1987, p. 246 - 254

9) D. J. C. Bures, Trans Amer. Math. Soc. **135**, 119, 1969

10) L. Dabrowski and A. Jadczyk, Quantum Statistical Holonomy. preprint, Trieste 1988

11) A. Uhlmann, Ann. Phys. (Leipzig) **46**, 63, 1989

12) M. Hübner: Explicit Computation of the Bures distance for Density Matrices. NTZ-preprint 21/91, Leipzig 1991

13) A. Uhlmann, Lett. Math. Phys. **21**, 229, 1991

14) G. Rudolph: A connection form governing parallel transport along 2×2 density matrices. Leipzig - Wroclaw - Seminar, Leipzig 1990.

15) J. Dittmann, G. Rudolph: A class of connections governing parallel transport along density matrices. Leipzig, NTZ-preprint 21/1991.

16) J. Dittmann, G. Rudolph: On a connection governing parallel transport along 2 x 2 -density matrices. To appear.

17) P. M. Alberti: A study of pairs of positive linear forms, algebraic transition probabilitiy, and geometric phase over noncommutative operator algebras. Leipzig, NTZ preprint 29/1991

Mathematical Physics Studies

Publications:

1. F.A.E. Pirani, D.C. Robinson and W.F. Shadwick: *Local Jet Bundle Formulation of Bäcklund Transformations*. 1979   ISBN 90-277-1036-8

2. W.O. Amrein: *Non-Relativistic Quantum Dynamics*. 1981
    ISBN 90-277-1324-3

3. M. Cahen, M. de Wilde, L. Lemaire and L. Vanhecke (eds.): *Differential Geometry and Mathematical Physics*. Lectures given at the Meetings of the Belgian Contact Group on Differential Geometry held at Liège, May 2–3, 1980 and at Leuven, February 6–8, 1981. 1983   ISBN 90-277-1508-4 (pb)

4. A.O. Barut (ed.): *Quantum Theory, Groups, Fields and Particles*. 1983
    ISBN 90-277-1552-1

5. G. Lindblad: *Non-Equilibrium Entropy and Irreversibility*. 1983
    ISBN 90-277-1640-4

6. S. Sternberg (ed.): *Differential Geometric Methods in Mathematical Physics*. 1984   ISBN 90-277-1781-8

7. J.P. Jurzak: *Unbounded Non-Commutative Integration*. 1985
    ISBN 90-277-1815-6

8. C. Fronsdal (ed.): *Essays on Supersymmetry*. 1986   ISBN 90-277-2207-2

9. V.N. Popov and V.S. Yarunin: *Collective Effects in Quantum Statistics of Radiation and Matter*. 1988   ISBN 90-277-2735-X

10. M. Cahen and M. Flato (eds.): *Quantum Theories and Geometry*. 1988
    ISBN 90-277-2803-8

11. Bernard Prum and Jean Claude Fort: *Processes on a Lattice and Gibbs Measures*. 1991   ISBN 0-7923-1069-1

12. A. Boutet de Monvel, Petre Dita, Gheorghe Nenciu and Radu Purice (eds.): *Recent Developments in Quantum Mechanics*. 1991   ISBN 0-7923-1148-5

13. R. Gielerak, J. Lukierski and Z. Popowicz (eds.): *Groups and Related Topics*. Proceedings of the First Max Born Symposium. 1992   ISBN 0-7923-1924-9

Kluwer Academic Publishers – Dordrecht / Boston / London

The manufacturer's authorised representative in the EU is Springer
Nature Customer Service Centre GmbH, Europaplatz 3, 69115 Heidelberg,
Germany. If you have any concerns regarding our products, please
contact ProductSafety@springernature.com

Printed and bound by CPI Group (UK) Ltd, Croydon, CR0 4YY
24/04/2026
02096308-0018